Applied Regression and ANOVA Using SAS

Patricia F. Moodie

Dallas E. Johnson

CRC Press
Taylor & Francis Group
Boca Raton London New York

CRC Press is an imprint of the
Taylor & Francis Group, an **informa** business
A CHAPMAN & HALL BOOK

First edition published 2021
by CRC Press
6000 Broken Sound Parkway NW, Suite 300, Boca Raton, FL 33487-2742

and by CRC Press
4 Park Square, Milton Park, Abingdon, Oxon, OX14 4RN

© 2022 John P. Hoffman

CRC Press is an imprint of Taylor & Francis Group, LLC

Library of Congress Cataloging-in-Publication Data

Names: Moodie, Patricia, author. | Johnson, Dallas E., 1938- author.
Title: Applied regression and ANOVA using SAS / Patricia Moodie, Dallas Johnson.
Description: First edition. | Boca Raton : Taylor & Francis Group, 2022. | Includes bibliographical references and index. | Summary: "Applied Regression and ANOVA Using SAS® has been written specifically for non-statisticians and applied statisticians who are primarily interested in what their data are revealing. Interpretation of results are key throughout this intermediate-level applied statistics book. The authors introduce each method by discussing its characteristic features, reasons for its use, and its underlying assumptions. They then guide readers in applying each method by suggesting a step-by-step approach while providing annotated SAS programs to implement these steps. Those unfamiliar with SAS software will find this book helpful as SAS programming basics are covered in the first chapter. Subsequent chapters give programming details on a need-to-know basis. Experienced as well as entry-level SAS users will find the book useful in applying linear regression and ANOVA methods, as explanations of SAS statements and options chosen for specific methods are provided. Features: Statistical concepts presented in words without matrix algebra and calculus Numerous SAS programs, including examples which require minimum programming effort to produce high resolution publication-ready graphics Practical advice on interpreting results in light of relatively recent views on threshold p-values, multiple testing, simultaneous confidence intervals, confounding adjustment, bootstrapping, and predictor variable selection Suggestions of alternative approaches when a method's ideal inference conditions are unreasonable for one's data This book is invaluable for non-statisticians and applied statisticians who analyze and interpret real-world data. It could be used in a graduate level course for non-statistical disciplines as well as in an applied undergraduate course in statistics or biostatistics"-- Provided by publisher.
Identifiers: LCCN 2021051446 (print) | LCCN 2021051447 (ebook) | ISBN 9781439869512 (hardback) | ISBN 9781032244662 (paperback) | ISBN 9780429107368 (ebook)
Subjects: LCSH: Regression analysis. | SAS (Computer program language)
Classification: LCC QA278.2 .M656 2022 (print) | LCC QA278.2 (ebook) | DDC 519.5/36--dc23/eng/20211204
LC record available at https://lccn.loc.gov/2021051446
LC ebook record available at https://lccn.loc.gov/2021051447

ISBN: 978-1-439-86951-2 (hbk)
ISBN: 978-1-032-24466-2 (pbk)
ISBN: 978-0-429-10736-8 (ebk)

DOI: 10.1201/9780429107368

Typeset in CMR10
by KnowledgeWorks Global Ltd.

To Ric, my support, wellspring of my inspiration and encouragement, and to the rest of my family who also continually inspire me by their dedication to what is important in life: the late Gordon and Eleanor Andrews, Zoe Moodie, Erica Moodie, Jon Wakefield, Dave Stephens, Eleanor and Eric Wakefield, Gordie and Jamie Stephens.

To my parents, Chet and Dorothy Johnson, for giving me the opportunity to pursue a college degree in mathematics education, and to my wife, Erma, for the support that she has provided during our 60 plus years of marriage.

FIGURE 1
Regression inspires art – a pendant designed and crafted by Ric Moodie, photo by Brian Rudolf, Empire Photography.

Contents

Preface

The spotlight in this book is on linear regression and ANOVA using SAS®, a popular software package for statistical analysis and graphics. We have written this book specifically for those who want to understand in depth how these methods can shed light on research questions. We think that this book will be helpful to researchers and applied statisticians because we address in detail complex issues that are confronted when data do not conform to classical textbook examples. Many of these issues are usually discussed only in books that assume a knowledge of matrix algebra, an assumption we have deliberately avoided. As our goal is to be statistically and scientifically rigorous without being mathematically challenging, we do not present any concepts or explanations in terms of matrix algebra or calculus, assuming our readers do not necessarily have a formal knowledge of these subjects. We do assume a familiarity with topics covered in an introductory statistics course such as descriptive statistics, hypothesis testing, and confidence intervals.

We do not assume any prior knowledge of SAS and provide SAS programs that illustrate examples of the statistical methods we present along with detailed explanations of these programs so they can be easily adapted for a researcher's own data. We place special emphasis on ODS graphics that are available in SAS Version 9.3 and in later versions. ODS graphics are high resolution, publication-ready graphics that are produced with virtually no effort on the part of the SAS user. Very simple SAS commands can generate these impressive graphs.

We have also placed major emphasis on a step-by-step approach to checking the assumptions (ideal inference conditions) of a statistical method before drawing conclusions from the results of that method. Alternative approaches are suggested for situations where ideal inference conditions for making valid conclusions are not reasonably satisfied. As each topic is considered, realistic examples are given, along with corresponding SAS programs and output which illustrate the analyses and graphics under discussion. Other important concerns are also discussed and evaluated using SAS programs. For example, are there outliers in the data? Are there any influential data points? Are there too many explanatory variables in the model? Are there too few explanatory variables in the model? Are there relationships among any of the explanatory variables that could lead to flawed interpretations? How can one deal with unordered categorical (qualitative) explanatory variables? What are the consequences of multiple comparisons? We have devoted an entire chapter to multiple testing and simultaneous confidence intervals, as understanding these topics is important when drawing certain conclusions from linear regression and ANOVA results.

In presenting material and discussing these issues, we have kept in mind the busy schedules and heavy workloads of our readers and therefore we often have distilled information in point form and in short subsections to allow easy navigation of the topics without short-changing the information being presented. To those individuals who are equation-averse, we offer our apologies as there are equations in this book. We thought that seeing these equations may make it easier to understand why evaluating ideal inference conditions is so important before rushing ahead to interpret the results of an analysis. However, we always give fair warning with section or subsection heading titles when an equation is about to appear on the page and so if you are reading this book for casual interest and not for a credit course, you might like to skip over these sections and subsections.

The issues addressed in this book are greatly influenced by our experiences in teaching, consulting, and collaborating for more than 40 years. These experiences have given us insight into the concerns of many researchers, data analysts, applied statisticians, and students. It is their concerns and challenges encountered in analysing real life data that have been at the forefront when we were writing this book. We hope our emphasis on interpretation of examples will help readers have a better understanding and a fuller appreciation of linear regression, ANOVA, and related graphics as accessible and useful research tools for discovery.

Acknowledgements

We would like to acknowledge P. F. M.'s support from the Department of Mathematics and Statistics at the University of Winnipeg. We would also like to acknowledge Shelley Sessoms at SAS Institute Inc. who facilitated P. F. M.'s access to SAS® software in Montreal during 2011–2012 via the SAS Author Assistance Software Program. We are most grateful as well to Ric Moodie for BibTeX support, to Erica and Zoe Moodie for LaTeX advice, and to Dave Stephens, who contributed Figures 2.2 and 2.3 in Chapter 2.

Finally we are greatly indebted to CRC Press, in particular to Rob Calver, for logistical, production, technical, and editorial expertise. The feedback we received from CRC Press reviewers was invaluable.

Thank you all!

Author Biography

Patricia F. Moodie is a Research Scholar in the Department of Mathematics and Statistics at the University of Winnipeg, Manitoba, Canada. Prior to that she was Head of Biostatistics in the Computer Department for Health Sciences in the College of Medicine, University of Manitoba, an adjunct lecturer in Biometry in the Department of Social and Preventive Medicine at the University of Manitoba, and a biostatistician in the Epidemiology and Biostatistics Department at the Manitoba Cancer Treatment and Research Foundation. Her statistical consulting and collaboration for over three decades as well as her substantive background in the biomedical sciences have made her appreciate the challenges in analysing and interpreting real-life data. She received a BSc (Hons) in Biology at Memorial University of Newfoundland, an MSc in Zoology at the University of Alberta, and an MS in Biostatistics at the University of Illinois at Chicago. She has been an enthusiastic SAS user since 1980.

Dallas E. Johnson, Professor Emeritus in the Department of Statistics, Kansas State University, has published extensively in the areas of linear models, multiplicative interaction models, experimental design, and messy data analysis. He is the author of *Applied Multivariate Methods for Data Analysts* and co-author with George A. Milliken of the following books: *Analysis of Messy Data, Vol. I – Designed Experiments, Vol. II – Nonreplicated Experiments, Vol. III – Analysis of Covariance, and Vol. I – Designed Experiments 2nd Edition*. An active presenter of short courses, and a statistical consultant for over 50 years, he was the recipient of ASA's award for Excellence in Statistical Consulting in 2010. He received his B.S. degree in Mathematics Education, Kearney State College, a M.A.T. degree in Mathematics, Colorado State University, a M.S. degree in Mathematics, Western Michigan University, and a Ph.D. degree in Statistics, Colorado State University. He has been a SAS user and mentor since 1976.

1

Review of Some Basic Statistical Ideas

Although we assume our readers have had an introductory statistics course, we will briefly review some important statistical ideas, to which we will refer throughout the book, just in case any of this material has been forgotten.

1.1 Introducing Regression Analysis

Definition 1.1. *Regression analysis* is a statistical methodology which investigates whether variables are related to each other and if they are related, to what extent are they related.

Researchers can use a regression analysis for:

- prediction of one variable by one or more other variables

- investigating the processes underlying their research by understanding how variables are related

There are two types of variables in a regression analysis – *a response variable* and *a predictor variable*.

Definition 1.2. A *response* variable is a variable you designate to be predicted from knowledge of other variables in the regression analysis.

A *response* variable is also referred to as an **outcome** variable, a **dependent** variable, or a Y variable.

Definition 1.3. A *predictor* variable is the variable that predicts the response variable in the regression analysis, provided a relationship truly exists between the response and predictor variable.

A predictor variable is also called an **explanatory** variable, an **independent** variable, *a regressor*, or an X variable.

We will use these alternate terms for the response and predictor variables interchangeably throughout this book to reflect the diversity of their usage in the literature.

Two of the ways by which various regression methods can be distinguished are:

- the number of response variables and predictor variables considered in the analysis

- the classification of the response and predictor variables (Section 1.2)

Although there are other important ways to distinguish the various regression methods that we will present, we will defer discussion of these until later chapters where we present these methods in detail.

DOI: 10.1201/9780429107368-1

1.2   Classification of Variables

1.2.1   Quantitative vs. Qualitative Variables

One important way a variable can be classified is whether it is *quantitative* or *qualitative*.

Definition 1.4. A *quantitative* variable is a variable that has values that are measured or counted, for example, body temperature, number of live offspring.

Definition 1.5. A *qualitative* variable is a variable that has values that are qualities not measurements, for example, hair colour, gender, place of birth.

A quantitative variable can be further classified by whether it is *continuous* or *discrete*.

Definition 1.6. A *continuous* variable is a quantitative variable that can have values with decimal points that make sense. Or to state this more formally, a continuous variable is a variable that can take on any numeric value from an interval of real numbers.

Weight in grams is an example of a continuous variable because values such as 1g, 2g or any value in between such as 1.00001g and 1.2436g are possible. A variable is usually considered to be continuous even though it may be rounded to a given level of accuracy when collecting the data.

Definition 1.7. A *discrete* variable is a quantitative variable which cannot have an infinite number of meaningful numeric values on the scale used to measure it. Values with decimal points are not meaningful.

Number of live births is an example of a discrete variable, because 2.5 births is not a meaningful possible value for this variable.

1.2.2   Scale of Measurement Classification

Variables are also classified by the scale of measurement, viz., *nominal, ordinal, interval,* or *ratio scale.*

Definition 1.8. A *nominal* variable has values that represent categories which have no relation to any sort of ranking.

Eye colour is a nominal variable because no specific eye colour category (blue, brown, grey) is considered to be better (higher rank) than another.

Definition 1.9. An *ordinal* variable has values that represent categories that can be ranked but where the difference between adjacent ranks is not the same.

For example, level of pain with categories low, medium, high is an ordinal scaled variable because one cannot say that the difference between low and medium pain level is the same as the difference between medium and high pain level.

Another example of an ordinal scaled variable is Apgar score, a score used to assess the health of a newborn child to determine quickly whether the newborn needs immediate medical care. Possible values for an Apgar score range from zero to 10, with lower values indicating poor cardiorespiratory and neurologic functioning and with higher values indicating good functioning. It is an ordinal scaled variable because the difference between Apgar scores of 9 and 10 does not have the same clinical implications as the difference between the scores 0 and 1.

Definition 1.10. An *interval* variable has values that represent categories that can be ranked and the difference between the adjacent categories is consistent and meaningful but there is no natural value of zero on this scale. Therefore it is reasonable to calculate differences between two values of an interval scaled variable but one should never calculate ratios between two such values.

An example of an interval scaled variable is temperature measured in Celsius or Fahrenheit. There is no natural zero for temperature. The temperature at which water freezes depends on the scale being used. Thus it does not make sense to calculate a ratio 30C/10C and say 30C is three times as hot as 10C.

Definition 1.11. A *ratio* variable has a natural value of zero as well as all the other attributes of an interval variable.

Examples of a ratio scaled variable are height and weight where a zero value has a natural meaning and does not depend on scale. Hence, if an object weighed twice as much as another object, it would weigh twice as much, regardless of whether pounds or kilograms were used.

1.2.3 Overview of Variable Classification and the Methods to be Presented in this Book

The methods to be presented in this book all involve modelling a single response variable. This response variable should be quantitative, typically continuous, and measured on an interval or ratio scale (Definitions 1.10 and 1.11 respectively). Table 1.1 summarizes how these methods are distinguished on the basis of the number and classification of predictor variables represented in the analysis. As can be seen from this table, predictor variable classification is an important factor to be considered at the outset when you are initially choosing a statistical method. Other factors which also should be considered when choosing a method will be subsequently discussed as each method is described in detail.

TABLE 1.1
Characterization of methods based on the number and classification of predictor variables.

Method	Chapters	Number and Classification of Predictor Variables Represented in the Analysis
Simple Linear Regression	2-4	One quantitative predictor variable, typically continuous
Multiple Linear Regression	5-9	More than one predictor variable, where at least one is quantitative
One-way analysis of variance	11	A single predictor variable that is represented as a qualitative variable in the analysis
Analysis of Covariance	13	At least one predictor variable that is represented as a qualitative variable plus at least one quantitative predictor variable

1.3   Probability Distributions

You may recall from introductory statistics, a *probability distribution for discrete data* gives all the possible values of a random variable in a given population along with the probability for each value. The probability for each value is the relative frequency of that value in the population. A *probability distribution* for continuous data gives the probability of observing a given value of a random variable within an interval of values *a* and *b*. We will not discuss how these probabilities for continuous data are obtained, as they are based on integrating a function using calculus and we are not assuming any knowledge of calculus in this book. Most authors drop the "probability" and refer to probability distributions simply as distributions.

1.3.1   The Normal Distribution

A well known example of a probability distribution for a continuous random variable is the normal distribution on which many confidence interval estimates and hypothesis tests discussed in this book are based (Figure 1.1). The normal distribution is completely specified by two parameters, its mean and its standard deviation. A key property of the normal distribution is that it is perfectly symmetrical about its mean. The normal distribution is sometimes called the Gaussian distribution (named after Carl Friedrich Gauss) to dispel the notion that such a distribution is typical (the norm) and that any other distribution is abnormal.

FIGURE 1.1
A normal distribution in sterling and gold – a pendant designed and crafted by Ric Moodie, photo by Brian Rudolf, Empire Photography.

The Central Limit Theorem

The central limit theorem provides some justification for using the normal distribution in hypothesis tests and confidence interval estimation involving means of populations, when the original observations in the these populations have either a non-normal or an unknown distribution. For example, although a population of birth weights has a skewed (i.e., a non-normal) distribution, if a sufficiently large sample is drawn at random from this population, then it is reasonable to apply a Student's t-test which is based on normal theory in a hypothesis test regarding the population mean birth weight.

A formal statement of this theorem is that when the sample size n is sufficiently large, the means of samples of size n drawn at random from a population will tend to be approximately normally distributed, even if the individual observations in the population are not normally distributed. How large a sample size is sufficiently large? There is no precise formula that is applicable to all research data. However, a general guideline is that the more the original distribution departs from a normal distribution, the larger the sample size should be.

1.4   Statistical Inference

Definition 1.12. *Statistical inference* refers to the process of making a conclusion about a particular population based on what is observed in a random sample from that population.

It is useful to distinguish between a target and a sampled population in a discussion of statistical inference.

Definition 1.13. A *target population* is the population about which one wants to draw conclusions.

Definition 1.14. A *sampled population* is a population from which a sample is drawn for the research study.

Statistical Generalization Based on Random Sampling

One can make a straightforward statistical generalization from a sample to a target population when a sample has been obtained at random from the target population. Recall from introductory statistics that random sampling requires that every member in the target population be uniquely identified, typically by an identification number. The researcher then applies a formal random process to select individuals for the sample such that each member of the population has an equal probability of being selected. The random process for selecting individuals could be as simple as drawing numbers out of a hat. However, a computer algorithm which generates pseudo-random numbers is commonly used, especially when the target population is large.

However, generalizing from a sample to a target population is not so straightforward if:

(i)   a nonrandom sample has been obtained from the target population

(ii) a sample (either random or nonrandom) has been drawn from a population other than the target population

(iii) there are missing data in the sample that are not missing completely at random as will be discussed in Section 1.5

Example of sampling scenario (i) nonrandom sampling from the target population: Researchers want to investigate a fish population in a particular lake in 2018, i.e., their target population is the fish population in this particular lake in 2018. They obtain a sample from this population by catching fish in a net set in this lake. This is a nonrandom sample from the target population. Obtaining a random sample of live lake fish in this situation is impossible as it would have required that all the individual fish of that species in the lake be uniquely identified before a random selection could be made. Therefore, as these researchers have not obtained a random sample from their target population, they cannot statistically generalize the results observed in their sample (fish caught in the net) to the entire lake population. Nets are known to be highly selective as to what part of a fish population they capture, as they select by size, shape, and behaviour of fish in the lake. Thus, every individual in the researchers' target population did not have an equal probability of being included in their sample.

Example of sampling scenario (ii) sampling from a population other than the target population: Researchers want to investigate oral contraceptive use in the general population of females under 35 who reside in the province of Manitoba, Canada in 2018. They use a sample of convenience, volunteers, under the age of 35 who are medical students at the University of Manitoba. To generalize conclusions from their sample to their target population, these researchers would have to justify that what they observed in their sample of medical student volunteers would apply to their target population. Such a justification would require the researchers' collecting information regarding factors considered to be associated with oral contraceptive use and indicate whether these factors are comparable in their sample and the target population. If these factors are not comparable, a statistical modelling technique (e.g., analysis of covariance, Chapter 13) might be considered to adjust for differences. However, conclusions based on results from a random sample from the researchers' target population would have been much more straightforward and convincing.

It is always extremely important to report in research methods a detailed description (who, where, when) of the population sampled, for example, a population of seniors, greater or equal to 65 years, resident in Manitoba, Canada in 2018. It is also important to report the details of how the sample was obtained from this population as well as details involving any missing data.

1.5   Missing Data

Rubin (1976) proposed the following three categories of missing data:

- *missing completely at random* (MCAR), which occurs when the reason that data values are missing is not related to their values or to the values of any other relevant variable in the data set so the sample with missing data can be considered a random sample from the complete data set: e.g., several data values are missing because the technician dropped some test tubes.

- *missing at random* (MAR), which means there might be systematic (i.e., nonrandom) differences between the missing and observed values but these can be explained by other observed variables in the data set. For example, suppose clinicians using routine health records for their research data have missing data on blood pressure. They are concerned that people with missing blood pressure data might have lower blood pressure and be younger on average than those who have blood pressure reported. Older people or people with cardiovascular disease are more likely to have blood pressure recorded as part of their standard care but young healthy people are not. In other words there might be a systematic difference between the blood pressure values of the missing and observed data. However, this difference can be explained by other observed variables in the data set, viz., age and presence of cardiovascular disease.

- *missing not at random* (MNAR), which means that the reason an observation is missing on a given variable is related to the value of that variable and cannot be explained by other observed variables in the data. For example, in a study of adolescent smoking, data on the daily number of cigarettes smoked may be missing in those heavy smokers who are concerned they will get into trouble if they report on the survey questionnaire how many cigarettes they smoke. However, no other variables recorded by the researchers in the data set can be related to this possible cause of missingness.

Procedures for handling missing data, which are beyond the scope of this book, are determined by these "missingness" categories. References for missing data procedures include van Buuren (2012) and Berglund and Heeringa (2014). The MI procedure (multiple imputation) in SAS implements methods to handle missing data.

1.6 Estimating Population Parameters Using Sample Data

Definition 1.15. In statistical terminology *a parameter* is defined as a summary quantity (e.g., a mean) for a population.

Definition 1.16. A *statistic* is defined as a summary quantity (e.g., a mean) for a sample. Typically, researchers do not investigate an entire population but rather just a sample from the population. They therefore cannot compute a value for a population parameter. Instead they must estimate the value of a population parameter by using the value of a corresponding statistic computed from their data. For example, the value of the sample mean is used as a **point estimate** of the mean of the population from which the sample is drawn.

Definition 1.17. A *point estimate* of a population parameter is defined as a single value computed from a sample which estimates a population parameter.

Some Notation

We will mostly use the "hat" symbol notation throughout this book to identify a sample statistic, indicating that the statistic is an estimator of the corresponding population parameter. For example, $\widehat{\text{VAR}}(Y)$ is the notation we will use for the variance of Y sample statistic, which is an estimator of the population parameter $\text{VAR}(Y)$. However, in the case of means, we will follow the widely used convention of using the Greek symbol μ to refer to a population mean and the symbol $\bar{X}$ to refer to a sample mean.

Confidence Interval Estimation

Confidence interval estimation is another way of gaining information about the value of a population parameter based on sample data from that population. A confidence interval for a population parameter is an interval consisting of two endpoints that are estimated by a statistical procedure in which there is a specified level of confidence. The level of confidence is based on the coverage probability of the procedure, i.e., the probability based on the proportion of times that many intervals estimated by this procedure will include (cover) the population parameter, provided all assumptions of the procedure are met.

Typically 90%, 95%, or 99% confidence intervals are considered, with 95% being the most common. It is important to remember that the level of confidence associated with a confidence interval is a characteristic of the statistical procedure that produced the confidence interval. It is not the level of confidence associated with any one particular confidence interval generated by that statistical procedure. When we construct a 95% confidence interval using our sample data, there is not 95% probability that the interval we compute contains the true population parameter. The 95% probability refers to the probability that many, many intervals produced by the statistical procedure will contain the true population parameter provided all the assumptions of the procedure were satisfied. The 95% probability does not refer to any single confidence interval generated by the procedure. Therefore, we do not know whether or not the particular 95% confidence interval we construct based on our sample data contains the population parameter, as our confidence interval could belong to the 5% of those intervals generated by the procedure that does not contain the population parameter.

The width of a confidence interval can indicate how accurately the value of the sample statistic is estimating the value of its corresponding population parameter, assuming the assumptions of the interval-producing procedure are reasonably satisfied. For example, the width of a confidence interval for a researcher's sampled population mean μ can indicate how accurately the value of a sample mean is estimating the mean of the population from which the sample is drawn.

In subsequent chapters we will discuss confidence intervals for various population parameters in detail. In our discussions we will emphasize the underpinning assumptions (ideal inference conditions) and how to evaluate these assumptions. When one or more of these assumptions are not met, the real level of confidence of the procedure is unknown. We cannot be sure of the true coverage probability of the procedure as it will depend on the extent to which assumptions are violated and on the robustness of the procedure. In Chapter 14 bootstrapping is discussed as an approach for estimating an approximate confidence interval for a population parameter when assumptions of traditional methods are not satisfied.

1.7   Basic Steps of Hypothesis Testing

The following basic steps of hypothesis testing can be used to evaluate a research question regarding one or more populations when only sample data are available:

Step (1) Write down the specific research question you want to investigate. Include such details as what you want for the response (outcome) variable and what you want for the predictor (explanatory) variable(s).

Step (2) Translate this research question into a statement of a test hypothesis and an alternative hypothesis. A test hypothesis is typically referred to as a null hypothesis, i.e., the hypothesis a researcher wishes to nullify or refute.

For example, if you were interested in investigating whether there was **any** difference in the mean birth weight of populations A and B, then you would want to carry out a two-sided test and therefore your null hypothesis and alternative hypotheses would be

$$H_0 : \mu_A = \mu_B$$
$$H_A : \mu_A \neq \mu_B$$

Or equivalently these null and alternatives hypotheses can be written as

$$H_0 : \mu_A - \mu_B = 0$$
$$H_A : \mu_A - \mu_B \neq 0$$

This is called a two-sided test because the alternative hypothesis does not specify whether the difference between the two population means $\mu_A - \mu_B$ is positive or negative. It merely states the two population means are unequal or equivalently the difference between them is not zero.

This is in contrast to a one-sided test where the alternative hypothesis of the test specifies a direction for the difference. Examples of the null and alternative hypotheses for a one-sided test are

$$H_0 : \mu_A \leq \mu_B$$
$$H_A : \mu_A > \mu_B$$

and

$$H_0 : \mu_A \geq \mu_B$$
$$H_A : \mu_A < \mu_B$$

Step (3) On the basis of Steps (1) and (2), choose an appropriate hypothesis test.

Step (4) Check if the hypothesis test's assumptions (ideal inference conditions) are reasonable for your data and only proceed to Step (5) if they are reasonable. We will discuss possible courses of action for various situations where ideal inference conditions are not reasonable in Chapter 14.

Step (5) Calculate the value of your *observed test statistic*, i.e., the value of the test statistic calculated using your sample data (for example, calculate the value of the t-test statistic for your data, if you are doing a Student's t-test). SAS will do this step for you and will include this value as part of your output.

Step (6) Find the *observed p-value* associated with your observed test statistic (Definition 1.18). SAS also carries out this step and typically includes the observed p-value as part of your computer output.

Definition 1.18. The *observed p-value* of a hypothesis test, is **the probability that the observed test statistic** has a particular value or values more unlikely, **if** the observed test statistic were to belong to a specific probability distribution. This probability distribution assumes that the specified null hypothesis is true and that all ideal inference conditions of the particular hypothesis test are satisfied.

The phrase "or values more unlikely" in this definition for the observed p-value may be confusing to some readers. Why not just give the probability of **only** the value of the test statistic you observed for your sample? Why bother with the other values of the test statistic you didn't observe which are more unlikely when the null hypothesis is true. The reason is when the test statistic is a continuous variable (the situation encountered for methods presented in this book) there are an infinite number of values possible and thus the probability of any one of these possible values is so extremely small that it is equivalent to zero. However, it is possible to give the probability of a set of values within an interval for a continuous variable. In this case where the continuous variable is the test statistic of a hypothesis test, it makes sense to assign a probability to the set of values of the test statistic within the interval that includes the value of the observed test statistic as well as values of the test statistic that are more unlikely when the null hypothesis is true. This probability is what we call the p-value of the hypothesis test.

Step (7) As an additional step in hypothesis testing we recommend computing a point estimate and confidence interval estimate for an *effect size* that is relevant to the research question being evaluated.

Definition 1.19. An **effect size** is defined as a measure which indicates the strength of a relationship between variables.

The difference between two population means is an example of an effect size and the difference between two sample means from these populations is a point estimate of this effect size (i.e., an effect size statistic). The term "effect" in this context is meant in its broadest sense in that "effect" is referring to a change in a variable that is either caused by or associated with another variable.

Examples of effect size confidence intervals will be given in subsequent chapters.

Step (8) Step (8) Assuming ideal inference conditions of the hypothesis test were found to be reasonable for your data in Step (4), evaluate your research hypothesis by considering numerous relevant factors, including:

- substantive expertise

- details about the study's implementation, e.g., sample size, whether the sample was randomly selected from the target population (Section 1.4), whether the study was observational, a quasi-experiment, or a controlled randomized trial (Section 1.12)

- whether there is measurement error (Definition 2.1) in the outcome or predictor variable or both

- the extent of random variability among the observations on the outcome variable in the data

- whether the ideal inference conditions of the hypothesis test applied to the data are reasonable (Chapter 3 introduces general techniques for evaluating ideal inference conditions)

- whether there are any outliers in the data (Sections 3.2, 3.13)

- whether the observed test statistic is based on a few influential observations (Section 3.3)

- whether there are missing data and if so how they were handled (Section 1.5)

- the *robustness* of the hypothesis test applied (Definition 1.23, Section 1.11)

- details regarding other statistical analyses applied in your study related to the research question, i.e., multiple testing (Chapter 12)

- a correctly interpreted confidence interval estimate for an effect size relevant to the research question (see Section 1.6 for general comments regarding confidence intervals, Section 1.10 for caveats regarding interpretation of effect size confidence intervals, and Greenland et al. (2016) for common misinterpretations of confidence intervals)

- a correctly interpreted observed p-value (based on Definition 1.18) from the hypothesis test investigating the research question (Section 1.8 discusses caveats for interpreting a single p-value from a hypothesis test.)

We end this section with a quote from early pioneers of hypothesis testing, Neyman and Pearson (1928) who wrote:

The sum total of the reasons which will weigh with the investigator in accepting or rejecting the hypothesis can very rarely be expressed in numerical terms. All that is possible for him[1] is to balance the results of a mathematical summary, formed upon certain assumptions, against other less precise impressions based upon a priori or a' posteriori considerations. The tests themselves give no final verdict, but as tools help the worker who is using them to form his final decision.

1.8   Why an Observed p-Value Should Be Interpreted with Caution

Examination of equations for various test statistics reveals that factors such as the sample size and an estimate of random variability in the data are used in the computation of a test statistic. Therefore the observed value of the test statistic for the data and its associated p-value are related to these factors. For this reason, a large p-value, say $P = 0.80$, does not necessarily indicate that the null hypothesis is likely to be true. In fact, it could be that the null hypothesis is actually false, and the large observed p-value of 0.80 associated with the observed test statistic may be a consequence of small sample size or large random variability among observations on the outcome variable in the study or both.

It is always important to consider how well a study was designed when assessing the usefulness of an observed p-value in providing information about the compatibility of the observed test statistic with the null hypothesis. Suppose, a study was poorly designed. For example, suppose random sampling was not employed and a non-representative sample was drawn from the target population. The p-value associated with the observed test statistic from this study may not accurately indicate compatibility of the observed test statistic with the null hypothesis regarding the target population. This is because the observed test statistic was computed using biased sample data that were not representative of the target population.

It is also important to evaluate whether the assumptions (ideal inference conditions) of the hypothesis test are reasonable for the data when assessing the usefulness of an observed p-value in providing information about the compatibility of the observed test statistic with the null hypothesis. When the ideal inference conditions of the test are not met by the data, the observed (nominal) p-value for the test statistic may not be correct.

Even if ideal inference conditions are met, a low p-value from a hypothesis test does not necessarily imply that the test has detected a meaningful result with regards to the substantive nature of the research. Consider a study investigating whether Drug A was better than Drug B in reducing high systolic blood pressure. The null hypothesis and alternative hypotheses specified for the statistical test were

$$H_0 \; \mu_A = \mu_B$$
$$H_A \; \mu_A \neq \mu_B$$

---

[1] In 1928, unfortunately gender-neutral references to a worker were rare or possibly nonexistent.

Suppose this researcher obtained an observed p-value of 0.0001. This low observed p-value does not necessarily indicate that there is a large difference between the mean systolic blood pressure in the group getting Drug A and the group getting Drug B, even when ideal inference conditions are met. The low p-value could be a result of a huge sample size and/or low variability of blood pressure within the two groups. It could be that, although the test statistic was associated with a low p-value, the actual difference in mean systolic blood pressure between the two groups may, in fact, be really small, say 1.00 mm Hg, and not a meaningful difference from a biomedical viewpoint. However, if the researcher reports a confidence interval estimate for the relevant effect size, viz., the mean difference between the sampled population means, information regarding biomedical significance may be evaluated (but see caveats regarding confidence intervals in Section 1.10).

Previous and concurrent studies that are related to the research question of a study can provide important information about the null and alternative hypothesis. However, the computation of the test statistic of one hypothesis test does not take these other sources of information into account. Neither does the computation of the test statistic take into account any information about how well the study has been designed or about how reasonable the ideal inference conditions of the test are for the researcher's data. As noted in Step (8), Section 1.7, many factors have to be considered when drawing a conclusion about a research question evaluated by a hypothesis test. Thus, any substantive decisions, which are based solely on the p-value of the observed test statistic from a hypothesis test, without consideration of these other factors, may be unsound.

This section underscores some of the reasons why:

- a large p-value should never be interpreted as unequivocal evidence that the null hypothesis is true

- a small p-value cannot be interpreted as unequivocal evidence that the null hypothesis is false and the alternative hypothesis is true

An observed p-value may be interpreted as a measure of the conformity of the data with the statistical model underlying the hypothesis test, where one of the assumptions of the statistical model is the null hypothesis. A large p-value suggests that there is conformity of the data with the assumptions of the statistical model whereas a small p-value indicates the data is not in accord with these assumptions. As Greenland et al. (2016) pointed out, a small p-value does not tell us which particular assumption(s) of the statistical model is (are) unlikely. A small p-value does not necessarily indicate that the assumption of the null hypothesis is unlikely. A large p-value implies only that the data are not unusual under the assumptions of the statistical model. A large p-value does not imply that any of the model assumptions including the null hypothesis are correct.

For a discussion of "the dirty dozen" common misinterpretations of a single p-value, see Goodman (2008). Greenland et al. (2016) expanded on this topic in their guide to 14 common misinterpretations of a single p-value.

1.9   A World Beyond 0.05?

Too many incorrect research decisions have been made based solely on the dichotomy of whether an observed p-value from a hypothesis test exceeds or does not exceed a pre-specified threshold value, typically 0.05. Traditionally when an observed p-value was greater than 0.05 the test result was declared "not statistically significant" and when an observed p-value was less than 0.05 the test result was declared "statistically significant". Unfortunately a number of researchers have equated:

- the term "not statistically significant" with proof that the null hypothesis is true

- the term "statistically significant" with proof that the null hypothesis is false, the alternative hypothesis true, implying that a meaningful substantive test result has been detected

It is not difficult to see how incorrect decisions can be made with this threshold approach of p-values when one considers the caveats regarding p-values discussed in Section 1.8. In the American Statistical Association (ASA) statement on statistical significance and p-values, Wasserstein and Lazar (2016) wrote, "The widespread use of 'statistical significance' (generally interpreted as 'p ≤ 0.05') as a license for making a claim of a scientific finding (or implied truth) leads to considerable distortion of the scientific process".

A special issue of the *American Statistician* in 2019 (73:sup1) was devoted entirely to the perils of threshold p-values. Wasserstein et al. (2019) in the issue's editorial entitled *Moving to a World Beyond "p < 0.05"* wrote: "We conclude, based on our review of the articles in this special issue and the broader literature, that it is time to stop using the term 'statistically significant' entirely. Nor should variants such as 'significantly different', 'p < 0.05', and 'nonsignificant' survive, whether expressed in words, by asterisks, in a table, or in some other way".

In 2021 the ASA President's task force on statistical significance and replicability stated (Benjamini et al., 2021) that "Comparing p-values to a significance level can be useful, though p-values themselves provide valuable information. P-values and statistical significance should be understood as assessments of observations or effects relative to sampling variation, and not necessarily as measures of practical significance. If thresholds are deemed necessary as a part of decision-making, they should be explicitly defined based on study goals, considering the consequences of incorrect decisions".

While the 2021 task force statement and the editorial in the 2019 special issue of the *American Statistician* indicate controversy concerning the usefulness of threshold p-values, there is nevertheless widespread agreement within the statistical community that properly interpreting p-values with "thoughtfulness" (Wasserstein et al., 2019) is of paramount importance. To quote the 2021 task force (Benjamini et al., 2021): "In summary, p-values and significance tests, when properly applied and interpreted, increase the rigor of the conclusions drawn from data". In this book as we introduce each statistical method, we will emphasize its proper application and interpretation, in anticipation statistical thoughtfulness will flourish!

1.10 Caveats Regarding a Confidence Interval Estimate for an Effect Size

When interpreting a confidence interval estimate for effect size, it is important to bear in mind:

- Even if all statistical assumptions of the interval-producing procedure are met, when you construct a confidence interval for an effect size in your study, say a 95% confidence interval, there is not 95% probability that the particular interval you obtain based on your data will contain (cover) the true population effect size.

- If all statistical assumptions of an interval-producing procedure are not met, we cannot be sure of the true coverage probability of the procedure as it will depend on the extent to which these assumptions are violated and the robustness of the procedure.

- The width of a confidence interval is inversely related to sample size, that is the larger the sample size the narrower the confidence interval.

Suppose the effect size being considered is the population mean difference between two groups (μ_A minus μ_B) and the 95% confidence interval estimate for this effect size is (-6.22, 4.88). This interval estimate suggests zero is a plausible value for the population mean difference as zero lies between -6.22 and 4.88. However this interval estimate also suggests that any of the other values that lie between -6.22 and 4.88 are also plausible. For example -5.01 is a plausible value for the population mean difference suggesting that $\mu_A < \mu_B$ or 4.80 is a plausible value for the population mean difference suggesting that $\mu_A > \mu_B$. Thus we would conclude from this confidence interval that there is insufficient evidence in the data to draw a conclusion regarding the difference between the population means. We cannot conclude that there is no difference in the populations means from this confidence interval. Perhaps if we had larger sample sizes in the two groups we might have ended up with a 95% confidence interval estimate for the population mean difference that was not so wide and did not include zero, e.g., (1.64, 3.04) instead of (-6.22, 4.88). Thus, sample size must be taken into consideration when interpreting a confidence interval estimate for an effect size.

- The width of a confidence interval is related to the random variability among observations on the outcome variable in the data: the greater the random variability the wider the confidence interval.

Thus, reducing random variability through the design of a study, e.g., holding nuisance variables (Definition 1.24) constant and/or through appropriate statistical modelling, e.g., analysis of covariance (Chapter 13) will result in a narrower confidence interval estimate for a study's effect size.

- The width of a confidence interval is related to the percentage level of confidence interval, i.e., the greater the percentage level of confidence interval, the wider the interval.

If you construct a 99% confidence interval for an effect size based on your study data, it will be wider than if you construct a 95% confidence interval. Thus the range of plausible values for the effect size suggested by a 99% vs. a 95% confidence interval for the same data differ. Consider the situation where a 99% confidence interval for a population mean difference between two groups includes zero whereas the narrower 95% confidence interval applied to the same data excludes zero. In this situation, the 99% confidence interval does not preclude the possibility that zero may be a plausible value for the difference between the population means whereas the 95% confidence interval obtained for these same data which excludes zero suggests that the population means may be different. In other words, in this example different possibilities are suggested for the same research data depending on whether the researchers report a 99% vs. a 95% confidence interval.

1.11 Type I and Type II Errors, Power, and Robustness

1.11.1 Type I and Type II errors

Type I and *Type II* errors are two possible types of errors you can make in drawing a conclusion about a null hypothesis.

Type I Error

Definition 1.20. *Type I error* is the mistake you make when you conclude a null hypothesis is false when it is actually true.

For example, suppose your null hypothesis is that the mean birth weight of population A is equal to the mean birth weight of population B and the alternative hypothesis is the two population means are not equal, i.e.,

$$H_0 : \mu_A = \mu_B$$
$$H_A : \mu_A \neq \mu_B$$

You would commit a Type I error if you make the mistake of concluding that the mean birth weight of population A is not equal to the mean birth weight of population B, although these two population means are actually equal to one another.

The probability of Type I error is usually denoted by the Greek letter alpha, i.e., α.

Type II Error

Definition 1.21. *Type II error* is the mistake you make in **not** rejecting a null hypothesis when it is actually false. Consider the previous example where

$$H_0 : \mu_A = \mu_B$$
$$H_A : \mu_A \neq \mu_B$$

You would commit a Type II error if you did not reject the null hypothesis, when the null hypothesis is false and the alternative hypothesis is true, i.e., these two population means are actually different.

The **probability** of a Type II error is usually denoted by the Greek letter beta, i.e., β. The researcher does not directly set the value of β. The value of β is a function of a number of factors, including the effect size, the sample size, and the variance of the population(s) sampled.

Power of a Hypothesis Test

Definition 1.22. The *power* of a hypothesis test is the probability of rejecting the null hypothesis when it is not true, i.e., power is the probability of **not committing a Type II error** and is given by

$$\text{Power} = 1 - \beta$$

where β is the probability of a Type II error

Because power is a probability, the maximum power you can attain with a test is 1.00 and the minimum is 0. The calculation of the power of a hypothesis test requires specification of values for the parameters involved. Such values are almost always unknown, so typically in a power calculation one uses likely values (educated guesses) for such parameters in the target population(s).

Robustness of a Statistical Procedure

Definition 1.23. *Robustness* is a term relating to the extent that the result of a statistical procedure is affected by departures from the procedure's ideal inference conditions (assumptions).

If a procedure is said to be highly robust, its result is not affected very much by departures from the test's assumptions. Simulation research in statistics provides evidence about the robustness of various statistical procedures. Wicklin (2013) illustrates how this simulation research can be implemented in SAS.

1.12 Basic Types of Research Studies

It is important to distinguish between three basic types of research studies because the interpretation of statistical results is fundamentally different for each type of study. These three basic types are:

- observational studies

- controlled randomized trials, also known as randomized clinical trials in biomedical science

- quasi-experiments

All three study types ideally involve random sampling of subjects. However, a controlled randomized trial, as its name suggests, is distinguished from an observational or a quasi-experimental study because subjects are randomized, i.e., assigned at random to the "treatment" groups in a controlled randomized trial whereas subjects are not assigned at random in the other two study types. It is important to note that randomizing subjects to a treatment is carried out via a formal statistical procedure which typically involves first labelling every subject in the trial with some sort of identification number and then referring to a random numbers table or to a random numbers generator to determine which subject is assigned to which treatment group. When subjects are randomized to the treatment groups, it is possible to draw a cause and effect type of conclusion in a controlled randomized trial, provided the trial is well designed and an appropriate statistical analysis is applied.

To illustrate key differences among these types of research designs, we consider hypothetical studies that investigate calcium supplementation and heart attack risk (subsections 1.12.1 to 1.12.2). Calcium supplementation and heart attack risk is of interest because calcium supplements raise serum calcium levels, possibly accelerating arterial calcification, which can increase heart disease risk (Reid et al., 1986; Pletcher et al., 2004; Bolland et al., 2008).

1.12.1 Observational Studies

Suppose researchers conducted an observational study to investigate calcium supplementation and heart attack risk. They took a random sample of individuals from a specified population, say, women 65 years old or older, living in Winnipeg, Canada in 2018 and classified the individuals in the sample into one of two groups on the basis of whether they reported they are taking calcium supplements or not. The researchers then followed these individuals for five years and observed the heart attack rate in each group.

A difficulty with this type of study is that the subjects belonged to these groups (calcium supplementation, yes or no) before the study began. Therefore, the two groups may differ in ways other than whether or not they took calcium supplements. The researchers cannot rule out the possibility that the individuals in the calcium supplementation group have a different heart attack rate than those in the no calcium supplementation group because some other factor or combination of factors associated with taking a calcium supplement is also associated with having a heart attack. For example, it may be those people in the study who report they are taking calcium supplements are more likely to have a different socio-economic status, exercise less because they have osteoporosis, consume more saturated fat, smoke, etc., than the people who do not take calcium supplements. The researchers could record some of these *"nuisance" variables* (Definition 1.24) and try to control for them in a statistical analysis.

Definition 1.24. *A nuisance variable* is a variable that is associated with the outcome variable and the explanatory variable but is not of direct interest to the research question being evaluated (e.g., socioeconomic status could be a nuisance variable in the calcium supplementation study described above).

As we will discuss later, the size of the sample limits the number of nuisance variables that can be reasonably controlled for in a statistical analysis. A key point is that there may be other factors unknown to the researcher that are associated with heart attacks and calcium supplementation that have not been taken into account. Thus, no matter how well the researchers design this observational study, one may never be able to conclude that calcium supplementation **causes** a change in heart attack rate. From this observational study, one can conclude only that calcium supplementation may be **associated** with a change in heart attack rate.

Another difficulty with this observational study is that the researchers are only observing and not administering the "treatment" (explanatory) variable, calcium supplementation. As a consequence, there may be a lot of variation among the individuals in the calcium supplement group such as variation in: the total daily dose of calcium supplement; how many times a day the supplement is taken; whether it is taken consistently every day; whether it is taken with a meal; whether the supplement is calcium carbonate or calcium citrate. In later chapters we will discuss how variation within a treatment group can result in a researcher's not being able to detect a true population difference between treatment and control.

1.12.2 Controlled Randomized Trials

Example

An example of a controlled randomized trial investigating calcium supplementation and heart attack risk could be a study where researchers take a random sample from a specified population, say, women 65 years old or greater, living in Winnipeg, Canada in 2018 and then randomize each individual in the sample to one of two treatment groups, one group which is prescribed a dose of 500 mg of calcium citrate to be taken daily at the evening meal and a control group that is prescribed a placebo to be taken daily at the evening meal. These individuals are followed and the heart attack rate is recorded after five years. Because the individuals are assigned to either group at random, each individual in the study has an equal chance of being in either group. This means any differences occurring between the groups at the beginning of the study are random occurrences which can be taken into account in the statistical analysis. Thus, at the beginning of the study both treatment groups are statistically comparable and therefore a cause and effect conclusion should be possible with this type of study. For example, if one observes that the calcium supplementation group has a higher heart attack rate than the placebo group, then one may be able to conclude that calcium supplementation is the cause, provided the experiment was well designed in all aspects and an appropriate statistical analysis was applied. However, a cause and effect type of conclusion cannot be automatically be made in a controlled randomized trial, as difficulties may arise such as noncompliance to the treatment prescribed. For example, some individuals in the calcium supplementation and placebo groups may have sometimes forgotten to take their prescribed pill daily as directed by the protocol.

There is some debate as to how to handle those individuals who are randomized to a treatment but end up for one reason or another not getting the treatment. Should they be excluded from the statistical analysis as they didn't actually get the treatment? Excluding them from the analysis, however, may introduce bias and undo the beneficial effects of the randomization process. This could occur as the individuals excluded from the treatment group in the analysis may represent a type of individual that has not been excluded in the placebo group and this would mean that the two groups are no longer statistically comparable. A conventional approach to this problem is to include non-compliant individuals in the treatment group in the analysis, although they did not actually receive the treatment. This approach is known as "intention to treat". One argument for this approach is that with an "intention to treat" analysis, one is able to validly evaluate the effect of **prescribing** the treatment (or intervention program) even though one cannot not necessarily evaluate the effect of the treatment per se.

We will not discuss in depth potential difficulties that may arise in randomized clinical trials, that may include noncompliance, irregular compliance, withdrawal of subjects from treatment, loss of subjects to follow-up, and random baseline covariate imbalance, i.e., imbalance of subject prognostic characteristics occurring by chance among the treatment groups before treatment is applied. However, we would like to raise awareness that the details of a randomized trial have to be carefully reviewed before one can make a cause and effect type of conclusion.

1.12.3   Quasi-experiments

Example

An example of a quasi-experiment investigating calcium supplementation and heart attack risk could be a study that was designed exactly the same way as the controlled randomized trial described in subsection 1.12.2 except there was no randomization of the individuals to the treatment groups. The researchers obtained a random sampling from the population of women, 65 years old or greater, living in Winnipeg in 2018. The researchers wanted to study two treatment groups, one group to which they prescribed a dose of 500 mg of calcium citrate to be taken daily at the evening meal and the other group to which they prescribed a placebo to be taken daily at the evening meal. The individuals in these two groups were followed and the heart attack rate was recorded in each group after five years. However, unlike a randomized study, the researchers did not use a formal statistical procedure to assign the individuals at random to either the treatment or placebo. For example, suppose they obtained a list of 100 individuals who were randomly sampled from the population but they sorted this list in alphabetical order by surname for logistic expediency and assigned the first 50 individuals on the list to calcium supplementation and the remaining 50 to the placebo treatment. A cause and effect type of conclusion cannot be made from this study, no matter how well it was designed in every aspect other than lack of randomization to treatment. Without randomization, one cannot be sure that some other factor or combination of factors associated with alphabetical order of surname is also associated with heart attack rate.

1.13 Why Not Always Conduct a Controlled Randomized Trial?

Why you might ask wouldn't investigators always choose to conduct a controlled randomized trial to evaluate their research hypothesis? The answer is that randomizing individuals to the treatment groups is not always ethical, possible, or practical. For example, it would be unethical to assign an individual to a smoking group, because smoking has been shown in observational studies to be strongly associated with lung cancer as well as with many other diseases. An example where a controlled randomized trial is not possible would be if a researcher wanted to compare the effect of the sex of a fish on a particular outcome. In this situation it is not possible to assign at random an individual fish to the "treatment group" of being male or female. An example of an impractical controlled randomized trial could be where a researcher wanted to investigate the effect of winning a huge lottery on a particular outcome. In a controlled randomized trial this would involve the researcher's assigning individuals at random to one of two groups, (i) the group where each individual would not win or (ii) the group where each individual would win a huge lottery, which would have to be given by the researcher.

1.14 Importance of Screening for Data Entry Errors

Screening research data for data entry errors before applying a statistical analysis is essential. It eliminates the loss of valuable time which inevitably occurs if a data entry error is detected after or during the course of a statistical analysis. If a data entry error is not detected during the analysis, then faulty conclusions based on the analysis of erroneous data may result.

Most data processing outsourcing companies provide verification of data entry as part of their services. However, when a data set is small, it is often entered manually by researchers or students, so in-house screening for data entry errors becomes important. One way to accomplish this is to get a listing of the data input using the PRINT procedure in SAS and then check this listing against the original data records.

Sometimes the nature of data collection involves a measurement device that digitally records measurements which can be directly input into a computer thereby bypassing manual input and the concern of data entry errors. However, even when data are input digitally from a measurement device, it can still be useful to get a listing of at least a few cases of the data input into SAS so you can verify that your SAS program has correctly read in the data. Program 1.1, in Section 1.A, is an example of a SAS program that creates a SAS data file using data from Example 1.1 in Section 1.15 and lists the data input.

1.15 Example 1.1: Environmental Impact Study of a Mine Site

Fish often serve as sentinels of environmental contamination, somewhat analogous to a canary in a coal mine. The environmental impact of a mine site on a creek was investigated by examining tissue levels of metal contaminants in a species of fish that were sampled upstream and downstream from the site. The mine site is located in an isolated area with no roads leading into the site. Researchers had to fly in by helicopter to collect the samples, which was quite costly. Budgetary restrictions precluded a longitudinal study and resulted in small sample sizes collected during a single field trip.

One of the metal contaminants studied was arsenic and a question of interest was whether the level of arsenic in a fish's tissue was related to the total length of the fish. We will use a subset of the arsenic and total length data to illustrate an important first step in any data analysis, i.e., checking whether the data were correctly input into the statistical program. This can be accomplished by getting a listing of the data values read into the computer program and seeing if these data values agree with those on the original data records. The SAS Program 1.1 (Section 1.A of this chapter's appendix) inputs the data values for arsenic and total length of the fish and then lists the values that were input. Before explaining the details of Program 1.1, we first will go over some commonalities that apply to all SAS programs.

1.16   Some SAS Basics: Key Commonalities that Apply to all SAS Programs

This section gives some basic commonalities that apply to all SAS programs. This is by no means an exhaustive list, just some key points to serve as an initial introduction:

- A SAS statement (which consists of keywords, SAS names, special characters, and operators) either provides information or specifies an operation to perform in a SAS program.

- Every SAS statement must end in a semi-colon.
  This is one of the most important things to remember when writing a SAS program. When a semicolon is inadvertently left out, you will often get a strange error message that does not indicate you have left out a semicolon. So if you do get an inexplicable error message in your SAS log, always first check to see if you have left out a semicolon at the end of any of the statements in your SAS program prior to the place where the SAS log indicates an error has occurred.

- SAS is not case-sensitive in that upper and lower case are both acceptable in most SAS statements. However, the case you use for a variable name for the first occurrence of that variable in your program will be the case that SAS will use for that variable name in the program's output. In Program 1.1 we wanted our variables to be displayed in upper case in the program's output, we therefore used upper case when specifying those variables in the INPUT statement.

- Any number of blank lines may occur before or after a SAS statement or comment.

- More than one SAS statement may be written on a line. However, some individuals find writing one statement per line often results in fewer programming errors and facilitates detecting errors if they do occur.

- Comments can be written anywhere in a SAS program.
  A **line comment** begins with an asterisk * and ends with a semi-colon, for example:
  * This program lists the data of Example 1.1. ;

A **block comment** allows one or more lines to be written between the /* and */ symbols, for example:

/* Function of Program 1.1 is listing of data
This program was written by John Doe, January, 2018
Last modified January, 2020 */

Note: Unlike a line comment, a block comment does not require a semi-colon at its end.

We recommend using comments in a SAS program to document your steps. Using comments is especially important if your program is to be shared with others or if you plan to go back to the program at a later point in time.

- A **SAS data set**, as defined at sas.com, "is a SAS file stored in a SAS library that SAS creates and processes. A SAS data set contains data values that are organized as a table of observations (rows) and variables (columns) that can be processed by SAS software". An example of a SAS data set is given in Program 1.1.

- A typical SAS program consists of a **DATA step** and one or more **PROC step(s)**.

- A **DATA step** is a group of SAS statements used to read in the raw data, assign SAS variable names to the data, create new variables, select a subset of the data, write a report, write to an external file or create a SAS data set.

- A **DATA statement always begins a DATA step.** This statement provides a user-supplied name for the SAS data set created in the DATA step.

- A **PROC step** is a group of SAS statements that specifies the descriptive, statistical, and graphics analyses you would like to perform. The format of SAS statements in a PROC step will vary according to the procedure to be used.

- A **PROC statement always begins a PROC step.** This statement specifies the SAS procedure you would like to use. By default a SAS procedure will use the most recently created SAS data set as input. However if you would like to use a different data set as input other than the one most recently created, you can indicate this via the DATA=*dataset name* option in the PROC statement. Some programmers recommend using the DATA=*dataset name* option as standard practice and note that this option can be especially helpful in debugging complicated programs where more than one data set is created. As well as the DATA=*dataset name* option, other procedure options can be specified in the PROC statement.

- SAS assumes by default that the user-supplied name for a SAS variable or a SAS data set, will conform to the following rules:

    – The first character of a user-supplied name must be a letter of the Latin alphabet in upper or lower case (A, B, ..., Z, a,b, ...z) or the underscore character. Subsequent characters can be Latin alphabet letters, numeric digits (0, 1, ..., 9), or underscores.

    – No blanks or special characters except for the underscore can be embedded in a user-supplied name.

    – The length of a user-supplied SAS name is determined by what is being named in SAS. Many user-supplied names can be 32 characters long whereas others have a maximum character length of eight bytes. SAS online documentation (see Language Reference: Concepts) provides a list of the maximum length of user-supplied names for specific types of SAS language elements.

– The name can contain mixed-case letters. However, when processing a variable name SAS converts it to upper case so that the user-supplied variable names MINESET, mineset, Mineset all represent the same variable. Therefore you cannot represent different variables by using a different combination of upper and lowercase letters.

NOTE: Some of these rules for user-supplied names for SAS variables names and SAS data sets can be circumvented by respectively specifying VALIDVARNAME=ANY option and the VALIDMEMNAME=EXTEND option as documented at sas.com.

1.16.1   Temporary and Permanent SAS Data Sets

Temporary and permanent SAS data sets can be created by a SAS program according to the specifics of the operating environment. It is important to note that a temporary SAS data set exists only for the duration of your current SAS session in most operating environments. Therefore if you try to execute a program that uses a particular temporary SAS data set as input and you have exited the SAS session in which you created this temporary SAS data set, in most operating environments, you will encounter error messages and your current program will not run. To avoid this problem you will have to re-create the particular temporary SAS data set you currently require. One way to do this is by simply running, in your current SAS session, the original program that created this particular temporary SAS data set. Alternatively you can copy the DATA STEP statements and data from the original program that created the particular temporary SAS data set and pasting these statements and data at the beginning of your current program. If, however, you prefer to create a permanent SAS data set, consult the specifics for your particular operating environment at sas.com. For a helpful introduction to data file management such as creating permanent SAS data sets, and importing and exporting external files from other software (e.g., Excel), see Goad (2020).

1.17   Chapter Summary

This chapter:

- describes the characteristic features of a regression analysis

- defines variable classifications and characterizes the methods to be presented in this book on the basis of these classifications

- describes probability distributions, in particular the normal distribution

- discusses the following key concepts that are relevant to the interpretation of results from any statistical analysis

    – the distinction between a target population and a sampled population
    – generalization of results
    – hypothesis testing
    – why p-values have to be viewed with caution
    – caveats regarding confidence interval estimation

- – Type I and Type II errors
- – the power of a hypothesis test
- – robustness of a hypothesis test
- – missing data
- – the distinctions between an observational study, a quasi-experiment, and a controlled randomized trial

- emphasizes the importance of data editing before applying a statistical analysis

- describes some commonalities that apply to all SAS programs and illustrates with a simple SAS program

Appendix

1.A Program 1.1

```
title Tissue Level of Arsenic and Total Length of Fish;
data minesite;
input RECORD AS TL;
label RECORD=Record Number
      AS=Arsenic
      TL=Total Length;
datalines;
797     0.16     72
798     0.20     81
799     0.16     74
777     0.28     77
778     0.18     73
779     0.28     79
780     0.20     84
781     0.30     90
782     0.18     71
783     0.29     79
784     0.26     78
785     0.34     64
786     0.23     73
787     0.24     63
788      .       62
789     0.24     55
790     0.23     59
791     0.28     48
792     0.25     42
793     0.35     49
794      .       45
795      .       36
796      .       37
;
proc print data=minesite label;
run;
```

1.A.1 Explanation of SAS Statements in Program 1.1

The TITLE statement specifies the title line(s) you want for a SAS program and its output. A TITLE statement can be used anywhere in a SAS program.

data minesite; The DATA statement in Program 1.1 assigns the file name MINESITE to the temporary SAS data set that is created in this DATA step.

NOTE: Every SAS dataset, whether temporary or permanent, has a two-level name of the form *fileref.filename* where fileref is the name of the library where the file is stored. However, typically a temporary SAS data set is stored in the SAS library called WORK by default. Therefore a temporary SAS dataset is usually referenced only by a one-level name, the file name you assign (e.g., mineset in Program 1.1).

The INPUT statement specifies the names to be used to identify the input variables as well as where to find the values for these variables on each data record.

Program 1.1 uses the simplest form of data input known as **simple list input**. The INPUT statement for simple list input provides a list of user-supplied variable names corresponding to the order that the variables are found on the input record for each observation. The INPUT statement in Program 1.1 informs SAS that the first value encountered on each data record is the ID number of the fish, the second value on the data record is the value for the variable AS (the user-supplied name for the variable, arsenic tissue level) and the last value encountered on the data record is the value for the variable TL (the user-supplied name for the variable, total length of the fish).

When simple list input is used to read in your data the following restrictions apply:

- The names of the variables should be specified in the INPUT statement in the same order that appears for these variables in the input data records.

- The values for the variables are always separated by at least one blank (space) unless a delimiter other than a blank is specified as an option on an INFILE statement. (See Data Step Statements: Reference at sas.com for details regarding the INFILE statement.)

- Missing values in the data are identified by a period (.) unless a character other than a period is specified in a MISSING statement that typically appears in a DATA step. (See Global Statements at sas.com for details regarding the MISSING statement.)

- Character values do not have embedded blanks unless a delimiter other than a blank has been specified in the INFILE statement

- Character values for each variable are not longer than eight bytes unless specified in a previous LENGTH, ATTRIB, or INFORMAT statement. (See Data Step Statements: Reference at sas.com for details regarding these statements.)

- Data values are in standard numeric or character format. (See SAS Language Reference: Concepts at sas.com. for information about standard format.)

The INPUT statement also indicates whether a variable's possible values are numeric or character or both. SAS assumes by default that a variable has only numeric values. If a variable has character values or a mixture of numeric and character values you indicate this in the INPUT statement by following the variable name with a dollar sign ($). The variables in Program 1.1 have only numeric values.

A LABEL statement in a DATA step assigns a descriptive label to a variable which can be used instead of a variable name by certain procedures. As illustrated in Program 1.1, an equals sign (=) is required after the variable name and then everything which follows the equals sign is the label for that variable until the next variable is encountered.

A DATALINES or LINES statement informs SAS that data lines immediately follow. Program 1.1 is an example of reading in raw data from "instream data lines", i.e., from raw data which has been typed into the program itself. When a DATALINES statement is used, it is important not to type a semicolon at the end of any data line as SAS will interpret a data line with a semicolon as a command statement.

As illustrated in Program 1.1 when a DATALINES or LINES statement is used, a null statement (i.e., a single semicolon on a line) signals the end of the data lines and executes the DATA step.

The PROC PRINT statement requests the PRINT procedure. The DATA= option specifies that the data set named MINESITE is to be printed. The PRINT procedure offers an easy way to get a listing of your input data for the purpose of data screening.

The LABEL option in the PROC PRINT statement allows the possibility to create user-friendly output as this option specifies that variable labels rather than variables names should be used as column headings by the PRINT procedure.

In Program 1.1 by default the PRINT procedure will output a list of the values of all the variables found in data set called MINESITE. If a list of just some of the variables in the data set is wanted, you can indicate this by specifying those variables separated by at least one blank in the VAR statement. For example, if a listing of only the ID and TL variables were wanted, you would indicate this as follows:

```
proc print data=minesite;
var id tl ;
run;
```

If you don't want to print all the records in the data you can restrict the number of records printed by the (obs=*n*) option where *n* is the number of records you want printed. For example you can request that only the first five records from the data set MINESITE be printed as follows:
```
proc print data=minesite (obs=5);
```

The RUN statement in Program 1.1 tells SAS to execute the statements in the preceding PROC step.

1.B   The SAS Log

After you have run your SAS program, in addition to any output requested, a SAS log will be produced, which documents whether all aspects of your SAS program have been successfully executed and if not will give error message(s). These error messages are usually quite helpful. However, as previously mentioned, if an execution error has resulted from a semi-colon being omitted from the end of a SAS statement or a line comment, the error message can be confusing.

The SAS log also reports the number of observations and the number of variables in each data set created by your SAS program. It is always important to examine your SAS log before examining the results output by your program. In some situations, although results are output, they may be invalid as indicated by information reported in the SAS log. The SAS log resulting from submitting Program 1.1 is as follows:

SAS Log for Program 1.1

```
1      title Tissue Level of Arsenic and Total Length of Fish;
2      data minesite;
3      input RECORD AS TL;
4      label RECORD=Record Number
5            AS=Arsenic
6            TL=Total Length;
7      datalines;

NOTE: The data set WORK.MINESITE has 23 observations and 3 variables.
NOTE: DATA statement used (Total process time):
      real time            0.01 seconds
      cpu time             0.01 seconds


31   ;
32   proc print data=minesite label;
33   run;

NOTE: There were 23 observations read from the data set WORK.MINESITE.
NOTE: PROCEDURE PRINT used (Total process time):
      real time            0.46 seconds
      cpu time             0.20 seconds
```

1.C   Listing of Data in the SAS Data File MINESET

The following is a listing of the values of all the variables in the SAS data file MINESET obtained from Program 1.1. The first variable in the output list, Obs, is supplied by SAS and identifies each data record with a number indicating the order that data record was input in the data step of the program, e.g., the data record for the fish with ID=797 was input first. You can omit the OBS column by specifying the option NOOBS of the PROC PRINT statement, e.g., proc print data=mineset label noobs;

Tissue Level of Arsenic and Total Length of Fish

Obs	ID	Arsenic	Length
1	797	0.16	72
2	798	0.20	81
3	799	0.16	74
4	777	0.28	77
5	778	0.18	73
6	779	0.28	79
7	780	0.20	84
8	781	0.30	90
9	782	0.18	71
10	783	0.29	79
11	784	0.26	78

12	785	0.34	64
13	786	0.23	73
14	787	0.24	63
15	788	.	62
16	789	0.24	55
17	790	0.23	59
18	791	0.28	48
19	792	0.25	42
20	793	0.35	49
21	794	.	45
22	795	.	36
23	796	.	37

2

Introduction to Simple Linear Regression

2.1 Characteristic Features

Characteristic features of a simple linear regression model are:

- There is only one response (outcome, dependent, Y) variable in the model.

- The response variable is continuous.

- There is only one explanatory (predictor, independent, X) variable in the model.

- The explanatory variable is typically a continuous or interval variable.

- The model assumes that a linear relationship exists between the response variable and the explanatory variable in the model.

2.2 Why Use Simple Linear Regression?

You would want to use a simple linear regression to:

- **evaluate whether a single explanatory (predictor) variable in a simple model can explain any variation in a response (outcome) variable** in the target population, for example, to evaluate whether age (an explanatory variable) can explain variation in systolic blood pressure (a response variable) in the target population.

- **predict the value of a response variable for an individual from the target population based on the value of that individual's explanatory (predictor) variable**, for example, to predict the systolic blood pressure (SBP) of a 65-year-old individual from the target population

- **predict the average value of a response variable for individuals in the target population who have the same value for the explanatory variable**, for example, to predict the average SBP for individuals who are sixty-five years old in the target population

- **predict change in a response variable's value when the value of an explanatory variable is changed.** For example, suppose in a Japanese quail population, there is a positive linear relationship between daily body weight gain (the response variable) and the amount of food in grams given daily (the explanatory variable). One can estimate from a simple linear regression model how much the amount of food in grams given daily should be increased to achieve a desired weight gain in the quail.

DOI: 10.1201/9780429107368-2

2.3   Example 2.1: Systolic Blood Pressure and Age

This example investigates whether age is linearly predictive of SBP in men, i.e., whether it is reasonable to use a single straight line to model the relationship between SBP and AGE in men when SBP is the response (outcome) variable and AGE is the explanatory (predictor) variable. Figure 2.1 displays these data in a scatter plot along with a simple linear regression line estimated from these data. This scatter plot was generated by Program 2.1 (Section 2.A) using the REG procedure of the SAS/STAT software.

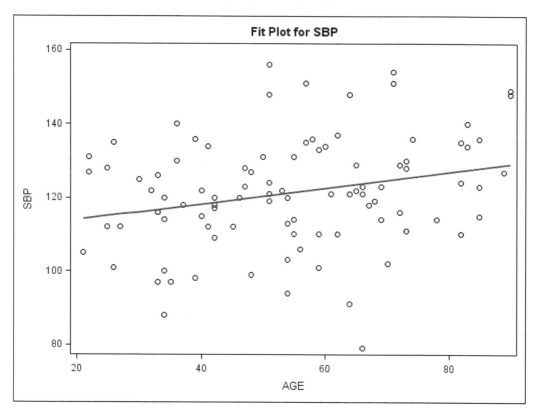

FIGURE 2.1
Scatter plot of observed data values for Example 2.1 with estimated simple linear regression line. SBP denotes systolic blood pressure.

2.4   Rationale Underlying a Simple Linear Regression Model

Typically, a response value of an individual cannot be perfectly predicted by a single explanatory variable. Consider the scatter plot in Figure 2.1, where SBP is the response (Y) variable and AGE is the predictor (X) variable. It is obvious from this scatter plot that an individual's age does not perfectly predict that individual's SBP because for each age, there is not one single SBP value but a number of different values. It is also obvious from Figure 2.1

that it is impossible to draw a single straight line that will pass through all the data points. Therefore, in simple linear regression we adopt a more plausible model which makes the assumption that **a single straight line can be drawn through the population means of the response variable, Y, at each distinct value of X**. This assumption, known as the **linearity assumption**, will be further discussed in Section 2.11.

2.5   An Equation for the Simple Linear Regression Model

An equation for the simple linear regression model is

$$\mu_{Y|X} = \beta_0 + \beta_1 X \tag{2.1}$$

where

$\mu_{Y|X}$ denotes the mean value of the response variable, Y, for all the individuals in the sampled population that have the same value of the explanatory variable, X. In other words, $\mu_{Y|X}$ denotes the population mean of Y given X. (The vertical bar in the notation $\mu_{Y|X}$ stands for the word given.)

β_0 is the Y intercept of the straight line that describes the linear relationship between $\mu_{Y|X}$ and the explanatory variable X in the sampled population.

β_1 is the slope of the straight line that describes the linear relationship between $\mu_{Y|X}$ and the explanatory variable X in the sampled population. Thus, β_1 is the change in the sampled population mean of Y for each unit change in X. Often β_1 is called the regression coefficient for X.

This simple linear regression model equation is an example of a model that is **linear in the parameters**. A model that is linear in the parameters has the basic form such that:

- Each explanatory variable in the model is multiplied by a single unknown parameter.

- There is at most one unknown parameter with no corresponding explanatory variable.

- All of the terms in the model are added together to give the value of the function. For example, in Equation 2.1, all the terms are added together to give the value of $\mu_{Y|X}$.

The simple linear regression model equation for the sampled population of the SBP example can be written as

$$\mu_{SBP|AGE} = \beta_0 + \beta_1 AGE \tag{2.2}$$

where

$\mu_{BP|AGE}$ denotes the mean SBP of all the individuals in the sampled population who have the same age

β_0 is the Y intercept of the straight line that describes the linear relationship between $\mu_{SBP|AGE}$ and AGE in the sampled population

β_1, the regression coefficient for AGE in the sampled population, is the change in mean SBP for each unit change in AGE in the sampled population

2.6   An Alternative Form of the Simple Linear Regression Model

A simple linear regression model for a sampled population can be alternatively expressed in terms of an individual's response value instead of $\mu_{Y|X}$ as in Equation 2.1. The equation for this alternative form of simple linear regression model is

$$Y_i = \beta_0 + \beta_1 X_i + \epsilon_i \qquad (2.3)$$

where ϵ_i is commonly referred to as the **model error** or the **error term** or the **error component** for individual i. In simple linear regression

$$\epsilon_i = Y_i - (\beta_0 + \beta_1 X_i) \qquad (2.4)$$

We can think of ϵ_i as the part (or component) of individual i's Y value that is not explained by the simple linear regression model.

For the SBP example (Example 2.1), this alternative form of the simple linear regression model equation can be written for individual i as

$$SBP_i = \beta_0 + \beta_1 AGE_i + \epsilon_i \qquad (2.5)$$

where

SBP_i is the observed value of SBP for individual i

β_0 is the Y intercept of the straight line that describes the linear relationship between $\mu_{SBP|AGE}$ and AGE in the sampled population

β_1, the regression coefficient for AGE in the sampled population, is the change in mean SBP for each unit change in AGE in the sampled population

AGE_i is the age of individual i in years

ϵ_i, the model error for individual i, is the part of individual i's SBP value that is not explained by the simple linear regression model which has AGE as an explanatory variable

2.7   How Simple Linear Regression is Used with Sample Data

The equations for predicting a mean and an individual Y value for a particular X value (Equations 2.1 and 2.3) describe population models. Typically, the values of the parameters of these models, β_0 and β_1, are unknown because research is usually conducted on only a sample from the sampled population and not on the entire population. You can nevertheless use simple linear regression when only sample data is available by:

- using the sample data to estimate values for the parameters of the model for the sampled population, viz., β_0 and β_1

- substituting these estimated values in the estimated simple linear regression model equations

Estimates of the sampled population parameters β_0 and β_1, denoted respectively by $\hat{\beta}_0$ and $\hat{\beta}_1$, can be obtained by applying **the methodology of ordinary least squares** to a sample from that population. Ordinary least squares (OLS) methodology can be applied without making any assumptions other than the model errors have a mean equal to zero.

The method consists of choosing values for the estimates of β_0 and β_1 according to the criterion that the sum of the squared vertical distances of the sample points (x,y) from the regression line defined by those estimated values for β_0 and β_1 is at a minimum. The estimates of β_0 and β_1 chosen by using this criterion are often referred to as the **least squares estimates** of β_0 and β_1.

The least squares estimates of β_0 and β_1 are given by the REG procedure. The following statements direct the SAS/STAT software to apply a simple linear regression analysis using the REG procedure, where Y is the response variable and X is the predictor variable in the model.

```
proc reg;
model y=x;
run;
```

The following statements from Program 2.1 (Section 2.A) direct the SAS/STAT software to use the REG procedure to apply a simple linear regression to the data of Example 2.1 where systolic blood pressure (sbp) is the response variable and age is the predictor variable.

```
proc reg data=one;
model sbp=age;
run;
```

The following results are obtained from the REG procedure for the least squares estimates of the sampled population parameters for the simple linear regression model applied to Example 2.1:

Parameter Estimates

Variable	DF	Parameter Estimate	Standard Error	t Value	Pr > |t|
Intercept	1	109.60348	4.72937	23.18	<.0001
age	1	0.21685	0.08279	2.62	0.0103

From the preceding parameter estimates for Example 2.1, we see that $\hat{\beta}_0$ is equal to 109.60 and $\hat{\beta}_1$, the regression coefficient for AGE, is equal to 0.217. We will not discuss the results of the standard errors and t-tests until Chapter 4 because first we should evaluate whether these standard errors and t-tests are valid based on model checking procedures discussed in Chapter 3.

Equations for the Least Squares Estimators, β_0 and β_1

The equations for the least squares estimators, β_0 and β_1, are derived from minimization methods in calculus.

The equation for the least squares estimator of β_1 is

$$\hat{\beta}_1 = \frac{\sum_{i=1}^{n}(X_i - \bar{X})(Y_i - \bar{Y})}{\sum_{i=1}^{n}(X_i - \bar{X})^2} \tag{2.6}$$

where

n is the size of the sample

X_i is the value of the explanatory variable for individual i

$\bar{X}$ is the sample mean of the X values

Y_i is the value of the response variable for individual i

$\bar{Y}$ is the sample mean of the Y values

The equation for the least squares estimator of β_0 is

$$\hat{\beta}_0 = \bar{Y} - \hat{\beta}_1 \bar{X} \tag{2.7}$$

2.7.1 Properties of Least Squares Estimators

It can be shown that OLS estimators are unbiased estimators of their respective parameters under the assumption that the model errors (the ϵ_i of Equation 2.4) have a mean of zero in the sampled population. Furthermore, it can be shown that a least squares estimator has **minimum variance** in estimating a population parameter as compared to all other **unbiased linear estimators of that parameter**, if the following assumptions are met in the sampled population:

- The model errors have a mean of zero.

- The model errors are uncorrelated.

- The model errors have equal variance at each X (i.e., the equal variances assumption, which will be discussed in Section 2.11).

In addition to the preceding assumptions being met, if the model errors also have a normal distribution, a least squares estimator has **minimum variance** in estimating a population parameter as compared to all other unbiased estimators of that parameter. Thus least squares estimators of β_0 and β_1 are optimal assuming all assumptions are met.

2.8 Prediction of a Y Value for an Individual having a Particular X Value

The **predicted value** from a simple linear regression model predicts the Y value of an individual having a particular X value, assuming that individual is from the sampled population. This predicted value is obtained by substituting the values of $\hat{\beta}_0$, $\hat{\beta}_1$, and the particular X value, say X=x_0, in the estimated simple linear regression equation, giving

$$\hat{Y}_{x_0} = \hat{\beta}_0 + \hat{\beta}_1 x_0 \tag{2.8}$$

where

$\hat{Y}_{x_0}$ denotes the predicted Y value for an individual having an X value equal to x_0, assuming that individual is from the sampled population

x_0 denotes any X value in the range of X values observed in the sample

For example, the predicted SBP of a 29-year-old male from the sampled population of Example 2.1, where $\hat{\beta}_0 = 109.60348$ and $\hat{\beta}_1 = 0.21685$, is given by

$$\hat{SBP}_{AGE=29} = 109.60348 + 0.21685(29)$$
$$= 115.89213$$

Note, although there are no 29-year-old individuals in the sample you can obtain a predicted SBP value for age 29 in the sampled population using the estimated simple linear regression equation because the age, 29, is in the range of X values observed in the sample where ages range from 21 to 90.

It is easy to obtain predicted values in SAS. One way is by using the P option in the MODEL statement of the REG procedure i.e.,

```
proc reg;
model y=x/p;
run;
```

The P option computes predicted values for all the individuals in the sample. If, however, you are interested in a predicted value for a certain value of the predictor variable not found in the sample, such as age 29 in Example 2.1, a predictor value can still be obtained using SAS. A simple way to accomplish this using the REG procedure is shown in Program 2.1 (Section 2.A).

2.9 Estimating the Mean Response in the Sampled Population Subgroup Having a Particular X Value

The mean of the Y values in the sampled population subgroup with a particular X value can also be estimated by the predicted value, which you can get by specifying the P OPTION in the MODEL statement of PROC REG (see Section 2.8). The estimated mean of the Y values at $X = x_0$ in the sampled population is denoted by $\hat{\mu}_{y|x_0}$ and is given by

$$\hat{\mu}_{y|x_0} = \hat{\beta}_0 + \hat{\beta}_1 \, x_0 \tag{2.9}$$

where x_0 denotes any X value in the range of X values observed in the sample

For example, the estimated mean SBP of 29-year-old males in the sampled population of Example 2.1 is

$$\hat{\mu}_{SBP|29} = 109.60348 + 0.21685(29)$$
$$= 115.89213$$

Note that the right hand side of Equations 2.9 and 2.8 are identical. As we have pointed out the predicted value from a simple linear regression is used both as an estimate of the sampled population mean of Y given X as well as an estimate of the Y value for an individual with a particular X value when that individual is from the sampled population. However, as will be discussed in Chapter 4, the population mean of Y given X will be always be estimated with greater accuracy than an individual Y value given X.

2.10 Assessing Accuracy with Prediction Intervals and Confidence Intervals

The accuracy of the predicted Y value for an individual having a particular X value (Section 2.8) can be assessed by a prediction interval. Similarly the accuracy of the estimate of the population mean of Y in the sampled population subgroup having a particular X value (Section 2.9) can be assessed by a confidence interval. We postpone a discussion of prediction and confidence intervals until Chapter 4 because the validity of both prediction and confidence intervals depend on ideal inference conditions (Section 2.11). In Chapter 3 we present methods which evaluate whether these ideal inference conditions are reasonable for a sampled population.

2.11   Ideal Inference Conditions for Simple Linear Regression

The ideal inference conditions for hypothesis tests, confidence intervals, and prediction intervals in simple linear regression are:

- **INDEPENDENCE OF ERRORS** The population model errors, i.e., the ϵ in Equation 2.4 are independent and therefore uncorrelated with any factor. Sometimes the design of the study in itself will inform you whether this assumption is tenable. For example, if there is more than one observation on any of the individuals in the study (say at monthly intervals), then the independence condition for simple linear regression is likely to be violated because the Y values for the same individual are likely to be correlated. Other examples where independence of errors are violated as a result of how the data were collected are given in Chapter 14. However, it may not always be obvious from the study design whether some or all of the model errors are dependent. The next chapter on model checking will describe methods for evaluating whether the independence of errors condition is violated.

- **LINEARITY** In the sampled population the means of Y at each distinct value of X lie on a single straight line. Figure 2.2 illustrates this ideal inference condition for a simple linear regression where the sampled population means of the Y variable (SBP in mmHg) at each distinct value of the X variable (age in years) lie on a single straight line. For the sake of simplicity, there are just four ages represented in this illustration, although this ideal inference condition applies to all values of X that are in the range of X values observed in the sample.

- **EQUALITY OF VARIANCES** The population model errors have the same variance for each value of X in the range of the X values observed in the sample. In simple linear regression this implies that the variances of the Y values in the sampled population are equal at each distinct value of X in the range of the X values observed in the sample. Figure 2.2 illustrates this ideal inference condition for a simple linear regression where the variances of the Y variable (say SBP) at each distinct value of the X variable (say age) are equal. Figure 2.3 illustrates a case where the ideal inference condition of equal variances is not met. It is obvious from Figure 2.3 that the variance of Y at $X = 85$ is much greater than the variances of Y at $X = 65$, $X = 45$, and $X = 25$. Only the variances of Y at $X = 45$ and $X = 25$ are equal.

  The equality of variances condition is sometimes referred to as the condition of **homogeneity of variances** or **constant variance** or **homoscedasticity**. The terms **heteroscedasticity** or **heterogeneity of variances** is used to refer to the situation where the variances of the groups at each distinct value of X in the sampled population are not all equal.

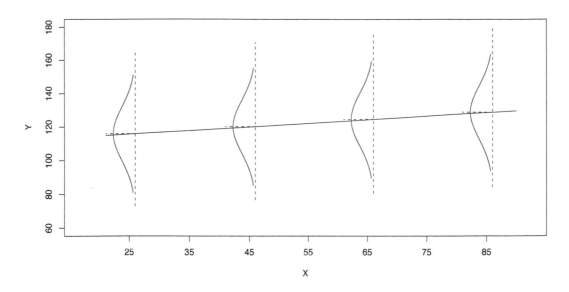

FIGURE 2.2
Illustration of a simple linear regression where the ideal inference conditions of linearity, equality of variances, and normality are met in the sampled population.

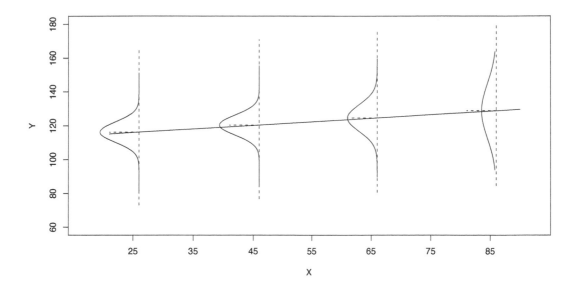

FIGURE 2.3
Illustration of a simple linear regression where the variance of Y at each X is unequal in the sampled population.

- **NORMALITY** The population model errors follow a normal distribution. This implies that in the sampled population, there is a normal distribution of Y values at each distinct value of X in the range of X values observed in the sample (Figure 2.2).

- **NO MEASUREMENT ERRORS**

  Definition 2.1. *Measurement error* is the error that occurs when the value recorded for an observation is not the actual value. An example of measurement error in X in a regression analysis is the situation where the predictor variable is body weight of an individual and the self-reported values for body weight used in the analysis are not always accurate because some individuals may underestimate or overestimate their actual body weight.

Chapter 3 describes methods for evaluating whether the first four of these ideal inference conditions are reasonable for the sampled population. Chapter 14 suggest alternative approaches for situations where these ideal inference conditions are violated.

2.12   Potential Consequences of Violations of Ideal Inference Conditions

If one or more ideal inference conditions for simple linear regression are violated the potential consequences are:

- $\hat{\beta}_1$, the OLS point estimate of the population regression coefficient, may not be accurate, i.e., not close in value to the true population parameter, β_1.

- The point estimate of the sampled population mean of Y for a given X estimated by Equation 2.9 may not be accurate.

- The prediction of an individual Y value for a given X value in the sampled population using Equation 2.8 may be not accurate.

- The results from hypothesis tests as well as confidence and prediction intervals related to simple linear regression (Chapter 4) may lead to erroneous conclusions about the relationship between the response variable and the explanatory variable.

2.12.1   How Concerned Should We Be About Violations of Ideal Inference Conditions?

Any violation of independence of errors is always a cause for concern. However, concern regarding the impact of violation of linearity, equal variances, and normality can be related to the degree of departure from these ideal inference conditions in the range of X observed in the sample. Expert knowledge in the research area and graphical methods discussed in Chapter 3 can be useful in assessing the extent of nonlinearity in the range of the observed X. Subsection 14.6.1 provides some approximate guidelines as to when unequal variance violations might cause serious harm in regression. Hypothesis tests as well as confidence intervals may not be affected by slight or even moderate departures from the normality assumption, especially if the sample size is large, due to the central limit theorem (Section 1.3.1). However, prediction intervals can be affected by departures from normality even if the sample size is large (van Belle et al., 2004). When there is a major departure from normality, tests of regression hypotheses as well as confidence and prediction intervals may result in invalid conclusions. The nature of the departure from normality may be of concern in that the mean of Y given X is not a good estimator of the centre of location for certain nonnormal distributions. For example, recall from introductory statistics the mean of a right skewed distribution is typically larger than the median and the mean of a left skewed distribution is typically less than the median.

Measurement error in the X values in simple linear regression produces a biased estimate of the regression coefficient for the X variable such that this regression coefficient is underestimated. However, if this measurement error is small relative to the range of variation of X in the individuals, then departure from this ideal inference condition has negligible consequences. Measurement error in the X values is also not a concern if the objective of the study is to investigate whether the X variable as measured, even if it is not completely accurate, can predict a response of interest. For example, consider the situation where the objective is to investigate whether an individual's self-measured BMI (body mass index), potentially measured with error (the X variable), can be used to predict the individual's SBP. In this situation the prediction of SBP by the "true value" of the X variable, i.e., the BMI, measured without any error by a health practitioner, is not relevant to the question the researcher wants to investigate. Apart from these two situations, measurement error in the X values is a cause for concern and should be initially addressed at the data collection phase. In those situations where measurement error in the X values is a problem and cannot be avoided, then methods dealing with this problem, which are beyond the scope of this book, should be considered. Graybill and Iyer (1994) have provided a helpful discussion of measurement error in X values for researchers.

2.13 What Researchers Need to Know about the Variance of Y Given X

The following are some key facts you need to know about the variance of Y given X:

- The variance of Y given X is the population variance of Y in a subgroup where the individuals have the same X value.

- It is extremely important to have a reliable estimate of the variance of Y given X in simple linear regression, as this estimate plays a critical role in hypothesis testing, in estimation of confidence and prediction intervals, and in some model checking methods. Incorrect conclusions may result if this estimate is unreliable.

- The estimate of the variance of Y given X is important because it is a measure of the extent that factors **other** than X (the predictor variable) affect the response variable, Y.

- A model-dependent estimator of the variance of Y given X is typically used in simple linear regression. This model-dependent estimator is often referred to as the **Mean Square Error** or alternately as the **Residual Mean Square** or **Error Mean Square**.

- The Mean Square Error is called a model-dependent estimator because it is valid only when the model is appropriate for the data. If linearity, independence of errors, and equality of variances (Section 2.11) do not hold for the sampled population, then the Mean Square Error is not a reliable estimator of the variance of Y given X and remedial measures may have to be considered (Sections 14.3, 14.5, 14.6).

2.13.1   An Equation for Mean Square Error, A Model-Dependent Estimator of the Variance of Y given X

The Mean Square Error is given by

$$\text{Mean Square Error} = \frac{\text{Sum of Squares Error}}{n - p} \tag{2.10}$$

where

n is the total sample size

p is the number of population parameters estimated in the model. In a simple linear regression model when β_0 and β_1 are estimated, $p=2$

Sum of Squares Error is the sum of squares of the *residuals* for all the observations in the researcher's sample, i.e.,

$$\text{Sum of Squares Error} = \sum_{i=1}^{n} e_i^2 \tag{2.11}$$

where e_i, the (raw) residual for individual i, is defined as the difference between the Y value observed for individual i and the Y value predicted for individual i by the regression model. That is

$$e_i = Y_i - \hat{Y}_i \tag{2.12}$$

where, in simple linear regression, $\hat{Y}$, the predicted value, is given by

$$\hat{Y}_i = \hat{\beta}_0 + \hat{\beta}_1 X_i \tag{2.13}$$

2.14   Fixed vs. Random X in Simple Linear Regression

Fixed X and **random X models** in simple linear regression are distinguished by the way a researcher has obtained the sample for the investigation. In a fixed X model, the values of the explanatory variable, X, in the sample are fixed by the researcher, i.e., the researcher decides in advance the particular values of X and the associated sample sizes that are to be in the sample and then chooses individuals for the study which have these X values. For example, in a study investigating SBP and age, a researcher had decided in advance to have individuals who are 20, 40, 60, and 80 years of age represented and then randomly selected 25 individuals in each of these four age groups to be in the study.

In contrast, in a random X model, the values of the explanatory variable X in the sample are obtained at random and not under the control of the researcher. For example, in a study investigating SBP and age, a researcher obtained a random sample of 100 individuals without knowing what their ages were at the time of sampling and thus did not control which ages and associated sample sizes were represented in the sample.

Fortunately, all simple linear regression topics covered in this chapter and subsequent chapters are applicable regardless of whether the X values in the study sample have been fixed by the researcher or have been obtained at random. The advantage of a fixed X model is that you have the opportunity to increase statistical power through controlling the sample sizes and range of the X values in the sample (see Section 4.12). The advantage of a random X model is that it allows certain parameters of the sampled population (mean of the X values, variance of X, etc.) as well as quantities that describe the joint distribution of X

and Y (correlation coefficient, etc.) to be estimated whereas these parameters and quantities typically cannot be estimated in the fixed X model. They could be estimated in the highly unlikely research scenario where each subgroup defined by every X value in the sampled population is represented in the sample and the relative sizes of subgroups are known in this population – not at all a realistic scenario.

2.15 Chapter Summary

This chapter:

- describes the characteristic features and underlying rationale of a simple (single predictor) linear regression model and lists some reasons why you would want to use this model

- discusses how you can obtain estimates of the parameters in a simple linear regression model (i.e., the Y intercept and slope) from sample data via the method of least squares

- shows how the Y intercept and slope estimates are used in estimating the mean of Y in a sampled population subgroup with a particular value of X and in estimating the value of Y for an individual from the sampled population with a particular X value

- illustrates how to obtain these estimates using the REG procedure in SAS/STAT software

- describes ideal conditions for simple linear regression and discusses potential difficulties when these ideal inference conditions are not met

- discusses why it is important to get a valid estimate of random variability within the X subgroups in simple linear regression

- compares random and fixed X models in simple linear regression

- provides an example of a SAS program that produces a high resolution scatter plot of raw data along with a estimated regression line for these data. This program also illustrates how to obtain a listing of predicted values from a simple linear regression, where these values are sorted by the X variable in ascending order.

Appendix

2.A   Program 2.1

Program 2.1

- applies a simple linear regression to Example 2.1 using the REG procedure

- generates a scatter plot of the raw data with the estimated regression line

- produces a listing of observed and predicted values of SBP sorted in ascending order by the X variable, AGE

```
data one;
title Simple Linear Regression of Systolic Blood Pressure (SBP) against Age;
input RECORD SBP AGE @@;
datalines;
1 124 82 2 114 34 3 115   40 4 121 51 5 120 34
6 128 25 7 94 54 8 120 54 9 148 64
10 118 67 11 120 42 12 149 90 13 106 56 14 114 69 15 79 66 16 128 73 17 123 66
18 112 25 19 148 51 20 131 22 21 102 70 22 122 65 23 135 57 24 112 41 25 99 48
26 112 27 27 100 34 28 134 83 29 123 69 30 140 36 31 110 62 32 117 42 33 121 66
34 127 89 35 114 78 36 154 71 37 131 50 38 135 26 39 103 54 40 116 72 41 136 85
42 88 34 43 120 46 44 118 37 45 151 57 46 122 40 47 134 60 48 136 58 49 123 47
50 115 85 51 134 41 52 98 39 53 119 68 54 129 65 55 151 71 56 122 53 57 114 55
58 156 51 59 124 51 60 127 22 61 91 64 62 131 55 63 125 30 64 97 35 65 122 32
66 128 47 67 148 90 68 133 59 69 126 33 70 112 45 71 121 61 72 111 73 73 127 48
74 136 39 75 110 82 76 135 82 77 140 83 78 101 59 79 130 73 80 110 59 81 116 33
82 129 72 83 105 21 84 110 55 85 113 54 86 123 85 87 130 36 88 121 64 89 118 42
90 97 33 91 109 42 92 116 33 93 136 74 94 101 26 95 137 62 96 119 51 97 . 29
;
run;
proc reg data=one  plots(only stats=none) =(fit(nolimits));
   model sbp=age/p;
   id record;
   output out=predvalues p=PREDICTED;
run;
proc sort data=predvalues ;
   by age;
run;
proc print data=predvalues;
run;
```

2.A.1 Explanation of SAS Statements in Program 2.1

The DATA, TITLE, COMMENTS, and DATALINES statements have already been explained for Program 1.1 (subsection 1.16).

The INPUT statement in Program 2.1 uses as a "double trailing at" symbol i.e.,@@. Placing @@ at the end of an INPUT statement directs SAS to keep reading on the same line according to the simple list format specified in the INPUT statement. (Simple list format is described in Section 1.A.1.) Thus, after reading in the data values for SBP and AGE for case with RECORD=1, SAS continues on this same line to read in the data values for the next cases viz., RECORD=2, SBP=114, AGE=34; RECORD=3, SBP=115, AGE=40, and so on until SAS inputs data values for cases on the last data line.

The last case input RECORD=97, with a missing value for SBP indicated by a single decimal point, "." and AGE=29, is a fictional case. We added this case because we wanted to get a predicted value for someone whose age is 29 and there were no actual cases with age 29 in the data. The REG procedure normally calculates predicted values only for values of X (AGE in this example) actually found in the data. However, adding this case will cause the REG procedure to calculate a predicted value for AGE 29. The estimation of the model parameters in the REG procedure will not be affected by adding this fictional case because the value of SBP is missing for this case.

A Note of Caution: You have to be very careful if you use the @@ symbol with simple list format as it tends to be quite error prone such that if you inadvertently leave out a data value for a case, then all the data values input for the cases from that point onwards will be incorrectly input. However, if it is convenient for you to use the @@ feature with simple list format in order to be able to input multiple cases per line, then we strongly recommend verifying the values for all your input variables by obtaining a listing of your input data using the PROC PRINT procedure, as has been illustrated in Program 1.1.

proc reg data=one plots(only stats=none) =(fit(nolimits)); This PROC REG statement requests the REG procedure be applied to the data set called "one". Specifying the SAS keyword **only** for the PLOTS option suppresses all default plots generated by PROC REG except the plot that is requested. In Program 2.1 the suboption (fit(nolimits)) requests that the REG procedure generate a scatter plot of the original data with the estimated regression line inserted and no confidence or prediction limits. We suppressed confidence and prediction limits as the default option for these limits in the REG procedure computes conventional confidence or prediction limits. The difficulty associated with conventional confidence limits for a regression line will be discussed in the chapter on multiple testing (Section 12.11). The suboption **stats=none** indicates that no model fit statistics should be inserted in the plot.

If the ONLY option is not specified all the default ODS plots available in the REG procedure would be output.

model sbp=age/p; This MODEL statement of the REG procedure specifies that the Y variable is SBP and the X variable is AGE. The keyword p after the forward slash in the MODEL statement results in predicted values being displayed in the output.

id record; When any of the MODEL statement options of the REG procedure are requested, the ID statement allows you to specify additional variables you want displayed in the output. In Program 2.1, the p option of the MODEL statement is requested and the ID statement **id record** specifies that the variable "record" will be output for each observation when its predicted value is displayed.

output out=predvalues p=PREDICTED; The OUTPUT statement creates a new data set named predvalues which consists of those quantities specified as options in the output statement as well as the data in the current data set. Unless an OUTPUT statement is used, quantities calculated by the REG procedure will not be available to subsequent procedures in the program as these quantities are not saved after the PROC REG step is finished executing. The general format of an OUTPUT statement in the REG procedure given in SAS documentation is

```
OUTPUT <OUT=SAS-data-set>< keyword=names> <...keyword=names>
```

Keywords for various quantities calculated by the REG procedure are given in SAS documentation. In this example we are just interested in predicted values which are identified by the keyword **p** in SAS.

Thus, the OUTPUT statement in Program 2.1, **output out=predvalues p=PREDICTED**, may be explained as follows:

out=predvalues requests that a new SAS dataset be created and it should be called "predvalues".

p=PREDICTED requests that predicted values for all the cases should be included in this new data set called "predvalues". In addition to the predicted values the new data set "predvalues" will automatically include the raw data from SAS dataset called "one" (the data set which the REG procedure used for the regression). The variable name we arbitrarily specify for the predicted values is "PREDICTED". We use upper case as we want this variable name in upper case in the output. Recall from Chapter 1 that the case used for the first occurrence of a variable name in a SAS program determines the case that SAS will use for that variable name in the program's output.

proc sort data=predvalues; This PROC SORT statement requests the SORT procedure be applied to the SAS dataset named "predvalues". As the OUT= option is not specified in the PROC SORT statement, the sorted data set output by this procedure will be placed in the data set "predvalues" overwriting the original order of the observations in "predvalues". If you do not want to replace the original order in the data set named "predvalues" you can use the OUT= option of the SORT procedure to specify the name for the new dataset in which the sorted data will be placed. For example, the statement

proc sort data=predvalues out=sortedpv; This PROC SORT statement specifies that the sorted data should be placed in the data set arbitrarily called "sortedpv".

The BY statement, **by age;**, specifies that the cases in this data set should be sorted by age in ascending order. Ascending order is the default option. If descending order is desired, it can be specified using the DESCENDING option in the BY statement, i.e., by age descending;

proc print data=predvalues; This PROC PRINT statement requests the PRINT procedure be applied to the SAS dataset named "predvalues".

In the PRINT procedure the VAR statement specifies which variables in the input data set should be printed. It also allows you to specify the order in which these variables should be printed. In Program 2.1 the VAR statement was omitted so the procedure printed the values for all the variables in the data set "predvalues". If, however, you wanted to print only the variables RECORD AGE SBP, you could specify this by adding a VAR statement after PROC PRINT e.g.,

    proc print data=predvalues; var record age sbp; run;

2.A.2   A Message in SAS Log for Program 2.1

The following message is given in the SAS log for Program 2.1:

```
NOTE: SAS went to a new line when INPUT statement reached past the
end of a line.
```

This message is no cause for concern in Program 2.1 because the trailing double at, @@, is used in the INPUT statement.

3

Model Checking in Simple Linear Regression

3.1 General Introduction

This chapter will discuss three important aspects of model checking which should be addressed before interpreting confidence intervals, prediction intervals, and hypothesis tests from a simple linear regression analysis. These model checking aspects involve assessing:

- **the model's ideal inference conditions** (Section 2.11)

- potential occurrence of **regression outliers** in the sample (Section 3.2)

- whether the estimated model has resulted largely from **the influence** of just one or a few data points in the sample (Section 3.3)

We begin the discussion by introducing the topic of regression outliers and influential cases in simple linear regression. We then present statistical tools, often called regression diagnostics, which can be used for model checking. We show how graphical displays of these various diagnostics can reveal information regarding ideal inference conditions, regression outliers, and potentially influential cases. As the entire model checking process can be somewhat overwhelming, we propose a step-by-step approach to integrate information from the various diagnostic methods described and provide a detailed example which illustrates this approach to model checking from start to finish. Although our focus in this chapter is on simple linear regression, the diagnostic tools presented can be used to assess any linear regression model, as will be discussed in later chapters.

3.2 Regression Outliers

Definition 3.1. A *regression outlier* is a data point (case) which does not support the linear regression relationship between the response and predictor variables that is shown by the majority of the other data points (cases) in the data set.

Example The data point (case) where $x=55$ and $y=130$ in Figure 3.1 is an example of a regression outlier. The word "case" used in the context of regression refers to "a particular observation on the response variable in combination with the associated values for the predictor variable(s)" (Cook and Weisberg, 1999). The word "case" is also used in statistics in the usual sense meaning "situation", for example "for the case of n independent samples". Usually the context of the sentence conveys the intended meaning.

A regression outlier should always be investigated to determine whether it is unusual as a result of some sort of error (measurement, recording, transmission, data entry, etc.) in

DOI: 10.1201/9780429107368-3

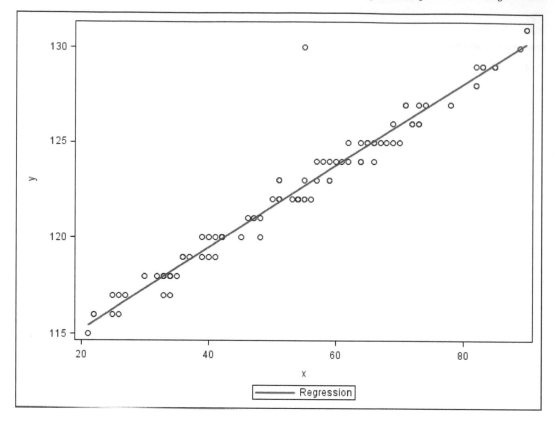

FIGURE 3.1
Scatter plot showing a regression outlier.

either its Y value or its X value or in both. If there has been an error, the researcher should correct it, and rerun the regression analysis on the corrected data. If it can be documented with concrete evidence that a regression outlier has resulted from some sort of error and it is impossible to correct the error at that stage of the research study, then that outlying erroneous data point should be eliminated from the data analysis. However, we do not recommend discarding a regression outlier in the absence of documented concrete evidence that the outlier is a result of an error. Although an outlier may be inconvenient in that it does not nicely fit with the hypothesized linear regression relationship, discarding such an outlier may result in throwing away important information about the entire sampled population. A simple first step that is often recommended is to analyse the data with and without the outlier included and compare the results. If materially similar results are obtained from the analyses with and without the outlier, a researcher may want to report this when publishing the results of the study. If materially different results are obtained from analyses with and without the outlier, other statistical approaches (Section 14.8) can be considered. What constitutes materially different results depends largely upon the goal of the research and upon expert opinion of the subject matter.

3.3   Influential Cases

Definition 3.2. An *influential case* is one which individually or together with a few other cases has a disproportionate influence on the estimated regression model as compared to the other cases in the sample.

Section 3.15 discusses measures which quantify the amount of influence a single case has on an estimated regression model. We postpone a discussion of these measures of influence until Section 3.15, as the accuracy of these measures depends on the ideal inference conditions of linearity, independence of errors, and homogeneity of variance. Therefore we first need to discuss evaluation of ideal inference conditions before considering measures of influence. In this section we introduce the concept of influence by graphical examples.

Examples The case where $x=100$ and $y=120$ in Figure 3.2 is an example of an influential case. This case, through the process of least squares estimation, has an influence on the estimate of the slope of the regression line such that the line is pulled closer to this case. This case is influential because it is a regression outlier and because it has an X value that is not close to the mean of the X values in the sample.

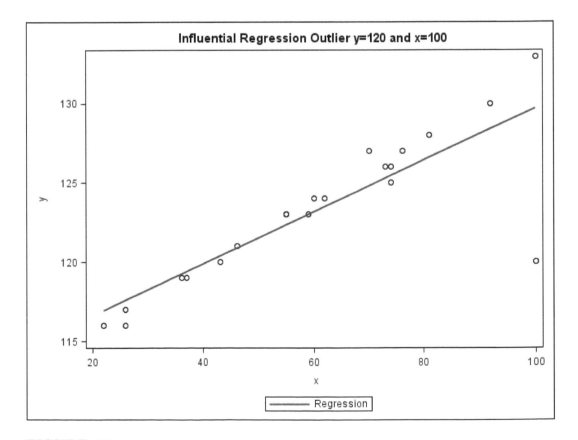

FIGURE 3.2
Scatter plot illustrating a regression outlier that has influence on the least squares slope estimate.

The case $x=55$ and $y=130$ in Figure 3.1 is an example of a regression outlier which does not have a disproportionate influence on the least squares estimate of the slope because the X value of this case is close to the mean of the X values in the sample. However, this case merits investigation for its influence on the least squares estimate of the Y intercept and also for its possible influence on the Mean Square Error for the analysis.

An influential case is not "bad" merely because it has a greater impact than the others on the estimation of the parameters of the regression model. However, it is important to identify an influential case. A researcher should investigate it fully to understand the implication of using an estimated model where just one or few cases disproportionately influence the estimation of the model's parameters.

3.4   Residuals as Diagnostic Tools for Model Checking

Residuals are typically used as diagnostic tools in model checking. They can reveal regression outliers and influential cases as well as provide information about whether a model's ideal inference conditions are reasonable for the sampled population We will consider three types of residuals in this chapter:

raw residuals (Section 3.6)

internally studentized residuals (Section 3.7)

studentized deleted residuals (Section 3.8)

3.5   Obtaining Residuals from the REG Procedure

3.5.1   Introduction

Raw, internally studentized, and studentized deleted residuals can be obtained by requesting the R and INFLUENCE options in the MODEL statement of the REG procedure. For example, the following SAS statements specify that a simple linear regression should be applied to the mine site study example described in Section 1.15. The MODEL statement of the REG procedure specifies the Y variable for the regression is AS (the arsenic level in the fish's tissue) and the X variable is TL (total body length of the fish). The R and INFLUENCE options of the MODEL statement direct SAS to output raw, internally studentized, and studentized deleted residuals from the regression analysis.

```
proc reg data=minesite;
model as=tl/r influence;
id record;
run;
```

3.5.2   Some Results from the REG Procedure for the Mine Site Example

The following Analysis of Variance table for the mine site example was generated by the preceding model statement for the REG procedure (subsection 3.5.1). An example of results for the first case, Record 797, in the mine site data, computed by specifying the R and INFLUENCE options in the MODEL statement are also reported in this subsection.

Analysis of Variance

Source	DF	Sum of Squares	Mean Square	F Value	Pr > F
Model	1	0.00319	0.00319	1.02	0.3273
Error	17	0.05329	0.00313		
Corrected Total	18	0.05647			

Output Statistics

Record Number	Dependent Variable	Predicted Value	Std Error Mean Predict	Residual	Std Error Residual	Student Residual	Cook's D
797	0.16	0.2417	0.0132	-0.0817	0.0544	-1.502	0.066

Output Statistics

Record Number	RStudent	Hat Diag H	Cov Ratio	DFFITS	DFBETAS Intercept	TL
797	-1.5646	0.0555	0.8991	-0.3792	0.0168	-0.0860

3.6 Raw (Unscaled) Residuals

Definition 3.3. The *raw residual* for case i is defined as the difference between the Y value observed for case i and the Y value predicted for case i by the regression model (Equation 2.12).

A *raw residual* is also referred to as an **ordinary residual**, an **unscaled residual**, or sometimes simply as a *residual*.

3.6.1 Example of a Raw Residual

As illustration, consider the case with Record=797 in subsection 3.5.2. Its raw residual, denoted as "Residual" in PROC REG's output, is −0.0817. The observed Y value, denoted as "Dependent Variable", and the predicted Y value for this case are respectively equal to 0.1600 and 0.2417. Thus the output value of the raw residual for Record 797 agrees with the following hand calculation:

$$\begin{aligned} e_{797} &= 0.1600 - 0.2417 \\ &= -0.0817 \end{aligned}$$

3.6.2   Estimated Standard Error (Standard Deviation) of a Raw Residual

Assuming the ideal inference conditions of linearity, independence of errors, and equality of variances hold in the sampled population, the estimated standard error of a raw residual for observation i is given by

$$\widehat{SE}(e_i) = \sqrt{(\text{Mean Square Error})\,(1 - h_i)} \qquad (3.1)$$

where

Mean Square Error $= \sum_{i=1}^{n} e_i^2/(n - p)$

where p is the number of estimated parameters in the model, e.g., $p = 2$ for simple linear regression.

h_i denotes the **hat (leverage) value** for case i where the hat value for case i in simple linear regression is a measure of how far away in X space, case i is from the mean of the X values of the cases in the sample. The range for a hat value is greater than or equal to zero and less than or equal to one. The farther the X value of case i is from the mean of the X values in the sample, the greater its hat value. Thus, cases with X values close to the sample mean of X have hat values near zero whereas cases with X values far away from the sample mean of X have hat values near one.

Example of an Estimated Standard Error of a Raw Residual

The estimated standard error of the raw residual for Record 797 generated by the REG procedure is 0.0544 (see "Output Statistics" in subsection 3.5.2). Also reported in subsection 3.5.2 is the Mean Square Error equal to 0.00313 for the regression analysis (see Analysis of Variance table) and the hat value ("Hat Diag H") for Record 797 equal to 0.0555 as reported in the "Output Statistics". Substituting these values in Equation 3.1, the standard error of the raw residual for Record 797 $= \sqrt{(0.00313)(1 - 0.0555)} = 0.0544$ which agrees with the standard error value for the raw residual of Record 797 reported in the "Output Statistics".

3.6.3   Difficulties Associated with Using Raw Residuals as a Diagnostic Tool

We do not recommend using raw residuals when evaluating ideal inference conditions because of the following difficulties:

- *Unequal Variances of Raw Residuals at each X* A key difficulty in using raw residuals for evaluating the ideal inference condition of equality of variances is that the variances of raw residuals at each value of X are not equal even when the ideal inference condition of equality of variances of Y at each X holds in the sampled population. This is because the variance of raw residuals at each X depends on how far away the X value is from the mean of the X values in the sample.

- *Lack of Independence among Raw Residuals* All raw residuals are correlated and hence lack independence, even if the ideal inference condition holds that the sampled population model errors are independent. However, this correlation decreases as the sample size increases and thus with large samples, lack of independence among the raw residuals usually is not a concern.

- *Possibility of Failing to Identify Influential Regression Outliers* A major difficulty can occur when a raw residual is used to identify a case as a regression outlier if the case is influential. In this situation the case can exert an influence on the least squares estimate of the regression model such that the estimated regression surface (line in the case of simple linear regression) will lie close to this influential case. Thus, this influential case will have a small raw residual, indicating it is not a regression outlier, although this case is markedly different from the majority of the cases in terms of the relationship between X and Y

- *Ambiguity in Interpretation of Q-Q Plots* We will expand on this difficulty associated with using raw residuals in Q-Q plots when we discuss these plots in subsection 3.9.3.

- *Supernormality of Residuals* Supernormality is the phenomenon, potentially occurring in small or moderate sample sizes, where raw residuals tend to be normally distributed even if the population model errors (the ϵ) are not. Thus, it might be concluded on the basis of a graphical display of raw residuals in a small or moderate-sized sample that the population errors are normally distributed at each X when in actuality they are not. This phenomenon of supernormality in small or moderate samples is a potential difficulty that is not restricted to raw residuals but also can occur in internally studentized residuals and studentized deleted residuals. Gnanadesikan (1997) and Andrews (1979) provide a more detailed discussion of supernormality.

3.7   Internally Studentized Residuals

Internally studentized residuals can be used for evaluating whether ideal inference conditions of a model are reasonable for a sampled population.

Definition 3.4. An *internally studentized residual* for case i is the raw residual for case i divided by its estimated standard error.

An *internally studentized residual* is also referred to simply as a *studentized residual*, as a *studentized raw residual*, or as a *standarized residual*.

Equation for an Internally Studentized Residual
The internally studentized residual for the ith case is given by

$$\text{STUDENT}_i = e_i / \widehat{\text{SE}}(e_i) \tag{3.2}$$

where
  STUDENT_i denotes the internally studentized residual for case i
  e_i denotes the raw residual for case i (Equation 2.12)
  $\widehat{\text{SE}}(e_i)$ is the estimated standard error of the raw residual for case i as given by Equation 3.1, under the assumption linearity, independence of errors, and equality of variances hold in the sampled population.

To illustrate application of Equation 3.2 we consider the internally studentized residual for RECORD 797 in the mine site example. The raw residual for Record 797 is -0.0817 and the estimated standard error for this raw residual is 0.0544 (subsection 3.5.2). Substituting these values into Equation 3.2 we obtain the value of the internally studentized residual for Record 797, as

$$\begin{aligned} \text{STUDENT}_{797} &= -0.0817/0.0544 \\ &= -1.502 \end{aligned} \tag{3.3}$$

which agrees with the value of the internally studentized residual given for Record 797 under the label "Student Residual" in the "Output Statistics" generated by the REG procedure as reported in subsection 3.5.2.

3.7.1   Advantages of Internally Studentized Residuals in Evaluating Ideal Inference Conditions

Internally studentized residuals rather than raw residuals are recommended for evaluating the ideal inference condition of equality of variances of the population errors at each X in the sampled population. The reason for this recommendation is that an internally studentized residual is a raw residual scaled (divided) by its standard error and thus the variance of an internally studentized residual does not depend on its corresponding X value.

Another advantage of using an internally studentized residual instead of a raw residual for evaluating ideal inference conditions is that an internally studentized residual is not affected by the scale of the response variable. If the ideal inference conditions for the model are met, then internally studentized residuals will have an ***approximate*** Student's t distribution with a mean near zero and a variance slightly larger than one. Thus if the ideal inference conditions for the model are met, any case with an absolute value of an internally studentized residual greater than two may possibly be a model outlier. It is important to be aware, as will be discussed in subsection 3.8.3, that there are situations where an internally studentized residual will not detect a regression outlier, although a studentized deleted residual will detect a regression outlier in these situations.

Internally studentized residuals would seem preferable for evaluating model assumptions regarding the ϵ (Myers, 1990) as these residuals are likely to be more representative of the ϵ than either raw residuals or studentized deleted residuals (Ryan, 2009).

3.8   Studentized Deleted Residuals

Studentized deleted residuals are often used to detect regression outliers in least squares linear regression.

Definition 3.5. A ***studentized deleted residual*** for case i is the studentized difference between the observed Y value and predicted Y value for case i when its predicted value is based on the regression equation that is estimated with case i deleted from the sample.

A ***studentized deleted residual*** is also referred to as a ***standardized deleted residual***, a ***standardized deletion residual*** or simply a ***deletion residual***.

Equations for a Studentized Deleted Residual

The studentized deleted residual for the ith case is given by

$$\text{RSTUDENT}_i = \frac{Y_i - \hat{Y}_{i(-i)}}{\widehat{\text{SE}}(Y_i - \hat{Y}_{i(-i)})} \tag{3.4}$$

where
  RSTUDENT_i denotes the studentized deleted residual for case i
  Y_i is the observed Y value for case i

$\hat{Y}_{i(-i)}$ is the predicted Y value for case i obtained from the regression model equation which has been estimated with case i excluded from the sample

$\widehat{SE}(Y_i - \hat{Y}_{i(-i)})$ is the estimated standard error of the difference $(Y_i - \hat{Y}_{i(-i)})$

The following expression for a studentized deleted residual which is algebraically equivalent to Equation 3.4 allows us to calculate a studentized deleted residual for case i without having to omit case i from the regression:

$$\text{RSTUDENT}_i = e_i \sqrt{\frac{n - p - 1}{(\text{Sum of Squares Error})(1 - h_i) - e_i^2}} \tag{3.5}$$

where

e_i is the raw residual for case i

$n =$ the sample size

$p =$ the number of parameters estimated in the regression model

Sum of Squares Error is the sum of squares of the raw residuals

h_i is the hat (leverage) value for case i.

As an example of Equation 3.5, we hand-calculated the value of RSTUDENT for Record 797 from the mine site example, where $n = 19$; the Sum of Squares Error for the regression= 0.05329, and the raw residual and the hat value for Record 797 are respectively equal to -0.0817 and -0.0555, as reported Section 3.5.2. Substituting these values in Equation 3.5, the hand calculation of the studentized deleted residual for Record 797 is -1.564056. The studentized deleted residual for Record 797, computed by the REG procedure under the label RSTUDENT reported in Section 3.5.2 is -1.5646. Rounding error accounts for the slight discrepancy between these two values.

3.8.1 Using a Studentized Deleted Residual to Identify a Regression Outlier

An approximate guideline for regression outlier identification is that a case with an absolute value of a studentized deleted residual greater than two may be a regression outlier (Myers, 1990; Fox, 2016). The basis for this rule of thumb is that when a model's ideal inference conditions are reasonable for the population sampled, studentized deleted residuals approximately follow a Student's t-distribution with $n - 1 - p$ degrees of freedom (p=the number of model parameters) and therefore fewer than 5% of studentized deleted residuals should have an absolute value greater than two.

3.8.2 Example

The studentized deleted residual for Record 797 in the mine site study example is -1.5646 as given in "Output Statistics" from Program 1.1 (Section 3.5.2). Before deciding whether this case is a regression outlier, we should first evaluate whether the ideal inference conditions are reasonable for the population sampled. If it turns out these inference conditions are reasonable then Record 797 would not be identified as a regression outlier, as the absolute value of its studentized deleted residual is less than two.

3.8.3   Advantages of Studentized Deleted Residuals in Regression Outlier Dectection

Studentized deleted residuals are typically used in outlier detection in least squares linear regression as sometimes regression outliers can be missed using raw residuals or internally studentized residuals. Consider the situation where a case is influential and the sample size is not large. Through the process of least squares estimation this influential case will influence the estimate of the regression model parameters so that the estimated regression line will be pulled close to this influential case. Thus this case will have a small raw residual or a small internally studentized residual that does not reveal the extent that this case differs from the rest of the cases (data points) in terms of the regression relationship between X and Y. However, this problem does not occur with the studentized deleted residual in this situation. This is because the studentized deleted residual is formed by the difference between the observed Y value for a case and the predicted Y value where the predicted Y value is based on the regression model which has been estimated with the current case excluded from the sample.

3.8.4   Caveat Regarding Studentized Deleted Residuals in Outlier Detection

Studentized deleted residuals are not ideal for detecting regression outliers when there are two or more influential outlying data points that lie close together. Methods to detect influential outliers that lie in a cluster have been discussed by Ryan (2009) and Clarke (2000) but are beyond the scope of this book.

3.9   Graphical Evaluation of Ideal Inference Conditions in Simple Linear Regression

3.9.1   Graphical Evaluation of Independence of Errors

Graphical displays can be used in an informal evaluation of independence of model errors in simple linear regression. For example, such displays can be useful in an observational study where cases have been non-randomly sampled with respect to one or more nuisance variables such as time (e.g., date of sampling) and/or space (e.g., distance from a transformer that distributes electricity). This informal evaluation can be carried out by plotting internally studentized residuals against a nuisance variable of concern. A random scatter of the data points about the line $Y=0$ along the horizontal axis in such a plot would suggest that the model errors are independent of the nuisance variable investigated whereas a systematic pattern of these data points in a plot would suggest that errors are not independent of the nuisance variable investigated.

Figure 3.3 is a scatter plot where internally studentized residuals are plotted against sequence number (seq) where seq indicates the order a case was entered into the study. This plot shows a random pattern of the internally studentized residuals about the baseline $Y=0$ suggesting that the model errors are independent of sequence number. Figure 3.4 is an example of a scatterplot which suggests a violation of independence of errors, as there is a systematic pattern of the data points. The data in this plot are a sample from a population in which internally studentized residuals and sequence number are negatively correlated.

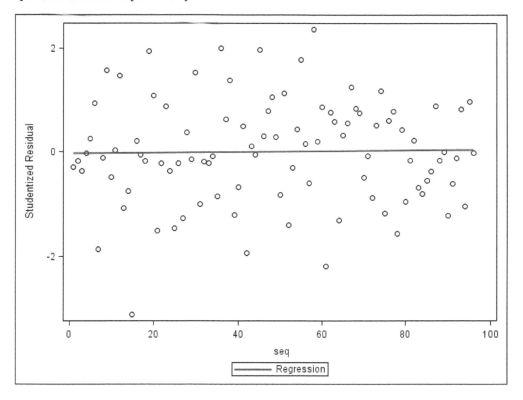

FIGURE 3.3
Scatter plot of internally studentized residuals against sequence number (seq) where errors are independent of sequence number.

3.9.2 Graphical Evaluation of Linearity and Equality of Variances

The ideal inference conditions of linearity and equality of variances can be informally evaluated by plotting internally studentized residuals against their corresponding X values in simple linear regression. Linearity is indicated if the residuals, centred about the line $Y=0$, occur in a horizontal band as illustrated in Figure 3.5. Equality of variances is indicated by the equal width of this horizontal band along the X axis as shown in Figure 3.5, i.e., no systematic increase or decrease in the size of the internally studentized residuals as X increases.

Figure 3.6 provides scatter plots for the same sample drawn from a population where there is a curvilinear relationship between the response and a single predictor variable. In the left hand panel of this figure where internally studentized residuals from the simple linear regression model are plotted against X, the curvature in the pattern of points clearly reveals a violation of the ideal inference condition of linearity. However, this linearity violation is not obvious in the right hand panel of Figure 3.6, where the actual (x,y) values of this sample are plotted and a regression line is fitted to these data. The scale imposed by the Y values in the right hand panel makes it difficult to detect the curvilinear relationship between X and Y in this sample. Moreover, when we consider deviations of actual data points from the fitted line, we are basing our evaluation of a model's fit on raw residuals (Section 3.6) which can be misleading, as the standard error of the residuals is not taken into account. This example underlines the importance of using a scatter plot of internally studentized residuals when assessing a departure from linearity in simple linear regression.

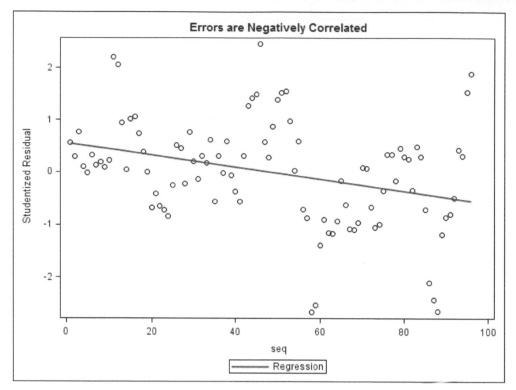

FIGURE 3.4
Scatter plot of internally studentized residuals from a simple linear regression model against sequence number (seq) for a sample from a population where the errors are negatively correlated with sequence number.

Figure 3.7 is an example of residuals from a simple linear regression model where there is a violation of the model's equality of variances assumption. In this figure, the funnel-shaped pattern that is widest at the right of the graph where the values of X are larger suggests the model errors increase with increasing X. In contrast, a funnel-shaped pattern that is widest at the left of the graph where the values of X are smaller suggests the model errors decrease with increasing X.

The linearity and equality of variances assumptions for a simple linear regression model can also be informally evaluated by plotting internally studentized residuals from the model against their corresponding fitted (predicted) Y values. The same basic pattern will be displayed whether internally studentized residuals are plotted against corresponding X values or fitted Y values in simple linear regression when the estimated slope $(\hat{\beta}_1)$ is positive. When $(\hat{\beta}_1)$ is negative, these two plots are mirror images of each other. Thus, information for evaluating the ideal inference conditions of linearity and equality of variances may be obtained regardless of whether internally studentized residuals are plotted against observed X values or against fitted Y values in simple linear regression. This will not be true for curvilinear regression or multiple regression.

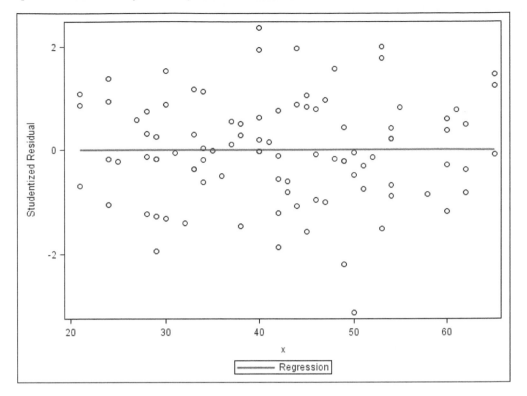

FIGURE 3.5
Scatter plot of internally studentized residuals from a simple linear regression model vs. X for a sample drawn from a population where ideal inference conditions of linearity and equality of variances for the model are met.

3.9.3   Graphical Evaluation of Normality

A *normal Q-Q plot* of residuals, sometimes referred to as a *normal probability plot of residuals*, can be used to evaluate whether there is a departure from the ideal inference condition of normality.

Definition 3.6. A *normal Q-Q plot of residuals* is a plot of Q1, the quantiles (percentiles) of the sample residuals against Q2, the expected corresponding quantiles of a normal distribution with the same mean and standard deviation.

Figures 3.8 and 3.9 illustrate how a Q-Q plot can be helpful in evaluating the ideal inference condition of normality:

• Figure 3.8 is a Q-Q plot of internally studentized residuals which illustrates the situation where the ideal inference condition of normality holds. The residual points from the sample in this normal Q-Q plot lie along the reference line. Figure 3.8 can be considered an idealized plot, not always seen in the real world, especially when the sample size is small.

• Figure 3.9 is a normal Q-Q plot of internally studentized residuals which illustrates a departure from normality because the distribution is skewed to the right. Recall a normal distribution is always perfectly symmetrical about its mean, i.e., not skewed to the left or right. A right skewed distribution (also referred to as a positively skewed distribution) is indicated in this plot where the normal quantiles are on the horizontal axis because

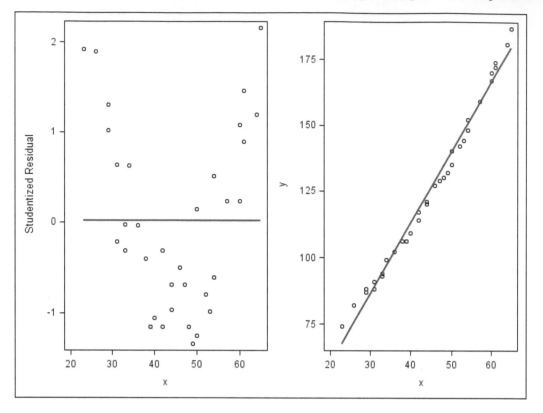

FIGURE 3.6
Scatter plots of internally studentized residuals from a simple linear regression model vs X and Y vs X for a sample drawn from a population where there is a curvilinear relationship between X and Y.

the plot has an approximate shape of a concave curve.[1] A possible example illustrating a right skewed distribution of Y given X could be in a population where subjects ranged in age from 0 to 65 years (average age 23 years), where Y is an individual's annual medical cost to a health plan and X is age of individual (Lumley et al., 2002). At each age there is a right skewed distribution of Y because at each age there were a few individuals whose annualized cost to the system was much greater than the others.

• Figure 3.10 is a normal Q-Q plot of internally studentized residuals which illustrates a distribution that is skewed to the left. In this plot where the normal quantiles are on the horizontal axis, a left skewed distribution (i.e., a negatively skewed distribution) is indicated because the plot has an approximate shape of a convex curve.[2] A possible example illustrating the distribution of Y given X which is left skewed could be where Y is an individual's age and X is year of sampling in a population where the birth rate is close to zero and there is no immigration or migration so very few babies are added to the population each year. At each year of sampling there is a left skewed distribution of age (the Y variable) because there are fewer very young individuals as compared to the many individuals who are older in the population.

---

[1]If normal quantiles are plotted on the vertical axis, a convex curve would indicate a right skewed distribution.

[2]If normal quantiles are plotted on the vertical axis, a concave curve would indicate a left skewed distribution.

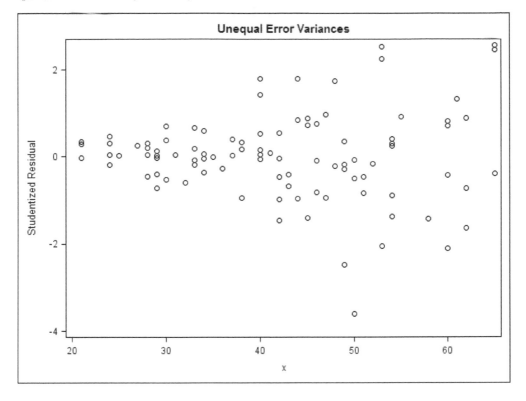

FIGURE 3.7
Scatter plot of internally studentized residuals from a simple linear regression model against X where the model's equality of variances assumption is violated.

- Figure 3.11 is a normal Q-Q plot of internally studentized residuals illustrating a distribution with long tails, i.e., a distribution where more values of a variable are spread out in the tails of the distribution than you would expect if the distribution were normal. A distribution with long tails is steepest at the top and bottom of a normal Q-Q plot. Examples of long-tailed distributions are a frequency distribution of book titles sold by an internet book seller (Brynjolfsson et al., 2010; Brynjolfsson et al., 2011) and a frequency of internet search terms (Wicklin, 2014).

Why Use Internally Studentized Residuals in Q-Q Plots?

Although a normal Q-Q plot of raw residuals can sometimes yield a similar pattern as a normal Q-Q plot of internally studentized residuals, it is preferable to use internally studentized residuals when evaluating normality. The difficulty with using raw residuals in a normal Q-Q plot is that raw residuals from the same regression have different standard deviations (standard errors) (Equation 3.1). This makes interpretation of a Q-Q plot less than straightforward. If there is a departure of the points from the reference line in the Q-Q plot, we cannot attribute the cause for this departure to a non-normal distribution of the raw residuals or to the fact that raw residuals have different standard deviations or to both of these factors. This difficulty does not occur when internally studentized residuals are used in a Q-Q plot, as they all have same expected standard deviation.

However, even when internally studentized residuals are used in a Q-Q plot, sampling variation as well as the phenomenon of supernormality of residuals in small to moderate

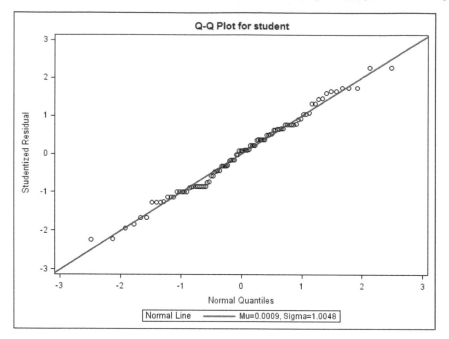

FIGURE 3.8

A normal Q-Q plot of internally studentized residuals for a sample drawn from a population where the ideal inference condition of normality is met.

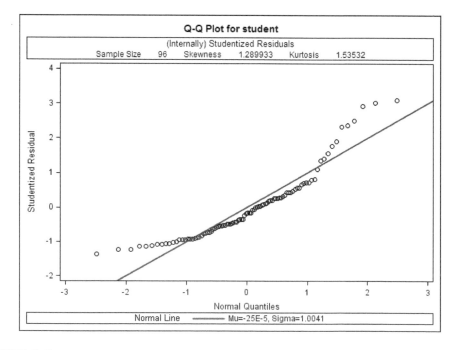

FIGURE 3.9

A normal Q-Q plot of internally studentized residuals for a sample drawn from a population where model errors are right skewed

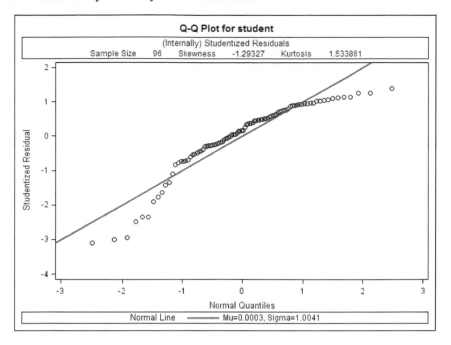

FIGURE 3.10
A normal Q-Q plot of internally studentized residuals for a sample drawn from a population where model errors are left skewed.

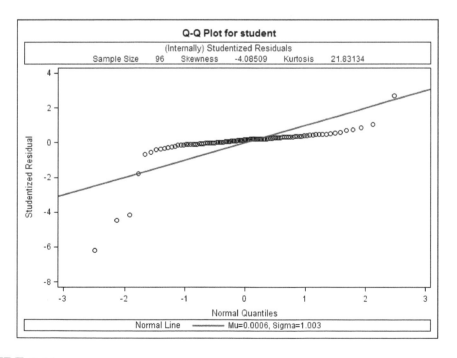

FIGURE 3.11
A normal Q-Q plot of internally studentized residuals for a sample drawn from a population where model errors have long tails.

samples make it difficult to reliably confirm that the model errors have a normal distribution in the sampled population. However, serious departures from normality, which have an important impact on substantive conclusions from hypothesis tests and estimated confidence and prediction intervals, can usually be detected, especially when the sample size is not too small.

3.10　Normality Tests and Gauging the Impact of Non-Normality

We do not recommend relying on results from statistical tests for normality to gauge the impact of non-normality on inference procedures (hypothesis tests, confidence intervals, and prediction intervals), which assume a normal distribution. When assessing the impact of non-normality in regression inference procedures, it is important to use judgement based on the combined information from expert opinion of the subject matter, graphical displays of the residuals, and statistical simulation studies that document the robustness of the regression inference procedure.

A key difficulty with normality tests is that they do not indicate the nature of departure from normality in the population, such as whether the distribution is skewed or multimodal. In a skewed distribution the population mean is no longer the centre of location for the data and thus investigating the mean of Y given X in regression would typically not make substantive sense; for example in a right-skewed distribution the mean is larger than the median. Investigating the mean of a population that has a multi-modal distribution (i.e., has multiple modes) can also typically lead to a misleading conclusion.

Another difficulty with normality tests is that the larger the sample size, the greater the power these tests will have to statistically detect small unimportant departures from normality. Generally statistical simulations have suggested that minor departures from normality will not affect conclusions made from regression inference procedures which assume a normal distribution. With small sample sizes, a potential problem with normality tests is that they will not have sufficient power to detect important departures from normality which affect substantive conclusions.

When sample sizes are small, investigators sometimes apply a test for normality to the residuals in the analysis. The difficulty here is that residuals based on least squares estimation in regression are not all independent from each other, which violates an assumption of the normality test and thus the p-value of the test will not be exact. The non-independence of residuals becomes less important with increasing sample sizes.

3.11　Homogeneity of Variance Tests and Gauging the Impact of Unequal Variances in Regression

Potential difficulties associated with using a homogeneity of variance test to gauge the impact of violation of the ideal inference condition of equal variances in regression include:

- A homogeneity of variance test may lack power to detect unequal population variances because of small sample sizes. For example, in simple linear regression such a test may lack power because there is insufficient independent replication at each X in the sample.

- A homogeneity of variance test does not give an effect size. Rejection of a test's null hypothesis of equal variances does not inform us about the magnitude of the inequality of variances and hence does not inform us about potential serious inference problems caused by the unequal variances. Suggestions as to how large the ratio of the largest to smallest variance might be before causing serious harm in regression are discussed in Section 14.6.1.

- A homogeneity of variance test may give misleading results if the homogeneity of variance test's ideal inference conditions are not reasonable for the data (e.g., Hartley's F_{max} test, subsection 11.12.2).

3.12   Screening for Outliers to Detect Possible Recording Errors in Simple Linear Regression

One or more errors may be present in data because of errors that occurred at the data collection phase of the research. For example, a research assistant may reverse digits and record a value as 92 instead of 29. One way to informally screen for possible recording errors is to identify and investigate any case in your data that has an unusual value, i.e., any case that is:

- a **Y-outlier**, i.e., a case that has an extreme Y value for the population sampled

- an **X-outlier**, i.e., a case that has an extreme X value for the population sampled

- a **contextual outlier**. We define a **contextual outlier** in simple linear regression as a case that has an unusual combination of an X and a Y value in the context of the population sampled

- a **regression outlier** (Definition 3.1, Section 3.2)

Examples of Outlier Types
Consider a scatter plot of data which was randomly sampled from a population of Montreal males in 2019. The Y variable of this scatter plot is height in feet and the X variable is age in years.

- A Y-outlier would be a case with a height of seven feet.

- An X-outlier would be a case with an age of 100 years.

- A contextual outlier would be a case that is five feet tall and six years of age.

An example of a regression outlier has been given Figure 3.1 (Section 3.2).

A scatter plot of Y against X, which includes the estimated simple regression line, may reveal these types of outliers. Outlying observations for the Y and the X variables can also be identified using the UNIVARIATE procedure to provide a listing of the unusual (extreme) values for these variables along with the identification number of the cases (e.g., Program 3.1A, Section 3.A).

As discussed in Section 3.8, studentized deleted residuals can be used to detect regression outliers (e.g., Step (9) Program 3.1B, Section 3.B). However, you would not want to employ studentized deleted residuals in the preliminary stage of outlier screening before ideal inference conditions of independence of errors, linearity, equality of variances, and normality have been evaluated. Studentized deleted residuals are useful as a diagnostic tool for regression outlier detection only if ideal inference conditions are reasonable for your data.

If investigation reveals that any Y-outliers, X-outliers, contextual, or regression outliers have resulted from a recording error, the error should corrected before proceeding to the subsequent model checking steps in Section 3.13.

Consider the situation where the value of a Y-outlier, an X-outlier, or a contextual outlier is so extreme that it would be considered impossible and therefore a result of a recording error, for example, a height of 160 feet, an age of 200 years, or a one-year-old who is five feet tall. If the investigator cannot correct the error in this situation, then the case with the impossible value should be removed from the analysis. However, we would like to emphasize the distinction here between an impossible and an improbable value. We do not recommend eliminating a case with an improbable value solely because it seems unlikely. In doing so one might be throwing away important information about a rare occurrence in the sampled population. There is always the danger when a case is excluded that misleading conclusions about the sampled population may be made due to selection bias. Exclusion of any case(s) from an analysis should always be reported in the description of the research study. Another course of action that may be appropriate when impossible values of Y-, X- or contextual outliers occur in the data is to consider such values as missing data (Section 1.5).

3.13   Overview of a Step-by-Step Approach for Checking Ideal Inference Conditions in Simple Linear Regression

Assessing ideal inference conditions, even in simple linear regression, can be a somewhat bewildering process as there are many diagnostic measures, graphical displays, and hypothesis tests readily available for this purpose. To simplify the process we propose the following nine-step approach:

Step (1) Screen for any data input errors and correct any errors found – the first essential step of any data analysis (Section 1.14).

Step (2) Screen your data for possible recording errors by identifying and investigating outliers (Section 3.12).

Step (3) Confirm that the response variable is continuous.

If the response variable is continuous, proceed to Step (4).

If the response variable is quantitative discrete where there are many values; for example, if the response variable is number of diatoms in a mussel's stomach where the sample values range from 50 to 16,800 diatoms, then proceed to Step 4, because a simple linear regression model may still be adequate for the data, although the ideal inference condition of normality does not hold.

If the response variable is nominal categorical (qualitative), ordinal categorical, or quantitative discrete, where there are only a few values (for example, number of nests where the only values are 0,1,2,3 nests), then stop at this step and consider categorical data analysis instead of simple linear regression. For a comprehensive reference for categorical data analysis using SAS, see Stokes et al. (2012).

Step (4). Review the details of the sampling design for your study to see if there are any obvious aspects of the design that would suggest that simple linear regression is inappropriate and that a more complicated model is required. Examples of sampling designs where a more complicated model would be required (such as repeated measures on the same individual at monthly intervals) are discussed in Chapter 14. If there is an obvious aspect of your sampling design that would require a different model with a more complicated error structure, then stop at this step. If, on the other hand, there is nothing immediately obvious about your sampling design that would make your simple linear regression model inappropriate, then proceed to Step (5).

Step (5) Apply a simple linear regression to the data to obtain regression diagnostics.

Step (6) Informally evaluate the ideal inference conditions of linearity and equality of variances in the sampled population.
Examine a plot of internally studentized residuals against their corresponding X values (or equivalently in simple linear regression, a plot of internally studentized residuals against the corresponding fitted Y values) for violations of the ideal inference conditions of linearity and equality of variances. See subsection 3.9.2 for details. If the ideal inferences conditions of linearity, and equality of variances do not look reasonable, see a discussion of other approaches in Chapter 14.

Note Regarding Omitting Step (7): You can omit Step (7) which further investigates independence of errors using internally studentized residuals and proceed directly to Step (8) if your investigation was a well-designed controlled randomized experiment (Section 1.12) in which each subject was randomized (i.e., each subject was assigned at random using a formal randomization procedure) to the "treatment" groups in the experiment (e.g., each individual was randomly assigned to a drug or placebo group) or equivalently the "treatment" was randomized to each individual in the study. In a well-designed controlled randomized experiment, all aspects related to measuring the response variable in the study have been either held constant (e.g., the same technician measured the response variable of a study subject at the same time on the same date) or, if not held constant, these aspects have been randomized in the study (e.g., the same technician measured the response variable of the study subjects at different times on different dates but the date and time of measurement were assigned at random to the subjects via a formal randomization procedure).

Step (7) Use graphical displays of internally studentized residuals to informally evaluate whether the ideal inference condition of independence of errors has been violated for the simple linear regression model due to the errors being systematically related to the explanatory variable in the model or to any other variable not included in the model.

Violation of independence of errors can occur, for example, when:

- the sampling units (e.g., individual mice) for the study have been non-randomly sampled with respect to a variable such as time, space, or some other key variable related to the details of an observational study
- the measurement of the response variable has not been carried out in a constant manner in an observational study or in a quasi-experiment (i.e., a non-randomized experiment)

You can informally evaluate independence of errors in these situations by examining separate plots of internally studentized residuals against the explanatory variable in your model as well as against any time, space, or other key variable related to the details of the sampling and measurement of the cases. If the internally studentized residuals appear to be exhibiting a systematic (i.e., nonrandom) pattern in any of these plots (subsection 3.9.1), then this suggests that the model errors are not independent of the variable in the plot. You should therefore stop checking the appropriateness of your simple linear regression model at this step and consider other modelling approaches discussed in Chapter 14.

Step (8) Evaluate the ideal inference condition of normality.
Utilize graphical displays (subsection 3.9.3) and expert prior knowledge of the research area to evaluate this ideal inference condition.

Step (9) Identify and investigate regression outliers using studentized deleted residuals.
Studentized deleted residuals can be appropriately used for regression outlier detection if previous steps have suggested that the ideal inference conditions are reasonable. Investigate any cases with an absolute value of a studentized deleted residual (RSTUDENT) greater than a recommended cutoff value, often 2. A listing of the values of RSTUDENT and the corresponding identification numbers for the cases can be readily obtained by specifying the INFLUENCE option in the MODEL statement of the REG procedure. However, instead of examining the entire listing of RSTUDENT values to find those cases with absolute values that exceed a recommended cutoff, it is easier if you get SAS to check if there are any such cases and if so print out their RSTUDENT values as well as their values for variables of interest that might shed light on why those cases are regression outliers (as will be illustrated in Program 3.1B, STEP 9).

A plot of RSTUDENT values against leverage values can be helpful to a researcher, especially if the sample size is large, as this plot can reveal at a glance whether any cases that are regression outliers are also X outliers as indicated by their leverage values. This plot can be readily obtained from the REG procedure and will be further discussed in subsection 3.B.1.

If investigation of a regression outlier reveals a recording error, correct the error and then go back to Step (5) and apply the simple linear regression analysis on the revised data set. As discussed in Section 3.2, if specific facts, not merely speculation, can be forwarded as evidence that a regression outlier has resulted from a recording error, which is impossible to correct at that stage in the research, then the regression outlier may be deleted from the data set. If this is the situation, go back to Step (5) and apply the simple linear regression analysis on this revised data set where any cases with recording errors have been removed. However, whenever one or more cases are removed from an analysis, it is always important to indicate the removal and the justification for the removal in any report about the research study.

3.14   Example of Evaluating Ideal Inference Conditions

As an example of this nine-step approach in practice, we consider data from the mine site example (Example 1.1, Section 1.15). One of the questions of interest in this study was whether tissue level of arsenic (AS) in a fish was related to the fish's total body length (TL). The researchers decided that a simple linear regression model where AS is the response variable and TL the predictor variable might shed light on this question. However, first they wanted to know if the ideal inference conditions for simple linear regression would be reasonable for their sampled population. The results and conclusions from applying the nine-step approach for checking ideal inference conditions for this regression are as follows:

Step (1) Screen for any data input errors. This was carried out in Section 1.14 and no data input errors were found.

Step (2) Screen your data for possible recording errors by identifying and investigating outliers. Figure 3.12, a scatter plot of arsenic against total length (from the SGPLOT procedure in Program 3.1A, Section 3.A) did not reveal any obvious Y-, X-, contextual, or regression outliers.

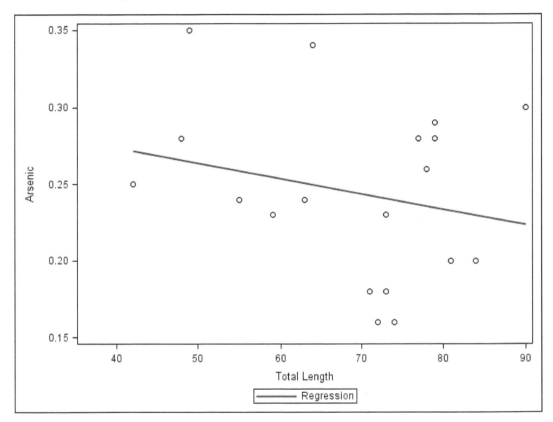

FIGURE 3.12
Scatter plot of arsenic vs. total length for Example 1.1.

Unusual (extreme) values for arsenic and total length reported by the UNIVARIATE procedure of Program 3.1A confirmed there are no obvious Y- or X- outliers as shown:

```
                The UNIVARIATE Procedure
                Variable:  AS  (Arsenic)

                    Extreme Observations

        --------Lowest--------        --------Highest--------

    Value     RECORD       Obs      Value     RECORD       Obs
```

0.16	799	3	0.28	791	18
0.16	797	1	0.29	783	10
0.18	782	9	0.30	781	8
0.18	778	5	0.34	785	12
0.20	780	7	0.35	793	20

Variable: TL   (Total Length)

Extreme Observations

---------Lowest--------			--------Highest--------		
Value	RECORD	Obs	Value	RECORD	Obs
36	795	22	79	779	6
37	796	23	79	783	10
42	792	19	81	798	2
45	794	21	84	780	7
48	791	18	90	781	8

Step (3) Confirm that the response variable is continuous. The response variable, the level of arsenic in a fish's tissue, is a continuous variable.

Step (4). Review the details of the sampling design for your study to see if there are any obvious aspects of the design that would suggest that simple linear regression is inappropriate and that a more complicated model is required. The details of the sampling design do not indicate any obvious reason why a simple linear regression would be inappropriate. There were no repeated measurements. Each data point is a measurement from a different individual. All the fish were collected from the same site at the same time in the same manner so there were no systematic differences in time, space, or sampling method associated with the collection.

Step (5) Apply a simple linear regression to the data to obtain regression diagnostics. This was carried out using the REG procedure in Step 5 of Program 3.1B (Section 3.B).

Step (6) Informally evaluate the ideal inference conditions of linearity and equality of variances in the sampled population. Figure 3.13, the plot of the internally studentized residuals against X, the total length, does not suggest that linearity has been violated as the pattern of points do not exhibit a curvilinear shape. Moreover, Figure 3.13 suggests the ideal inference of equality of variances is reasonable as the spread of the data points in the sample does not appear to markedly increase or decrease as total length increases.

Step (7) Use graphical displays of internally studentized residuals to informally evaluate whether the ideal inference condition of independence of errors has been violated for the simple linear regression model due to the errors being systematically related to the explanatory variable in the model or to any other variable not included in the model. A plot of internally studentized residuals against X, the total length, was obtained to informally evaluate whether the population errors were independent of X (Step 7, Program 3.1B). The random scatter of the internally studentized residuals about the

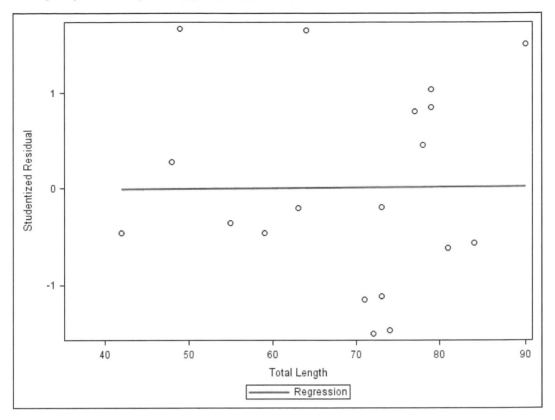

FIGURE 3.13
Internally studentized residuals against TL, total length for Example 1.1.

line $Y=0$ in this plot (Figure 3.13) suggests that the population residuals are uncorrelated with X. The following details of the study's implementation would suggest that violation of independence of errors is unlikely, viz., the preparation of the fish tissue samples for arsenic analysis and the measurement of total body length of the fish was carried out in the same way by the same experienced research assistant. Her expertise allayed any concern that her technique was associated with a learning curve over the time period when the tissue samples were prepared.

Step (8) Evaluate the ideal inference condition of normality. A normal Q-Q plot of internally studentized residuals was obtained from PROC UNIVARIATE (Step 8, Program 3.1B). Figure 3.14 suggests that the ideal inference condition of normality may be reasonable for these data because the plotted points fall on or near the straight line indicating reasonable agreement between the quantiles of the internally studentized residuals (obtained from applying the simple linear regression model to the data) and the quantiles of a normal distribution with a mean approximately equal to zero and a standard deviation approximately equal to one.

Step (9) Identify and investigate regression outliers using studentized deleted residuals.
As no serious violations of the ideal inference conditions of independence of errors, linearity, equality of variances, and normality were evident in the previous steps, studentized deleted residuals were used as diagnostics for regression outlier detection. The following

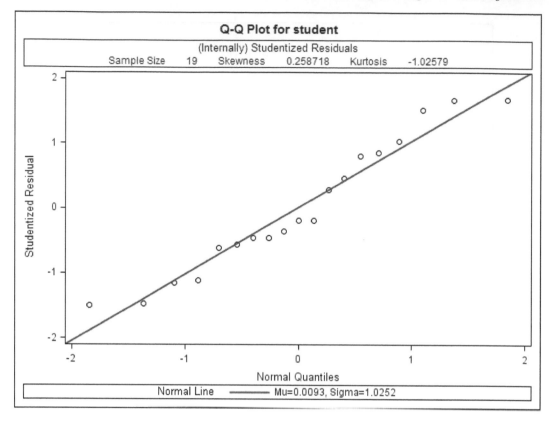

FIGURE 3.14
A normal Q-Q plot of internally studentized residuals for Example 1.1.

message in the SAS log from Program 3.1B confirms the fact that no cases (0 observations) were found that met the selection criterion of having an absolute value of RSTUDENT greater than 2.0.

Partial Output from SAS Log for Program 3.1B

```
*STEP 9 Identify regression outliers using studentized deleted residuals;
proc print data=diag; where abs(rstudent)>2.0;
run;
NOTE: No observations were selected from data set WORK.DIAG.
NOTE: There were 0 observations read from the data set WORK.DIAG.
      WHERE ABS(rstudent)>2;
```

3.15   Checking for Influential Cases

Another key aspect of model checking is the identification and investigation of influential cases. This aspect is important to researchers because of the substantive implication of using an estimated model where just one case or a few have a disproportionate influence on the estimation of the model's parameters. One approach used to assess the extent of influence that a single case in the sample has on the estimated regression model for the sampled population is to compare the estimated model with and without that case included in the analysis. This is the approach used by DFBETA, DFFITS, and COOK'S D, the measures of influence described in the following subsections. Program 3.1C (Section 3.C) illustrates how easily these measures can be obtained in SAS.

Before using these measures to identify influential cases you should first evaluate whether independence of errors, linearity, and equality of variances are reasonable for your data as these ideal inference conditions are underpinning these measures of influence. Unreliable values for these influence measures may result if these ideal inference conditions are not reasonable.

3.16   DFBETA: Influence on an Estimated Regression Coefficient

Definition 3.7. DFBETA is a measure of the influence that a case has on the estimation of a regression coefficient for a predictor in the model. It is the scaled difference in the estimated values of the β (the regression coefficient) for a given predictor estimated with and without the case included in the regression analysis. Here "DF" in the word DFBETA is meant to remind us of the word "difference" not "degrees of freedom".

3.16.1   General Equation for DFBETA

Let DFBETA$_{ij}$ indicate the DFBETA for the ith case and the jth predictor

$$\text{DFBETA}_{ij} = \frac{\hat{\beta}_j - \hat{\beta}_{j(-i)}}{\widehat{\text{SE}}^*(\hat{\beta}_j)} \tag{3.6}$$

where

$\hat{\beta}_j$ is the estimate of β_j when case i is included in the analysis

$\hat{\beta}_{j(-i)}$ is the estimate of β_j when case i is excluded in the analysis

$\widehat{\text{SE}}^*(\hat{\beta}_j)$ is a modified estimated standard error (standard deviation) of $\hat{\beta}_j$, modified in that one of the factors of this standard error, viz., the estimated common variance of Y at each X, is computed with case i excluded from the analysis

3.16.2   Equation for DFBETA for β_1 in Simple Linear Regression

Substituting $j = 1$ in Equation 3.6, the DFBETA for the ith case and the single predictor in simple linear regression is given by

$$\text{DFBETAS}_{i1} = \frac{\hat{\beta}_1 - \hat{\beta}_{1(-i)}}{\widehat{\text{SE}}^*(\hat{\beta}_1)} \tag{3.7}$$

where

$\hat{\beta}_1$ is the estimated value of β_1 when case i is included in the regression analysis

$\hat{\beta}_{1(-i)}$ is the estimated value of β_1 with case i excluded from the regression analysis

$\widehat{\text{SE}}^*(\hat{\beta}_1)$ is given by

$$\widehat{\text{SE}}^*(\hat{\beta}_1) = \sqrt{\text{Mean Square Error}_{(-i)}} \sqrt{c_{jj}} \tag{3.8}$$

where

Mean Square Error$_{(-i)}$ is used as an estimate of the common variance of Y at each X with case i excluded under the assumption that the ideal inference conditions of linearity, independence of error, and equality of variances are met.

c_{jj} is equal to $\sum(X_i - \bar{X})^2$ in simple linear regression

3.16.3   Interpretation of DFBETA

Assuming the ideal inference conditions of linearity, independence of error, and equality of variances are reasonable

  (i)   The plus or minus sign of a DFBETA value for case i and a given predictor indicates whether the estimated value of the regression coefficient for that predictor is respectively increased or decreased when case i is excluded from the analysis.

 (ii)   The size of the absolute value of DFBETA for case i and a given predictor indicates the extent of influence that case i has on the estimation of the regression coefficient for that predictor.

The larger the absolute value of a DFBETA value for case i and a given predictor, the greater the influence case i has on the estimation of the regression coefficient for that predictor. However, the size of the absolute value of DFBETA is inversely related to sample size, that is, with larger sample sizes, you will tend to get smaller values of DFBETAS. Therefore, Belsley et al. (1980) recommended the size-adjusted guideline that an absolute value of DFBETA greater than $2/\sqrt{n}$ for an observation indicates it is influential and warrants investigation. Their rationale for recommending this cutoff value is that DFBETA is a Student's t-like diagnostic so as a first approximation, one might want to investigate any case that has a DFBETA greater than two. However, as it is highly unlikely that removal of a single case would change the estimate of a beta as much as two in a large sample, Belsley et al. (1980) proposed this size adjusted cutoff of $2/\sqrt{n}$. Kutner et al. (2005) recommended that an individual case can be considered influential if the absolute value of its DFBETA is greater than one for small to medium sample sizes. However, see comment in 3.19 regarding guidelines for DFBETAS and other diagnostic measures of influence.

3.17   DFFITS: Influence of a Case on its Own Predicted Value

Definition 3.8. DFFITS is a measure of the influence that a case has on the estimation of its own predicted (fitted) value.

3.17.1   Equation for DFFITS

DFFITS for case i is given by

$$\text{DFFITS}_i = \frac{\hat{Y}_i - \hat{Y}_{i(-i)}}{\widehat{SE}^*(\hat{Y}_i)} \tag{3.9}$$

where

$\hat{Y}_i$ is the predicted value for case i obtained from the estimated regression model with case i included in the analysis

$\hat{Y}_{i(-i)}$ is the predicted value for case i obtained from the estimated regression model with case i excluded from the analysis

$\widehat{SE}^*(\hat{Y}_i)$ is a modified estimated standard error (standard deviation) of $\hat{Y}_i$, modified in that one of the factors of this standard error, viz., the estimated common variance of Y at each X, is computed with case i excluded from the analysis

$$\widehat{SE}^*(\hat{Y}_i) = \sqrt{\text{Mean Square Error}_{(-i)}}\sqrt{h_i} \qquad (3.10)$$

where

Mean Square Error$_{(-i)}$ is used as an estimate of the common variance of Y at each X with case i excluded

h_i denotes the hat (leverage) value for case i. Recall the leverage value for case i in simple linear regression is a measure of how far away in X space, case i is from $\bar{X}$

3.17.2 Interpretation of DFFITS

Assuming the ideal inference conditions of linearity, independence of error, and equality of variances are reasonable, the larger the absolute value of DFFITS for a case, the greater the influence that case has in determining its own predicted value. There is no agreed upon general guideline for identifying influence using DFFITS. Belsley et al. (1980) suggested a case could be considered influential, meriting further investigation, if that case has an absolute value of DFFITS greater than $2\sqrt{p/n}$ where n is the sample size and p is the number of estimated parameters in the regression model, i.e., the number of terms on the right hand side of the regression model equation. In simple linear regression $p = 2$. Belsley et al. (1980) proposed this cutoff value to adjust for the sample size and for the fact that the absolute value of DFFITS increases as the number of terms in the model increases. However, Staudte and Sheather (1990) suggested using $1.5\sqrt{p/n}$ for a cutoff value as they often found cases meriting investigation with DFFITS values less than the cutoff value suggested by Belsley et al. (1980). Kutner et al. (2005) suggested using one as a cutoff value for "small to medium" data sets and the cutoff value of Belsley et al. (1980) for "large" data sets.

3.18 Cook's Distance: Influence of a Case on All Predicted Values in the Sample

Definition 3.9. Cook's Distance (Cook's D) is a measure of the influence a case has on the estimation of the predicted (fitted) values of **all** the cases in the sample.

3.18.1 Equation for Cook's D

Cook's Distance for case i is given by

$$D_i = \frac{\sum_{j=1}^{n}(\hat{Y}_j - \hat{Y}_{j(-i)})^2}{(p)\,\text{Mean Square Error}} \qquad (3.11)$$

where
D_i is Cook's distance for case i

$\hat{Y}_j$ is the predicted value for case j obtained from the estimated regression model with case i included in the analysis

$\hat{Y}_{j(-i)}$ is the predicted value for case j obtained from the estimated regression model with case i excluded from the analysis

p denotes the number of estimated parameters in the regression model. In a simple linear regression model $p=2$

Mean Square Error of the fitted model is obtained using all the observations

An equivalent equation (Cook and Weisberg, 1999, p. 360), which expresses Cook's distance for the ith case as a function of the internally studentized residual (STUDENT) and leverage (h) for case i as well as the number of estimated parameters in the model is

$$D_i = \frac{(\text{STUDENT}_i)^2}{p} \frac{h_i}{1 - h_i} \tag{3.12}$$

3.18.2   Interpretation of Cook's Distance

Cook and Weisberg (1999) emphatically stated that a Cook's D value is not a test statistic for a hypothesis test which investigates whether a case has statistically significant influence. Rather they stated, the value of Cook's D for a case is a benchmark which can be used as an aid in detecting an influential case. In this regard, they suggested that "it is generally useful to study cases that have $D_i > 0.5$ and that it is always important to study cases with $D_i > 1$".

3.19   Caveats Regarding Cut-Off Values for Measures of Influence

The cut-off values given for DFBETAS, DFFITS, and Cook's distance should not be used to form hard and fast rules for detecting influence. These cut-off values can serve as an approximate guide but can never be used as a substitute for a researcher's judgement. Of cardinal importance to researchers is whether a substantive conclusion is changed after a particular observation has been removed from the analysis.

3.20   Detecting Multiple Influential Cases That Occur in a Cluster

When two or more influential cases (data points) lie close together, the approach for detecting influence described in the previous subsections where the effect of removing a single case is examined, may not be useful. This is because when multiple influential cases lie in a cluster and any one of the cases is removed from the analysis, the estimated model will not change much as the remaining influential case(s) in the analysis will pull the regression towards itself (or the cluster) through the process of least squares estimation. Methods have been suggested that address this problem (Belsley et al., 1980; Ryan, 2009) but as these methods require knowledge of matrix algebra we will not discuss them. However, researchers should be aware that there are methods that address this problem and that they can seek help from a professional statistician if they are concerned about the possibility of multiple influential cases that lie in a cluster in their data set.

3.21 Checking for Potentially Influential Cases in Example 1.1

As previous steps suggest that the ideal inference conditions of independence of errors, linearity, and equality of variance appear to be reasonable for Example 1.1 (Section 3.14), we therefore used DFFITS, DFBETAS, and Cook's D values to check if there were any influential cases in this example using Program 3.1C (Section 3.C). Program 3.1C enables detection of potentially influential cases as it:

- computes and generates graphical displays of DFFITS, DFBETAS, and Cook's D via the REG procedure

- calculates size-adjusted cut-off values for DFBETAS (Belsley et al., 1980) and for DFFITS (Belsley et al., 1980; Staudte and Sheather, 1990) as outlined in subsections 3.16.3 and 3.17.2

- lists cases that may merit investigation because they have absolute values which exceed suggested thresholds for Cook's D and DFFITS

3.21.1 DFBETAS Results for Example 1.1

Figure 3.15 from Program 3.1C reveals that two cases, RECORDS 781 and 793, have absolute values of DFBETAS that exceed 0.45883 (the size-adjusted threshold suggested by Belsley et al. (1980) but do not exceed the threshold value of one suggested by Kutner et al. (2005) for small to medium sample sizes.

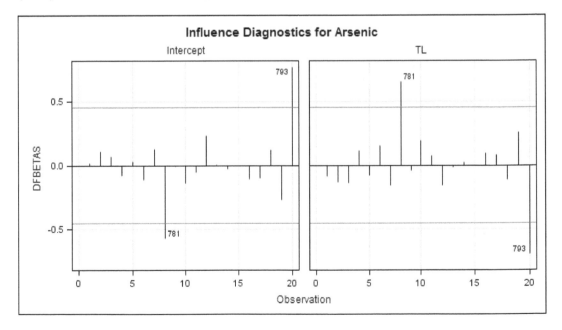

FIGURE 3.15
Values of DFBETAS for Example 1.1.

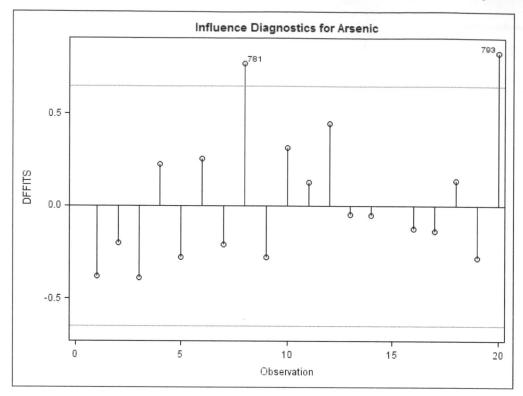

FIGURE 3.16
DFFITS values for Example 1.1.

3.21.2   DFFITS Results for Example 1.1

Figure 3.16 and the list of cases output by Program 3.1C show that RECORDS 781 and 793 have absolute values of DFFITS values that exceed Staudte and Sheather's (1990) size-adjusted cutoff value of 0.48665 and also Belsley et al.'s (1980) size-adjusted cutoff value of 0.64889.

The list of cases with DFFITS absolute values that exceed size-adjusted cutoff values for the mine site data are as follows:

```
    Cases with Abs. DFFITS Value > Belsley et al. 1980 Cutoff Value

Obs RECORD  AS  predicted TL rawresid student rstudent   cookd    dffits

  8   781  0.30  0.22362  90 0.076383 1.51830  1.58429 0.27485 0.77364
 20   793  0.35  0.26485  49 0.085149 1.67907  1.78353 0.30846 0.83430

Cases with Abs. DFFITS Value > Staudte and Sheather 1990 Cutoff Value

Obs RECORD  AS  predicted TL rawresid student rstudent   cookd    dffits

  8   781  0 30  0 22362  90 0 076383 1 51830  1 58429 0 27485 0 77364
 20   793  0.35  0.26485  49 0.085149 1.67907  1.78353 0.30846 0.83430
```

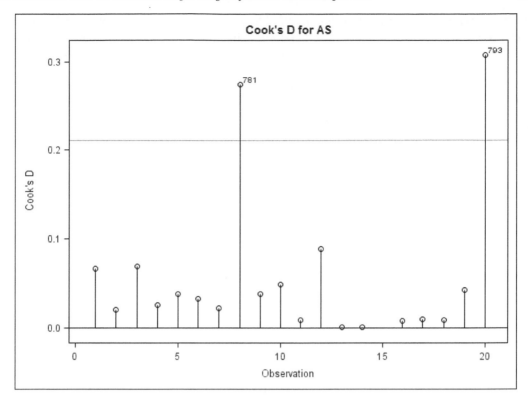

FIGURE 3.17
Cook's D values for Example 1.1.

3.21.3 Cook's D Results for Example 1.1

None of the cases have a Cook's D greater than 0.5, the cutoff value suggested by Cook and Weisberg (1999) for identifying cases which might merit investigation because of their influence on all the predicted values in the sample (Figure 3.17). The following partial output from the log file for Program 3.1C also provides this information:

```
proc print data=three; where cookd gt 0.5 ;
title Cases with COOKD Value > 0.5 ;
run;
NOTE: No observations were selected from data set WORK.THREE.
NOTE: There were 0 observations read from the data set WORK.THREE.
      WHERE cookd>0.5;
```

3.22 Discussion and Conclusion Regarding Influence in Example 1.1

Although cases 781 and 793 have absolute values of DFFITS that exceed suggested cutoff thresholds, the Cook's D values of these cases are less than 0.5, which on the basis of Cook et al.'s guideline (1999) suggests that neither of these cases singly has an influence on **all** the predicted values predicted in the sample. Moreover, the studentized deleted residual for each case was not greater than 2. The possibility that the values of the studentized deleted residual and Cook's D for cases 781 and 793 were misleadingly small due to these cases' masking each other's influential effect (subsection 3.20) does not appear to be a problem as these cases do not lie close together in x,y space, i.e., an (x,y) of (49,0.30) for case 781 and an (x,y) of (90,0.35) for case 793. Investigation of cases 781 and 793 did not reveal anything particularly unusual except case 781 had the largest total length in the sample. The conclusion of the regression analysis was not changed when the analysis was rerun with case 781 deleted from the data set. Likewise there was no change in this conclusion in a separate analysis when the analysis was rerun with case 793 deleted from the data set. This evidence suggests that the substantive conclusion of the regression analysis is not disproportionately attributable to the influence of either case 781 or case 793. Consequently neither case will be removed for the final analysis.

3.23   Summary of Model Checking for Example 1.1

No evidence of serious departures from ideal inference conditions was found (Section 3.14). No case in the sample appeared to be a regression outlier as none had a studentized deleted residual greater than two, a recommended cut-off value. Evaluation of potentially influential cases (Section 3.21) did not raise concern that the conclusions from the regression analysis are a result from just one or a few cases exerting a disproportionate influence on the precision or on the values estimated for the regression parameters. In summary, on the basis of the model checking methods presented in this chapter, we conclude that it is reasonable to use the hypothesis tests and confidence and prediction interval estimates in the next chapter to shed light on the researcher's question – is arsenic in a fish's tissue linearly related to the fish's body length.

3.24   Chapter Summary

This chapter:

- describes three key aspects of model checking: evaluation of ideal inference conditions, identification of regression outliers, and identification of influential cases

- discusses why these three key aspects of model checking need to be addressed before interpreting results from regression hypothesis tests, confidence intervals, and prediction intervals

- describes three types of residuals that can be used as diagnostic tools and discusses their advantages and disadvantages for model checking

- illustrates how graphical displays can informally evaluate the ideal inference conditions of independence of errors, linearity, equality of variances, and normality

- discusses potential difficulties regarding the use of hypothesis tests for evaluating the ideal inference conditions of normality and equality of variances

- proposes a nine-step approach for evaluating ideal inference conditions in simple linear regression and provides an illustrative example along with SAS programs that implement this approach

- describes measures of influence, quantities that can be used to identify influential cases, and discusses guidelines for the interpretation of these measures

- provides an illustrative example and a SAS program that enables identification of potentially influential cases

Appendix

3.A   Program 3.1A

Using the **temporary** SAS data set MINESITE which is created by Program 1.1 (Section 1.A), Program 3.1A checks for possible input and recording errors in the mine site data of Example 1.1. Program 3.1A screens for errors in the data by searching for outliers, i.e., this program implements Step (2) of evaluating ideal inference conditions in simple linear regression (Section 3.13). To implement this step the program generates a scatter plot of the data with a simple linear regression line inserted. The program also uses the UNIVARIATE procedure to search for extreme values for the X (total length) and Y (arsenic) variables in the data file.

Please note:

If you attempt to execute Program 3.1A after having logged out from the SAS session in which you executed Program 1.1, you will get error messages in most operating environments. One way to avoid this is by executing Program 1.1 in your current SAS session prior to running Program 3.1A or alternatively by copying the DATA step statements and data from Program 1.1 and pasting them at the beginning of Program 3.1A.

```
* Program 3.1A;
/*The following plot can be examined for contextual,Y- ,X-,
and regression outliers*/

proc sgplot data=minesite ;
   reg x=tl y=as;
run;
proc univariate data=minesite nextrval=3;
   var as tl;
   id record;
   ods exclude TestsForLocation;
run;
```

3.A.1   Explanation of SAS Statements in Program 3.1A Not Encountered in Previous Programs

proc sgplot data=minesite ; specifies that the SAS data set MINESITE is the input data for the SGPLOT procedure.

reg x=tl y=as; This REG statement requests that a simple linear regression model where the X and Y variables are respectively Tl and AS be fit to the data and that the data and regression line be displayed.

proc univariate data=minesite nextrval=3; This PROC UNIVARIATE statement requests that the UNIVARIATE procedure be applied to the data set called MINESITE. This PROC statement also requests the NEXTRVAL=n option which is a request to have the procedure display for each variable specified in the VAR statement the n high and n lowest distinct values and their corresponding frequencies in the MINESITE data set. The "Extreme Observations" table that is output by the UNIVARIATE procedure by default displays the observations (records) that have the five highest and five lowest but not necessarily distinct values in the data set (i.e., duplicate values may be displayed). The NEXTROBS=n option in the PROC UNIVARIATE statement can be used to specify a different number of extreme observations to be displayed other than the default number of five. Specifying $n=0$ for the NEXTROBS option suppresses the table. The "Extreme Observations" table provides identification numbers for the observations in its list. Both the "Extreme Observations" table and that of extreme values generated by the NEXTRVAL option can be useful in data screening.

var as tl; This **VAR statement** specifies the variables "as" and "tl" are to be analysed.
id record; This **ID statement** specifies that the corresponding values of the variable "record" should be included along with the extreme observations reported.

ods exclude TestsForLocation; excludes Tests For Location, i.e., tests of the null hypothesis that the population mean is equal to a specified value (zero by default). These tests are rarely useful in screening for data errors.

3.B   Program 3.1B

Program 3.1B implements Steps 5–9 of the evaluation of ideal inference conditions for the simple linear regression (Section 3.13) applied to the **temporary** SAS data set MINESITE created by Program 1.1 (Section 1.A).
Please note:
If you attempt to execute Program 3.1B after having logged out from the SAS session in which you executed Program 1.1, you will get error messages in most operating environments. One way to avoid this is by executing Program 1.1 in your current SAS session prior to running Program 3.1B or alternatively by copying the DATA step statements and data from Program 1.1 and pasting them at the beginning of Program 3.1B.

```
*Program 3.1B;
*STEP 5;
/* The REG procedure is used to fit a simple linear
 regression model to obtain regression diagnostics*/

proc reg data=minesite plots (only label)= RStudentByLeverage;
   id record;
   model as=tl/r influence ;
   output out=diag r=r rstudent=rstudent
   student=student p=predicted h=leverage;
run;

*STEPS 6 and 7;
```

```
/*The SGPLOT procedure is used to generate
a plot of internally studentized residuals vs. tl (X) for STEPS 6 and 7*/
proc sgplot data=diag;
    title;
    reg x=tl y=student;
run;


* STEP 8 EVALUATE NORMALITY;
* The UNIVARIATE procedure is used to generate
 a Q-Q normal plot for internally studentized residuals;

proc univariate data =diag;
    var student;
    qqplot student /normal (mu=est sigma=est);
    inset n='Sample Size' skewness kurtosis/
    pos=tm header='(Internally) Studentized Residuals';
run;


*STEP 9;
/*CHECK IF THERE ARE ANY REGRESSION OUTLIERS BASED ON RSTUDENT VALUES*/
/*Get a listing of variables of interest for cases identified as
 regression outliers based on a specified cutoff value for
 the absolute value of the studentized deleted residual (rstudent)*/
proc print data=diag;
    where abs(rstudent) gt 2.0;
run;
```

3.B.1   Explanation of SAS statements in Program 3.1B

proc reg data=minesite plots (only label)= RStudentByLeverage;
This PROC statement specifies that the temporary SAS data set MINESITE should be used as input. Specifying the keyword ONLY for the PLOTS option suppresses all default plots generated by the REG procedure with the exception of the plot of studentized deleted residuals by leverage values (RStudentByLeverage). Program 3.1B requests this option, as default plots from the REG procedure currently do not involve internally studentized residuals (denoted as STUDENT) which are used for Steps 6–8. However, SAS is typically adding new default plots so graphical displays of internally studentized residuals may be offered at a later date.

model as=tl/r influence ; This MODEL statement specifies AS to be the Y variable and TL to be the X variable in the simple linear regression model. The R option of the MODEL statement requests that predicted values and raw, internally studentized, and studentized deleted residuals be computed. It is not necessary to request the P option when the R option is requested. The INFLUENCE option is required along with the R option to obtain a listing of studentized deleted residuals.

The following OUTPUT statement
output out=diag r=r rstudent=rstudent student=student p=predicted h=leverage;

specifies that rstudent (the studentized deleted residuals), student (the internally studentized residuals), and the h (leverage) for each case should be saved in a new SAS data set called DIAG along with the values of variables for the records that were in the original data set MINESITE that was input in the REG procedure. We will need this data set DIAG in subsequent procedures in the program. The name a user specifies after the = sign for each of the quantities, i.e., rstudent=rstudent, student=student p=predicted, h=leverage is the name that these quantities will be given in the new data set DIAG. To keep things simple in Program 3.1B we specified the same names as the keywords except we renamed p as "predicted" and h as "leverage". (However, in more complex analyses where a data analyst specifies more than one model in the same REG procedure, the renaming of the keywords in the OUTPUT statement for each model is essential.)

SAS statements for the SGPLOT procedure have been previously explained.
Explanation of the QQPLOT statement **qqplot student /normal (mu=est sigma=est);** in the UNIVARIATE procedure

This QQPLOT statement requests that a normal Q-Q plot be constructed for the internally studentized residuals (student). The normal options mu=est and sigma=est request that a reference line be added to the plot corresponding to the estimated values of the mean μ_0 and the standard deviation σ_0 which are respectively the sample mean and the sample standard deviation. Agreement of the pattern of the data points with this specified reference line suggests that the normal distribution with parameters μ_0 and σ_0 is a good fit to the data.

The following INSET statement
inset n='Sample Size' skewness kurtosis /pos=tm header='(Internally) Studentized Residuals'; inserts in the QQ plot the sample size and shape statistics (skewness and kurtosis estimates) for the internally studentized residuals from the regression in this example. Shape statistics are not very reliable when the sample size is small as in this example. However, we have included this INSET statement in the program as you may have a larger sample and knowing the estimates of skewness and kurtosis may be helpful when evaluating normality. Recall from introductory statistics, the skewness and kurtosis are equal to zero in a normally distributed population. The options after the forward slash of the INSET statement, i.e, **pos=tm header='(Internally) Studentized Residuals'** give the position (top margin) and title of the box to be inserted in the plot. For other position options see SAS online documentation regarding the INSET statement for the UNIVARIATE procedure.

*STEP 9 Identify regression outliers using studentized deleted residuals (rstudent);
proc print data=diag; where abs(rstudent) gt 2.0;
The preceding WHERE statement improves the efficiency of the program and facilitates the identification of regression outliers because the PRINT procedure reads in and prints out only those cases that have an absolute value of rstudent greater than 2.

3.C   Program 3.1C

Using the **temporary** SAS data set MINESITE created by Program 1.1 (Section 1.A), Program 3.1C facilitates identification of potentially influential cases in the simple linear regression applied to the mine site data. This program computes and generates graphical displays of Cook's *D*, DFFITS, and DFBETAS values and calculates recommended cutoff values for DFBETAS and DFFITS that are appropriate for the sample size ($n=19$) and the number of estimated parameters ($p=2$) for Example 1.1.

Please note:

If you attempt to execute Program 3.1C after having logged out from the SAS session in which you executed Program 1.1, you will get error messages in most operating environments. One way to avoid this is by executing Program 1.1 in your current SAS session prior to running Program 3.1C or alternatively by copying the DATA step statements and data from Program 1.1 and pasting them at the beginning of Program 3.1C.

```
*Program 3.1C;
/*Apply a simple linear regression analysis to generate plots
of Cook's D, DFFITS, DFBETAS*/
proc reg data=minesite plots (only label)= (cooksd dffits dfbetas );
    id record;
    model as=tl;
    output out=diag r=rawresid student=student rstudent=rstudent
    p=predicted cookd=cookd dffits=dffits h=leverage;
run;
data three; set diag ;
    absdffits=abs(dffits);
run;
proc print data=three;
    where cookd gt 0.5 ;
    title Cases with COOKD Value > 0.5 ;
run;
proc print data=three;
    where absdffits gt 1.00;
    title Cases with DFFITS Absolute Value > 1.00;
run;


/*CALCULATE SIZE-ADJUSTED CUT-OFF VALUES
for DFBETAS: Belsley et al. 1980
for DFFITS Belsley et al .1980,Staudte and Sheather 1990*/
data cutvalues;
    /*parm=number of estimated parameters in the model
    n=sample size*/
    *dffitscutbels is cutoff value recommended by Belsley 1980;
    /*dffitscutss is cutoff value recommended
    by Staudte and Sheather 1990*/
    parm=2;
    n=19;
    nminusp=n-parm;
    dfbetacut=2/(sqrt(n));
    dffitscutbels=2*(sqrt(parm/n));
    dffitscutss=1.5*(sqrt(parm/n));
run;
proc print data=cutvalues;
    title Size-Adjusted Cutoff Values for DFBETAS and DFFITS;
run;
```

```
data influence;
   if _n_=1 then set cutvalues;
   set diag;
   absdffits=abs(dffits);
run;
proc print data=influence;
   where cookd gt 0.5 ;
   title Cases with COOKD Value > 0.5 ;
   var record as predicted tl  rawresid student
   rstudent cookd dffits;
run;
proc print data=influence;
   where absdffits gt 1.00;
   title Cases with DFFITS Absolute Value > 1.00;
   var record as predicted tl rawresid student
   rstudent cookd dffits;
run;
proc print data=influence;
   where absdffits gt dffitscutbels;
   title Cases with Abs. DFFITS Value > Belsley et al 1980 Cutoff Value;
   var record as predicted tl rawresid
   student rstudent cookd dffits;
run;
proc print data=influence;
   where absdffits gt dffitscutss;
   title Cases with Abs. DFFITS Value > Staudte and Sheather 1990
   Cutoff Value;
   var record as predicted tl rawresid student
   rstudent cookd dffits;
run;
```

3.C.1 Explanation of Program 3.1C

In Program 3.1C, new SAS statements and options that have not been encountered in previous programs are the following:

proc reg data=minesite plots (only label)= (cooksd dffits dfbetas); This PLOTS= option of the PROC REG statement directs SAS to generate only plots of Cook's distance, DFFITS, and DFBETAS.

The LABEL option of **plots (only label)= (cooksd dffits dfbetas)** followed by the ID statement, **id record;** results in SAS's labelling a case using the record identification number in the COOK's D, DFFITS, and DFBETAS plots to draw attention to cases that are considered influential (e.g., Figures 3.17, 3.16, and 3.15). SAS, citing Rawlings et al. (1998) as a reference, uses the following criteria for labelling a case, where p=the number of parameters estimated in the model and $n=$ the sample size:

- In the Cook's D plot, SAS labels a case if its Cook's D value is greater than $4/n$.

- In the DFFITS plot, SAS labels a case if its DFFITS value is greater than $2/\sqrt{p/n}$.

- In the DFBETAS plots, SAS labels a case if its DFBETAS value is greater than $2/\sqrt{n}$.

SAS also uses these criteria, following Rawlings et al. (1998), to draw reference lines on the relevant plots. The cut-off criteria for DFFITS and DFBETAS that SAS uses are typically recommended for labelling a case as influential. However, Cook and Weisberg (1999) suggested a different cut-off criterion for Cook's D than that used by SAS. Cook and Weisberg (1999) suggested it is always important to study cases with a Cook's D value greater than 1 and it is generally useful to study cases that have a Cook's D value greater than 0.5. Regardless of whether you agree with the criterion SAS uses in their Cook's D plot for labelling influential cases, this plot is nevertheless informative as it displays the distribution of the Cook's D values for all the cases in the sample.

model as=tl;
output out=diag r=rawresid student=student rstudent=rstudent p=predicted cookd=cookd dffits=dffits h=leverage;

The OUTPUT statement in Program 3.1C specifies the regression diagnostics and measures of influence to be saved in a new SAS data set arbitrarily called "diag" using the same names as the SAS keywords for these measures in the output statement, e.g., cookd=cookd, with the exception that the SAS keywords r, p, and h are respectively arbitrarily reassigned the names "rawresid", "predicted", and "leverage" in the data set "diag". Note the SAS keyword to reference a ODS plot of Cook's D values is **cooksd** in the PLOTS= option. However, using cooksd with a "s" instead of the keyword cookd in the OUTPUT statement will cause an error and no observations will be saved in the dataset "diag", which in turn will cause numerous errors for the rest of the program.

data cutvalues; This DATA step creates a SAS data set arbitrarily named "cutvalues". The size-adjusted cut-off values for DFBETAS (Belsley et al., 1980) and for DFFITS (Belsley et al., 1980; Staudte and Sheather, 1990) are computed in this data step for a simple linear regression ($p=2$) applied to a sample of 19 cases. These size-adjusted cut-off values are stored in "cutvalues". This data step can be easily modified to calculate these size-adjusted cut-off values for a linear regression model with a different number of estimated parameters and a different sample size by simply changing the values assigned to parm and n.

data influence; This DATA step creates a SAS data set arbitrarily named "influence". The size-adjusted cut-off values calculated in the previous data step as well as data from "diag" created in the REG procedure are saved in "influence" to be used in the subsequent PRINT procedures in this program.

The remaining code in this program uses the PRINT procedure to output separate lists of cases which have values of COOK'S D and DFFITS that exceed recommended cutoff values.

4

Interpreting a Simple Linear Regression Analysis

4.1 Introduction

The methods discussed in this chapter are valid only if there are no serious departures from the ideal inference conditions for a simple linear regression model. In Chapter 3 (Section 3.14) we did not detect any serious departures from the ideal inference conditions for the simple linear regression applied to the mine site example (Example 1.1), where the level of arsenic in a fish's tissue was the Y (outcome) variable and total length of the fish was the X (explanatory) variable. We will use this example to illustrate the methods in this chapter.

4.2 A Basic Question to Ask in Simple Linear Regression

One of the first questions to ask in a simple linear regression analysis is "Can the explanatory variable, X, provide any information about the outcome variable, Y?" One way we can evaluate this question is by investigating the null hypothesis that the population regression coefficient for X is equal to zero in the sampled population (i.e., investigating H_0: $\beta_1 = 0$). Recall β_1 is the unit change in Y per unit change in X. Therefore, if $\beta_1 = 0$, none of the change in Y is associated with a unit change in X. In other words, the explanatory variable, X, in the simple linear regression model, does not provide any information about Y.

Typically, we do not know the true value of β_1 in the sampled population. However, we can estimate β_1 using sample data from the population via ordinary least squares methodology (Section 2.7) and we denote this least squares estimate as $\hat{\beta}_1$. Just looking at the value of $\hat{\beta}_1$ alone will not tell us whether the population parameter β_1 is equal to zero. For example, the population parameter β_1 could be equal to zero although the value of $\hat{\beta}_1$ is, say, 2.3. This difference in the value of a population parameter and its corresponding sample estimate may be due to sampling variation. It is not at all surprising to find variation between a population parameter and its sample estimate because the sample consists of only some of the data from the population.

The following methods, which take sampling variation into account, can be applied to sample data to evaluate whether $\beta_1 = 0$ in the simple linear regression model for the sampled population:

- the Model F-test (Section 4.3)

- the t-test of $\beta_1 = 0$ vs. $\beta_1 \neq 0$ (Section 4.4)

- estimation of the endpoints (upper and lower limit) of a confidence interval for β_1 (Section 4.8)

These methods can be accessed in the REG procedure (e.g., Program 4.1, given in Section 4.A).

DOI: 10.1201/9780429107368-4

4.3   The Model F-Test

The Model F-test in simple linear regression tests the null hypothesis β_1 is equal to zero in the sampled population against the alternative hypothesis β_1 is not equal to zero in this population.

4.3.1   The Test Statistic for the Model F-Test

The test statistic for the Model F-test is given by

$$F = \frac{\text{Model Mean Square}}{\text{Mean Square Error}} \tag{4.1}$$

where in simple linear regression

Model Mean Square is a measure of the estimated variability in Y that is explained by the X variable in the simple linear regression model

Mean Square Error is a measure of estimated variability in Y that is not explained by X in the simple linear regression model, provided the ideal inference conditions of linearity, independence of error, and equality of variances of Y in the X-subgroups hold (as discussed in Section 2.13)

In simple linear regression the respective degrees of freedom (DF) associated with the numerator and denominator of the F-value are respectively equal to one, i.e., (p-1) and $(n - p)$, where p is the number of parameters estimated for the model and n is the sample size.

4.3.2   Mine Site Example

Program 4.1 (Section 4.A) applies a simple linear regression model to the mine site data (Section 1.15). As previously noted, results from the model checking methods in Chapter 3 indicate that the ideal inference conditions of independence of errors, linearity, equality of variances, and normality for simple linear regression appear to be reasonable for this example.

The following Analysis of Variance table from Program 4.1 reports that the observed F-value of the Model F-test in this simple linear regression is 1.02 with a p-value of 0.3273. The degrees of freedom (DF) associated with the numerator and denominator of this observed F-value are respectively equal to 1 and 17, i.e., (p-1) and (n-2) where $p = 2$ is the number of parameters in the model and $n = 19$ is the sample size. Section 4.5 discusses possible conclusions that may be considered when a p-value from a Model F-test such as 0.3273 is obtained in simple linear regression.

Analysis of Variance

Source	DF	Sum of Squares	Mean Square	F Value	Pr > F
Model	1	0.00319	0.00319	1.02	0.3273
Error	17	0.05329	0.00313		
Corrected Total	18	0.05647			

```
Root MSE              0.05599
Dependent Mean        0.24474
Coeff Var            22.87605
R-Square              0.0565
```

We will refer to the quantities (Root MSE, etc.) listed below the preceding Analysis of Variance table in later sections of this chapter.

4.4   The *t*-Test of $\beta_1 = 0$ vs. $\beta_1 \neq 0$ in the Sampled Population

The *t*-test of $\beta_1 = 0$ vs. $\beta_1 \neq 0$ is equivalent to the Model *F*-test in simple linear regression. The test statistic for the *t*-test of the null hypothesis $\beta_1 = 0$ vs. the alternative $\beta_1 \neq 0$ is given by

$$t = \frac{(\hat{\beta}_1 - 0)}{\widehat{SE}(\hat{\beta}_1)} \qquad (4.2)$$

where

  t is the observed *t*-value which can be shown to have a Student's *t*-distribution with *n*-2 degrees of freedom.

  $\hat{\beta}_1$ is the least squares estimate of the regression coefficient for X in the simple linear regression model for the sampled population.

  $\widehat{SE}(\hat{\beta}_1)$ is the estimated standard error of $\hat{\beta}_1$ and assuming the model's ideal inference conditions is given by

$$\widehat{SE}(\hat{\beta}_1) = \sqrt{\frac{\text{Mean Square Error}}{SS_X}} \qquad (4.3)$$

where

  Mean Square Error is the model dependent estimate of the population variance of Y at each X

SS_X is called the sum of squares of X and is given by

$$SS_X = \sum_{i=1}^{n}(X_i - \bar{X})^2 \qquad (4.4)$$

where

  X_i is the value of the X (explanatory) variable in the model for the ith case
  $\bar{X}$ is the mean of the X values in the sample

  The *t*-test of $\beta_1 = 0$ vs. $\beta_1 \neq 0$ is a two-sided test and the observed p-value for this two-sided test is computed by default by the REG procedure. If, however, researchers were interested in specifying a direction for the alternative hypothesis, e.g., suppose they were interested in whether or not β_1 is positive, the null hypothesis and alternative hypothesis for this one-sided test would be

$$H_0: \beta_1 \leq 0$$
$$H_A: \beta_1 > 0$$

and the test statistic given by Equation 4.2 would be the same. However, the probability for the one-sided test would be half of the value output by the REG procedure by default for the two-sided test. If researchers have applied a one-sided test it is important that they emphasize that the p-value they report is based on a one-sided test and that they provide substantive justification for applying a one-sided instead of a two-sided test.

4.4.1   Example

The following parameter estimates, obtained from Program 4.1, which applies a simple linear regression to the mine site example (subsection 4.3.2) shows that the least squares estimate for β_1, the regression coefficient for the explanatory variable, TL (total length of fish) is -0.00101 and its associated standard error is 0.00099721. Substituting these values in Equation 4.2 gives the observed t-value of -1.01 which is confirmed below. The p-value associated with the t-test of $\beta_1 = 0$ vs. $\beta_1 \neq 0$ (the two-sided test) is 0.3273.

Parameter Estimates

Variable	DF	Parameter Estimate	Standard Error	t Value	Pr > |t|
Intercept	1	0.31413	0.07000	4.49	0.0003
TL	1	-0.00101	0.00099721	-1.01	0.3273

It may be noted that $t^2 = (-1.01)^2 = 1.02$ is equal to the F-value in subsection 4.3.2 and that the p-values associated with the Model F-test and this two-sided test are identical and thus the same conclusion can be reached from either test.

4.5   Possible Interpretations of a Large p-Value from a Model F-Test or Equivalent t-Test of β_1 vs. $\beta_1 \neq 0$ in Simple Linear Regression

A large[1] p-value from a Model F-test or equivalent t-test of $\beta_1 = 0$ vs. $\beta_1 \neq 0$ in a simple linear regression model does not necessarily imply that X, the explanatory variable in the model, provides no information about Y, the response variable, in the sampled population.

It may be that X can actually provide some information about the Y variable in the sampled population but this relationship cannot be detected in a researcher's sample due to one or more of the following reasons:

- The sample investigated was not randomly obtained from the sampled population and does not give a representative picture of what is happening in the population.

- There is data missing not at random in the sample (Section 1.5).

- The sample size is too small.

---

[1]Recall labelling a p-value as large solely because it is greater than some threshold value such as 0.05 is not recommended (Section 1.9). The context of the research should always be taken into account when considering the size of a p-value.

- The range of X values is too narrow in the sample resulting in a small Model Mean Square and therefore a small observed F-value (and equivalent t-value) with a large p-value.

- There is a lot of variability among the Y values at each X value in the sample resulting in a large Mean Square Error and therefore a small observed F-value (and equivalent t-value) with a large p-value.

- One or a combination of the ideal inference conditions of the model is violated to such an extent that Mean Square Error does not provide a valid estimate of the population variance of Y given X.

- One or more additional explanatory variables need to be included in the linear model to reveal a relationship between X and Y, i.e., a multiple linear regression model is needed.

- A different model other than a linear regression model is required because the relationship between X and Y in the sampled population is not linear over all the values of X observed in the sample.

These numerous reasons as well as the cautionary remarks regarding p-values in Section 1.8 underscore why interpretation of a large p-value associated with the Model F-test (or equivalent t-test) in a simple linear regression model cannot be simplistic. These reasons also underscore the importance of using graphical displays in aiding interpretation.

4.6 Possible Interpretations of a Small p-Value from a Model F-Test or Equivalent t-test of $\beta_1 = 0$ vs. $\beta_1 \neq 0$ in Simple Linear Regression

A small p-value from the Model F-test or equivalent t-test of $\beta_1 = 0$ vs. $\beta_1 \neq 0$ in a simple linear regression model suggests (although does not unequivocally imply) that the X variable, within the range of X values observed in the sample, has some utility in explaining Y in the sampled population provided:

- The sample is representative of the sampled population, which can be achieved through random sampling. When random sampling is not possible, substantive knowledge of the research area is often used to justify extending conclusions from the non-random sample to the sampled population or to the population of interest.

- The ideal inference condition of independence of errors is not violated.

- The ideal inference conditions of linearity, equality of variances, and normality are reasonable for the sampled population.

However, a small p-value from the Model F-test or equivalent t-test of $\beta_1 = 0$ vs. $\beta_1 \neq 0$) does not imply:

- that Y is caused by X. The design of the study, not the statistical analysis, can enable researchers to conclude cause and effect (Section 1.12)

- that the simple linear regression model is the best way to describe the relationship between the X and Y variables in the sampled population. For example, it may be that a nonlinear model could be a better way to summarize the relationship between X and Y

4.7 Evaluating the Extent of the Usefulness of X for Explaining Y in a Simple Linear Regression Model

The Model F-test or the equivalent t-test of $\beta_1 = 0$ vs. $\beta_1 \neq 0$ evaluates only the very basic question "Is the X variable of any use at all in explaining Y in a simple linear regression model – yes or no?" These tests do not provide information about the **extent** of the usefulness of X as an explanatory variable in a simple linear regression model. This information may be investigated using other methods such as:

- estimating the endpoints (upper and lower limit) of a confidence interval for β_1, the regression coefficient for X (Section 4.8)

- calculating a measure called R^2 for the model (Section 4.9) or a measure called adjusted R^2 (Section 7.5)

- evaluating the prediction accuracy of the simple linear regression model

The prediction accuracy of a simple linear regression model can be evaluated via:

- the root mean square from the model (Section 4.10)

- the coefficient of variation (Section 4.11)

- estimating the endpoints of a confidence interval for the average Y of individuals having a particular X value that is in the range of X values observed in the sample data (Section 4.12)

- estimating the endpoints of a prediction interval for an individual's Y value when the individual has a particular X value that is in the range of X values observed in the sample data (Section 4.13)

- estimating a confidence band (i.e., estimating the endpoints of a set of **simultaneous** confidence intervals for the prediction of the average Y at each X, for all X values within a specified range of X observed in the sample (Section 12.13)

4.8 A Confidence Interval Estimate for β_1, the Regression Coefficient for X in Simple Linear Regression

The equation for estimating the upper and lower limits (endpoints) for a two-sided $100(1 - \alpha)\%$ confidence interval for β_1 in simple linear regression is given by

$$\hat{\beta}_1 \pm t_{1-\alpha/2;\ n-2}\widehat{SE}(\hat{\beta}_1) \tag{4.5}$$

where

$t_{1-\alpha/2;\ n-2}$ is the theoretical value from the Student's t distribution that is appropriate for obtaining a two-sided $100(1 - \alpha)\%$ confidence interval estimate for β_1 in a simple linear regression model. Thus the appropriate theoretical Student's t-value used in estimating a two-sided 95% confidence interval for β_1 in a simple linear regression model is $t_{.05/2,\ n-2}$.

$\widehat{SE}(\hat{\beta}_1)$ is the estimated standard error of $\hat{\beta}_1$ given by Equation 4.3

The term "two-sided" is often omitted when referring to a two-sided confidence interval as it is generally assumed that a confidence interval is two-sided unless it is specifically labelled as one-sided. We will consider only two-sided confidence intervals. Discussion of one-sided confidence bounds are typically found in introductory statistics textbooks.

4.8.1 Interpreting a 95% Confidence Interval Estimate for β_1

When we interpret a 95% confidence interval estimate for β_1 we make the assumption that the particular confidence interval obtained for the data does not belong to the 5% of the intervals that excludes the value of population parameter. We also assume that ideal inference conditions of the model are reasonable. With this in mind we consider the following scenarios in interpreting a two-sided 95% confidence interval estimate for β_1 which has been obtained using Equation 4.5.

When a 95% confidence interval estimate for β_1 includes zero, it does not necessarily imply that, in the range of X values observed in the sample, X, the explanatory variable in a simple linear regression model is useless as a predictor of Y for the sampled population. Nonzero values observed within the range of the 95% confidence interval estimate are also values for β_1 that should be considered and may suggest that X can provide some information about Y in the sampled population.

Moreover, it may be that a confidence interval would exclude zero if this interval were not so wide. A confidence interval for β_1 estimated from the sample data may be unduly wide due to one or more of the same reasons, outlined in Section 4.5, as to why an unduly large p-value may be observed for a Model F-test of $\beta_1 = 0$.

A 95% confidence interval estimate for β_1 which excludes zero suggests that, in the range of X values observed in the sample, X may be of some use in predicting Y in a simple linear regression model for the sampled population. How useful X is in predicting Y is suggested by the values for β_1 that lie within the range of the confidence interval estimate for β_1. Recall in simple linear regression β_1 is the change in Y for each unit change in X.

4.8.2 Example

The following "Parameter Estimates" results from Program 4.1 show that the estimated upper and lower limits for the 95% confidence interval for β_1, the regression coefficient for TL (the total length of a particular fish species) in the mine site example, are respectively -0.00311 and 0.00110. These limits were obtained by specifying the CLB option in the MODEL statement of the REG procedure as illustrated in Program 4.1.

Parameter Estimates

Variable	DF	95% Confidence Limits	
Intercept	1	0.16645	0.46181
TL	1	-0.00311	0.00110

We cannot say on the basis of this confidence interval estimate for TL, which includes both zero and nonzero values, whether or not, in the sampled population, total length of this fish species is of any use as a predictor of arsenic tissue level for the sampled population in a simple linear regression model.

It is interesting to note that the negative values for β_1 in this confidence interval estimate may not make sense based on expert knowledge of the research area. A negative value for β_1 suggests that larger individuals of this fish species would have lower levels of arsenic in their tissue which is somewhat unexpected, given the habitat of the sample as well as the range of total length of fish observed in the sample. Possibly the small sample size, resulting in a wide confidence interval estimate, may be a factor accounting for the negative values being included in this example's 95% confidence interval estimate for β_1.

4.9 R^2, the Coefficient of Determination

The coefficient of determination, often denoted as R^2, is a quantity that is sometimes used as an index of the extent that X and Y are related in a model.

In simple linear regression R^2 can be defined as the proportion of the total variation of the Y values about the sample mean that is explained by X, the predictor variable.

4.9.1   An Equation for R^2

$$R^2 = \frac{\text{Model Sum of Squares}}{\text{Total Sum of Squares}} \tag{4.6}$$

where

Model Sum of Squares is a measure of the variation of the observed Y values that is explained by the X variable in the model, i.e.,

$$\text{Model Sum of Squares} = \sum_{i=1}^{n} (\widehat{Y}_i - \bar{Y})^2 \tag{4.7}$$

Total Sum of Squares is a measure of the variation of the observed Y values about the overall sample mean of Y, i.e.,

$$\text{Total Sum of Squares} = \sum_{i=1}^{n} (Y_i - \bar{Y})^2 \tag{4.8}$$

When all of the variation in Y in the sample is explained by (or associated with) the X variable in the model, R^2 is equal to 1. When none of the variation in Y in the sample is explained by the X variable in the model, R^2 is equal to 0.

R^2 may also be interpreted in simple linear regression as:

- the square of the correlation between the X and Y values in the sample

- the square of the correlation between the observed Y and predicted Y values in the sample

In simple linear regression when cases are randomly sampled (i.e., X values not prespecified as in a fixed X model) and when the model's ideal inference conditions are met, R^2 is an estimate of ρ_{XY}^2, the square of true correlation coefficient between X and Y in the population sampled. Graybill and Iyer (1994, p. 192) have shown that ρ_{XY}^2 is a measure of how much better Y is predicted by the simple linear regression model $Y = \beta_0 + \beta_1 X$ than by the population mean of Y ignoring X

4.9.2   Some Issues Interpreting R^2

A researcher should be aware of the following issues when interpreting R^2 in simple linear regression:

- A large R^2 does not always indicate that the relationship between X and Y is linear. There are situations where the relationship between X and Y is nonlinear but you can still get a large R^2 (e.g., Kutner et al., 2005, p. 76; Montgomery et al., 2012, p. 36).

- An R^2 close to zero does not always indicate a weak relationship between X and Y, as there could be a strong relationship between X and Y but a simple linear regression model does not adequately reflect this relationship (e.g. Cook and Weisberg, 1999, p. 70).

- R^2 does not indicate whether X is an adequate predictor of Y. R^2 just informs us relatively how much better the simple linear regression model is for predicting Y as compared to the sample mean for predicting Y (e.g., Graybill and Iyer, 1994, p. 190).

- In simulations of simple linear regression fixed X models, Ranney and Thigpen (1981) showed that the expected value of R^2 can be increased by increasing the range of X.

- R^2 is biased in that it tends to overestimate the extent of the linear relationship between X and Y in the sampled population (Mickey et al., 2004).

4.9.3   Example

In the mine site study example, the R^2 value of 0.0565 (Section 4.3.2) tells us that total length of fish in the simple linear regression model explains only 0.0565 (only 5.65%) of the total variation of arsenic tissue level values in the fish sample. As this simple linear regression model is a random X model (total lengths are not preselected), assuming the model's ideal inference conditions hold, the R^2 value of 0.0565 suggests that predicting Y using the simple linear regression model would not be much better than predicting Y using the sample mean of Y and not considering X.

4.10   Root Mean Square Error

Root Mean Square is a measure of the average difference in the sample between the observed Y value and the Y value predicted by the model. Root mean square error, denoted as Root MSE, is given by

$$\text{Root MSE} = \sqrt{\text{Mean Square Error}} \qquad (4.9)$$

Root MSE is used as a summary measure for evaluating the adequacy of a model's prediction capability. Root MSE is measured in units of the response variable. For this reason, Root MSE is not recommended as a measure to compare models where the Y values are measured in different units.

The Root MSE from the simple linear regression model in mine site study example is 0.05599 micrograms^{-1} (subsection 4.3.2). The estimate of the standard deviation of the observations about the sample mean reported by the UNIVARIATE procedure in Program 3.1 A (Section 3.A) is 0.0560, which suggests that the average difference in predicting Y is not improved by using the estimated model to predict Y as compared with using the sample mean of Y to predict Y.

4.11   Coefficient of Variation for the Model

The coefficient of variation is a unit-less measure of a model's fit to the sample data, expressed as a percentage. It is calculated by dividing the Root MSE by the sample mean of Y and multiplying by 100.

$$\text{Coefficient of Variation} = \frac{\text{Root MSE}}{\bar{Y}} 100 \qquad (4.10)$$

The smaller the value of the coefficient of variation the better the fit of the model to the sample data. However, the coefficient of variation has a meaningful interpretation only if all the values of the Y variable are positive. The coefficient of variation for the mine site example is 22.88%, which is not suggestive of a close fit in the sample between the observed values and the values predicted by the model for arsenic level in the fish tissue.

4.12   Estimating a Confidence Interval for the Subpopulation Mean of Y Given a Specified X Value

Researchers may be interested in assessing how accurately the average Y can be predicted for a subgroup with a particular X value in the sampled population. This can be assessed by estimating the endpoints of a confidence interval for the mean of Y in the population subgroup having that particular X in the sampled population. The CLM option in the MODEL statement of the REG procedure estimates a $100(1 - \alpha)\%$ confidence interval for the mean of Y of each subgroup defined by the X value of every case in the sample. As illustrated in Program 4.1, the CLM option estimates a 95% confidence interval by default. Confidence interval estimates other than 95% can be also requested as illustrated in subsection 4.A.

Confidence interval estimates generated by the CLM option are not adjusted for multiplicity: In the context of confidence intervals, multiplicity is the issue that the greater the number of confidence intervals considered, the greater the coverage error probability, i.e., the greater the probability that one or more of these estimated confidence intervals will **not** include the true value of the parameter being considered. We will discuss adjusting confidence intervals for multiplicity in Section 12.11. A confidence interval that is **not adjusted for multiplicity** is referred to as a **conventional confidence interval** or a **standard confidence interval** or an **ordinary confidence interval** or sometimes **simply as a confidence interval**.

The results of the CLM option are given in the "Output Statistics" section of the REG procedure output under the column headings "Predicted Value" and "95% CL Mean".

4.12.1   Example

Suppose researchers were interested in how accurately the estimated simple linear regression model of Example 4.1 predicts the mean tissue level of arsenic in fish which have a total length of 72 mm in the sampled population. As reported in the "Output Statistics" of the REG procedure of Program 4.1 we find the predicted value for the fish identified by Record Number 797, which has a total length of 72 mm, is 0.2417, i.e., the predicted mean tissue level of arsenic in fish with a total length of 72 mm in the sampled population is 0.2417. The estimated endpoints for the 95% conventional confidence interval for this predicted mean are 0.2139 and 0.2695.

Output Statistics

Obs	Record Number	Dependent Variable	Predicted Value	Std Error Mean Predict	95% CL Mean	
1	797	0.16	0.2417	0.0132	0.2139	0.2695

4.12.2 Equation for Estimating the Endpoints of a 95% Conventional Confidence Interval for the Mean of Y for a Given X Value

Provided the ideal inference conditions of simple linear regression hold, the upper and lower limits for a 95% two-sided conventional confidence interval for the mean of Y in the population subgroup with a particular value of X, say x_0, are estimated by

$$\hat{\mu}_{y|x_0} \pm t_{.05/2,n-2}\widehat{SE}(\hat{\mu}_{y|x_0}) \tag{4.11}$$

where

$\hat{\mu}_{y|x_0}$ is the predicted mean of Y in the sampled population subgroup which has $X = x_0$

$t_{0.05/2,n-2}$ is from the t distribution with error degrees of freedom equal to n-2

$\widehat{SE}(\hat{\mu}_{y|x_0})$ is the estimated standard error of the mean of Y in the population subgroup which has $X = x_0$ and is given by

$$\widehat{SE}(\hat{\mu}_{y|x_0}) = \sqrt{\widehat{VAR}(Y|X)\left[\frac{1}{n} + \frac{(x_0 - \bar{X})^2}{SS_X}\right]} \tag{4.12}$$

The REG procedure assumes that the ideal conditions of linearity, independence of errors, and equality of variances hold and uses the Mean Square Error as an estimate of $VAR(Y|X)$ to compute $\widehat{SE}(\hat{\mu}_{y|x_0})$ (Equation 4.12).

SS_X is given by Equation 4.4.

It is important to note that the mean of Y for a given X in the sampled population can be estimated more accurately at those values of X that are closer to the sample mean of X. This is due to the term $(x_0 - \bar{X})^2$ in Equation 4.12 which makes the standard error of the estimate of the mean of Y given X larger when a particular X value, say x_0, is far from $\bar{X}$.

4.13 Estimating a Prediction Interval for an Individual Value of Y at a Particular X Value

Researchers may also be interested in assessing how accurately a predicted value from a simple linear regression model can predict a Y value for an individual with a particular X when that individual has been drawn at random from the sampled population. This can be assessed by estimating the endpoints of a prediction interval for the individual Y in the sampled population subgroup having that particular X value. It is called a prediction interval rather than a confidence interval because it is an interval for an individual Y value, rather than a population parameter. Typically a 95% prediction interval has been considered. Caution is advised in interpreting a prediction interval when the ideal inference condition of normality is suspect. As previously noted, the accuracy of a prediction interval can be degraded when there is departure from normality, even if the sample size is large. The CLI option of the MODEL statement of the REG procedure (as illustrated in Program 4.1) estimates the limits of a 95% conventional prediction interval for each individual Y value predicted in the sample. As prediction intervals estimated by the CLI option are not adjusted for multiplicity, the greater the number of prediction intervals that are considered, the greater the probability that one or more of these intervals will not include the true value under consideration. Prediction intervals are reported under the column headings "95% CL Predict" in the "Output Statistics" section of the REG procedure output.

4.13.1   Example

Suppose researchers were interested in how accurately the estimated simple linear regression model predicts the tissue level of arsenic of an **individual** fish having a total length of 72 mm in the sampled population. Results given in the "Output Statistics" section of the REG procedure in Program 4.1 show that the estimated upper and lower limits of the 95% conventional prediction interval for the tissue level of arsenic in an individual fish with a total length of 72 mm (the fish identified by Record Number 797) are respectively 0.1204 and 0.3631. The raw residual, −0.0817, i.e., observed arsenic tissue level minus the predicted arsenic tissue level for fish number 797 (Section 3.6) is also reported when the CLI option is requested.

```
                        Output Statistics

              Record
   Obs        Number        95% CL Predict              Residual

    1           797        0.1204      0.3631           -0.0817
```

Recall in Chapter 2, we stated that the predicted value from a simple linear regression is used both as an estimate of the mean of Y given X as well as an estimate of the individual Y value given X in the sampled population. It is important to note that the population mean of Y given X will always be estimated with greater accuracy than an individual Y value given X. Thus the confidence interval for the mean of Y given a particular X value will always be shorter than the prediction interval for an individual Y given that same particular X value.

4.13.2   Equation for Estimating the Endpoints of a 95% Conventional Prediction Interval for an Individual Y Given X

Provided the ideal inference conditions of simple linear regression normality hold, the upper and lower limits for a 95% two-sided conventional prediction interval for a value of Y at a specified X value for a single individual from the sampled population are estimated to be

$$\hat{y}_{x_0} \pm t_{.05/2,n-2}\widehat{SE}(\hat{y}_{x_0}) \tag{4.13}$$

where

$\hat{y}_{x_0}$ is the predicted value of an individual Y at $X = x_0$ in the study population

$t_{.05/2,n-2}$ is the appropriate theoretical value from the t-distribution with error degrees of freedom equal to $n-2$ for constructing a two-sided 95% conventional prediction interval for a simple linear regression

$\widehat{SE}(\hat{y}_{x_0})$ is the standard error of $\hat{y}_{x_0}$ and is given by

$$\widehat{SE}(\hat{y}_{x_0}) = \sqrt{\widehat{VAR}(Y|X)\left[1 + \frac{1}{n} + \frac{(x_0 - \bar{X})^2}{SS_X}\right]} \tag{4.14}$$

where

x_0 denotes any particular value of X in which a researcher is interested and which is found within the range of X values in the sample

$\bar{X}$ is the mean of the X values in the researcher's sample

SS_X is given by Equation 4.4

The REG procedure assumes the ideal inference conditions of linearity, independence of errors, and homogeneity of variances hold in the sampled population and uses the Mean Square Error as an estimate of $VAR(Y|X)$ in Equation 4.14.

Prediction of an individual Y value for a given X is more accurate at those values of X that are closer to the sample mean of X. This is due to the term $(x_0 - \bar{X})^2$ in Equation 4.14 which makes the standard error of the estimate of the individual Y given X larger, the farther the specified X value, x_0, is from the mean of X and hence the estimate is less precise.

4.14 Concluding Comments

Several important comments that pertain to inference methods presented in this chapter are:

- Extrapolation, i.e., drawing conclusions and making predictions about Y based on X values that lie outside the range of the X values observed in the sample, is not recommended in simple linear regression. Highly inaccurate predictions and misleading conclusions may result from extrapolation.

- Given that the ideal inference conditions are reasonable for the data, the power of a hypothesis test can be increased and the width of an estimated confidence and a prediction interval can be narrowed in simple linear regression by:

    (i)   increasing the sample size
    (ii) reducing the variability of Y at each X
    (iii) increasing the range of X values in the sample

4.15   Chapter Summary

This chapter:

- emphasizes that the inference methods described are not valid if there are any serious departures from ideal inference conditions

- discusses how you can answer the basic question "Is the X variable of any use at all in explaining Y variable in a simple linear regression model for a sampled population?" using the Model F-test or equivalent t-test

- discusses possible interpretations of a large vs. a small p-value from the Model F-test or equivalent t-test

- discusses how you can evaluate the extent that the X variable is useful in explaining Y in a simple linear regression model for the sampled population

- provides a SAS program illustrating methods discussed

- cautions against making predictions and drawing conclusions in simple linear regression beyond the range of X values observed in the sample

- outlines how the power of a hypothesis test can be increased and the width of a confidence and a prediction interval can be decreased in simple linear regression

Appendix

4.A   Program 4.1

Program 4.1 applies a simple linear regression analysis to the mine site data (Example 1.1, Section 1.15). In this regression the outcome variable is arsenic level in a fish's tissue and the explanatory variable is total length of the fish. Program 4.1 uses as input the **temporary** SAS data set MINESITE, created by Program 1.1.

Please note:

If you attempt to execute Program 4.1 after having logged out from the SAS session in which you executed Program 1.1, you will get error messages. One way to avoid this is by executing Program 1.1 in your current SAS session prior to running Program 4.1 or alternatively by copying the DATA step statements and data from Program 1.1 and pasting them at the beginning of Program 4.1.

```
*Program 4.1;
proc reg data= minesite plots(only stats=none) =(fit(nolimits));
   id record;
   model as=tl/clb clm cli;
quit;
```

Features in Program 4.1 Not Discussed in Previous Program Examples

The Model F-test and the equivalent t-test for $\beta_1 = 0$ are given by default in the REG procedure.

The CLB, CLM, and CLI OPTIONS specified after the forward slash on the MODEL statement of the REG procedure respectively request a confidence interval for β_1 in the sampled population, a conventional confidence interval for the subpopulation mean of Y given X in the sampled population, a conventional prediction interval for an individual value of Y for a given X in the sampled population. By default 95% confidence intervals are estimated. If a confidence interval estimate other than 95% is wanted, it can be requested with the "ALPHA=" option of the REG procedure's MODEL statement. For example, a 90% confidence interval estimate is requested by specifying alpha = 0.10 in the following MODEL statement:

```
model as = tl/clb clm cli alpha = 0.10;
```

5

Introduction to Multiple Linear Regression

5.1 Characteristic Features of a Multiple Linear Regression Model

Characteristics features of a multiple linear regression model are:

- There is only one response variable.

- The response variable is continuous.

- There are two or more explanatory variables in the model.

- The explanatory variables are often quantitative but with proper coding can also be qualitative.

- The model assumes that a linear relationship exists between the response variable and the terms associated with the explanatory variables in the model.

- Interaction effects, which will be discussed in Chapters 8 and 9, may be included in the model.

5.2 Why Use a Multiple Linear Regression Model?

You would want to use a multiple linear regression model to:

- **evaluate if and to what extent the explanatory variables in the model can explain the variation in a response variable** – for example, to evaluate if and to what extent the explanatory variables, age, body mass index (BMI), and sex assigned at birth can explain the response variable, diastolic blood pressure, in a linear regression model.

- **predict a value of a response variable for an individual, based on the values of the individual's explanatory variables, when that individual belongs to the population sampled** – for example, to predict the diastolic blood pressure of an individual on the basis of that individual's age, BMI, and sex assigned at birth.

- **predict the average value of the response variable for a group of individuals when each of the individuals in the group have the same values for the explanatory variables** – for example, to predict the average blood pressure for a group of individuals who have the same age, BMI, and sex assigned at birth (e.g., a group of 18-year-old individuals who were assigned male sex at birth with a BMI of 24).

DOI: 10.1201/9780429107368-5

- **investigate whether some variables can be replaced by others** – Sometimes it is expensive and inconvenient to measure one or more variables, so a pilot study can be set up to investigate whether other variables that are less costly and more convenient to measure can be used as replacement for the more costly variable(s) of interest (e.g., replacing the variable percentage body fat with BMI, which is easier to measure).

- **increase precision of prediction of the response variable as compared to a simple linear regression** – Often more precise prediction is achieved if there is more than one explanatory variable in the model.

- **increase the possibility of detecting a relationship between a response variable and an explanatory variable by adding additional explanatory variables to the model, thereby reducing previously unexplained random variability in the response variable.**

- **evaluate whether confounding is present in an observational study** – **Confounding** occurs in an observational study when the effect of an explanatory variable of interest (the study variable) on the response variable is **confounded** or mixed up with the effect of one or more "nuisance"[1] explanatory variables. When the principal objective of an investigation is to understand the relationship between a response variable and a study variable, it is important to identify which (if any) other explanatory variables are confounding the study variable's effect on the response. Confounding can be evaluated by comparing multiple regression models with and without explanatory variables that might be potential confounders.

- **adjust for a confounding effect of one or more "nuisance" variables** – One way to disentangle the effect of a study variable on a response variable from the effect of a certain "nuisance" variable(s) on the response variable is to include the "nuisance" variable(s) along with the study variable as explanatory variables in a multiple regression model.

- **investigate whether an *interaction effect* is present** – An interaction effect, which will be discussed extensively in Chapters 8 and 9, can have an important impact on the interpretation of a multiple linear regression model. Misleading conclusions may result if an interaction effect is ignored.

5.3   Example 5.1

Researchers wanted to investigate whether height of a male at age 18 years (ht18) could be predicted by his length at birth (birthl) and by the heights of his mother and father at age 18 (mht18, fht18 respectively). This example uses a subset of data given in Graybill and Iyer (1994).

Characteristics of a multiple linear regression model illustrated by this example are:

- There is a single response variable, height of a male at age 18.

- The response variable is continuous.

- There is more than one explanatory variable. The explanatory variables are birth length, mother's height at age 18, and father's height at age 18.

---
[1] Definition 1.24

5.4 Equation for a First Order Multiple Linear Regression Model

One way of writing a first order multiple linear regression model with k explanatory variables is

$$\mu_{Y|X_1,\ldots,X_k} = \beta_0 + \beta_1 X_1 + \cdots + \beta_k X_k \tag{5.1}$$

where

$\mu_{Y|X_1,\ldots,X_k}$ denotes the mean value of the response variable Y for all those individuals in the sampled population subgroup who have the same combination of values for the explanatory variables $X_1,\ldots,X_k$

β_0 is the intercept

$X_1,\ldots,X_k$ are the explanatory variables

β_j, $j = 1,\ldots,k$, is the expected change in $\mu_{Y|X_1,\ldots,X_k}$ per unit change in X_j when the other explanatory variables in the model are held at a constant value

The quantities $\beta_1,\ldots,\beta_k$ are referred to as **regression coefficients** or alternatively as **partial regression coefficients**.

This model is referred to as a **first order** model because the maximum power of the explanatory variables in the model is one.

5.5 Alternate Equation for a First Order Multiple Linear Regression Model

A equivalent formulation of a first order multiple linear regression model with k explanatory variables can be written in terms of an individual's response. For individual i, this equivalent formulation of the population model is written as

$$Y_i = \beta_0 + \beta_1 X_{i1} + \cdots + \beta_k X_{ik} + \epsilon_i \tag{5.2}$$

where

Y_i is the value of the response variable for individual i

X_{i1} is the value that individual i has for the explanatory variable X_1

$\vdots$

X_{ik} is the value that individual i has for the explanatory variable X_k

β_0 is the intercept

β_j, $j=1,\ldots,k$, is the expected change in Y per unit change in X_j when the other explanatory variables in the model, are held at a constant value

5.6 How Multiple Linear Regression is Used with Sample Data

The equations for predicting a mean and an individual response value for a particular combination of values for the explanatory variables (Equations 5.1 and 5.2) describe population models. Typically, the values of the parameters of these models, $\beta_0, \beta_1, \ldots, \beta_k$, are unknown because research is usually conducted on only a sample from the population and not on the entire population. You can nevertheless use multiple linear regression when only sample data is available by:

- using the sample data to estimate values for the parameters of the model for the population sampled

- substituting these estimated values for the parameters in the estimated multiple linear regression model equation (Equation 5.3)

Estimates of the sampled population parameters $\beta_0, \beta_1, \ldots, \beta_k$ denoted respectively as $\hat{\beta}_0, \hat{\beta}_1, \ldots, \hat{\beta}_k$ can be obtained by applying the methodology of ordinary least squares to sample data from the sampled population. The method of least squares chooses values for the estimates of the parameters in the model, according to the criterion that the sum of the squared differences between the observed values for Y in the sample and the values for Y predicted by the multiple linear regression model is minimized. Application of calculus methods to the principle of least squares provides equations for these least squares estimators of the model parameters. These equations are generally given in terms of matrices so they will not be given here.

The least squares estimates of the regression model parameters are always given in the output of the REG procedure. The following statements will fit a multiple linear regression model where Y is the response variable and $X1$, $X2$, $X3$, and $X4$ are the explanatory variables in the model.

```
proc reg;
model y=x1 x2 x3 x4;
run;
```

The following statements direct SAS to use the REG procedure to apply a multiple linear regression to Example 5.1:

```
proc reg;
model ht18=birthl mht18 fht18;
run;
```

The least squares estimates of the sampled population parameters for the preceding multiple linear regression model applied to Example 5.1 are as follows:

Parameter Estimates

Variable	DF	Parameter Estimate	Standard Error	t Value	Pr > |t|
Intercept	1	-78.38040	13.25051	-5.92	<.0001
birthl	1	1.31994	0.44712	2.95	0.0094
mht18	1	0.70482	0.16023	4.40	0.0004
fht18	1	1.10172	0.09903	11.12	<.0001

From the preceding display of least squares parameter estimates from the REG procedure we see that the intercept (β_0) is estimated to be -78.38 and the regression coefficients for the explanatory variables recorded in inches, viz., birth length, mother's height at age 18, and father's height at age 18 are respectively estimated to be 1.32, 0.70, and 1.10. We will postpone discussing the standard errors and t-test results for these parameter estimates until Chapter 7 as first we will evaluate whether these standard errors and test results are reasonable based on model checking procedures in Chapter 6.

5.7 Estimation of a Y Value for an Individual from the Sampled Population

The **predicted value** from a multiple linear regression model estimates the Y value for an individual having a particular combination of values for the explanatory variables, assuming that individual is from the sampled population.

5.7.1 Equation for Estimating a Y Value for an Individual from the Sampled Population

In multiple linear regression with k explanatory variables, the equation for the predicted Y value for an individual with values for the explanatory variables $X_1, \ldots, X_k$ respectively equal to $x_{i1}, \ldots, x_{ik}$ is

$$\hat{Y}_i = \hat{\beta}_0 + \hat{\beta}_1 x_{i1} + \cdots + \hat{\beta}_k x_{ik} \qquad (5.3)$$

where
$\hat{\beta}_0$ is the least squares estimated intercept
$\hat{Y}_i$ is the predicted Y value for individual i with values for the explanatory variables $X_1, \ldots, X_k$ respectively, equal to $x_{i,1}, \ldots, x_{ik}$
$\hat{\beta}_1, \ldots, \hat{\beta}_k$ respectively are least squares estimates of the sampled population regression coefficients, β_1 to β_k

5.7.2 Example

As an illustration, consider predicting the height of the first case (Record=1) in Example 5.1. The predicted value from the estimated multiple regression model for the height of a male at age 18 is 67.72 inches given his birth length is equal to 19.7 inches, his mother's height at age 18 is equal to 60.5 inches, and his father's height at age 18 is equal to 70.3 inches.

That is, for the first case,

$$\widehat{ht18} = -78.38 + 1.32(19.7) + 0.70(60.5) + 1.10(70.3)$$
$$= 67.72$$

where -78.38 is the estimated Y intercept and 1.32, 0.70, 1.10 are respectively the estimated regression coefficients for the explanatory variables (measured in inches): birth length, mother's height at age 18, and father's height at age 18 as reported in Section 5.6.

This predicted value of 67.72 is interpreted as follows – if an individual is drawn at random from the sampled population of males whose birth length is equal to 19.7 inches, whose mother's height at age 18 is equal to 60.5 inches, and whose father's height at age 18 is equal to 70.3 inches, then this individual's height at 18 years would be predicted to 67.72 inches.

It is quite easy in SAS to obtain the predicted Y value for an individual with a particular combination of values of the explanatory variables. One way is to use the P option in the MODEL statement of the REG procedure. Consider Example 5.1, the following statements give a predicted value for the height at age 18 of each individual in the sample.

```
proc reg;
model ht18= birthl mht18 fht18 /p;
run;
```

The REG procedure reports the predicted value for each individual in the "Output Statistics", e.g., the predicted value for the first case in the data set (Record=1) is 67.7153 as shown below. The "Residual", i.e., the difference between the observed value of the response (dependent) variable and the predicted value is also given in the "Output Statistics".

```
                       Output Statistics

                    Dependent     Predicted
       Obs   Record   Variable        Value     Residual

        1      1         67.2       67.7153      -0.5153
```

5.8 Using Multiple Linear Regression to Estimate the Mean Y Value in a Population Subgroup with Particular Values for the Explanatory Variables

The predicted value from the multiple regression also is an estimate of the **mean** of Y for the population subgroup defined by having a particular combination of values for the explanatory variables. For example, the predicted value of 67.7153 for Record 1 is also the estimated **mean** height in inches of males in the sampled population subgroup with birth length equal to 19.7 inches, mother's height at age 18 equal to 60.5 inches, and father's height at age 18 equal to 70.3 inches.

5.9 Assessing Accuracy of Predicted Values

The accuracy of the predicted Y value for an individual having a particular combination of values for the X variables in the model can be assessed by a prediction interval. Similarly the accuracy of the estimate of the population mean of Y in the sampled population subgroup having a particular combination of values for the X variables can be assessed by a confidence interval. We postpone a discussion of prediction and confidence intervals for multiple regression until Chapter 7 because the validity of these intervals depend on certain ideal inference conditions (Section 5.10). In Chapter 6 we present methods which evaluate whether these ideal inference conditions are reasonable for a sampled population.

5.10 Ideal Inference Conditions for a Multiple Linear Regression Model

Ideal inference conditions for a multiple linear regression model are:

• **RANDOM SAMPLING** (Section 1.4)

- **LINEARITY** In the sampled population, the mean value of the response variable, Y, for each specific combination of values for the explanatory variables $X_1, \ldots, X_k$ is a linear function of $\beta_0, \beta_1, \ldots, \beta_k$. That is, Equations 5.1 and 5.2 describe the relationship between the response and the explanatory variables.

- **INDEPENDENCE OF ERRORS** The ϵ_i, i.e., the model errors, are independent of each other.

- **EQUALITY OF VARIANCES** In the sampled population the variances of Y values are equal in all the subgroups defined by the same combination of values of the explanatory variables. This means that in the sampled population the variance of the model errors is the same in all the subgroups which have the same combination of values for their explanatory variables.

- **NORMALITY** In the sampled population the response variable, Y, has a normal distribution in each of the subgroups which have the same combination of values for their explanatory variables. This in turn implies that in the sampled population the model errors have a normal distribution in the subgroups which have the same combination of values for their explanatory variables.

- **NO MEASUREMENT ERRORS**

- **ABSENCE OF HARMFUL COLLINEARITY IN THE STUDY SAMPLE** which will be described in Section 5.12

If any violations of these ideal inference conditions occur, a multiple regression analysis may potentially yield:

- misleading conclusions about the relationship between the response and explanatory variables

- inaccurate predictions of the response variable

- misleading estimates of prediction and confidence intervals associated with the analysis

5.11 Measurement Error in Explanatory Variables in Multiple Linear Regression

In multiple regression, measurement error in an explanatory variable can produce a biased estimate of the regression coefficient for that variable such that the value of the regression coefficient can be too large or too small on average. If the measurement error in an explanatory variable is small relative to the range of variation of values for that variable, then departure from this ideal inference condition has negligible consequences. Measurement error in an explanatory variable is also not a concern if the objective of the study is to investigate whether explanatory variables, **as measured**, can predict a response of interest, even if these explanatory variables have not been measured with complete accuracy. For example, consider the situation where the objective is to investigate whether the explanatory variables, an individual's self-measured body mass index and an individual's self-reported minutes per day spent in moderate exercise can be used to predict the individual's systolic blood pressure. Both explanatory variables could be potentially measured with error.

However, in this study the prediction of systolic blood pressure by the "true value" of the explanatory variables is not relevant to the question the researcher wants to investigate. Apart from these two situations, measurement error in any of the explanatory variables is a cause for concern and should be initially addressed by researchers when planning a study.

5.12   Collinearity – Why Worry?

Definition 5.1 Collinearity in multiple regression, also sometimes called **multicollinearity**, refers to the situation in which one or more linear relationships exist among the explanatory variables that are included in a multiple regression model.

Collinearity can cause difficulties in multiple linear regression. When there is **perfect collinearity**, i.e., when an explanatory variable in the model can be perfectly predicted by a linear function of one or more of the other explanatory variables in the model, millions of multiple linear regression models comprised by these particular explanatory variables can fit the data equally well.[2] Littell et al. (2002, p. 21) provide an example of perfect collinearity and illustrate how the REG procedure flags this problem in the output. Perfect collinearity is also referred to as **complete collinearity** or **exact linear dependency**. If perfect collinearity is found, one of the explanatory variables involved should be dropped from the model. This does not create a loss of information as the dropped variable was completely redundant in the original analysis. In practice, perfect collinearity typically does not occur in multiple regression. However strong, but not perfect, collinearity can occur and cause harm as the regression coefficients for strongly collinear predictors, although unique, cannot be reliably interpreted. The following problems may ensue in the presence of strong collinearity:

• The estimated regression coefficients of the highly collinear explanatory variables are sensitive to small changes in the data values. That means that we cannot have confidence in the regression coefficients for collinear variables estimated from a researcher's sample data because another sample from the same population with just slightly different values could yield radically different estimates for these regression coefficients. An illustrative example can be found in Graybill and Iyer (1994).

• The estimated regression coefficients of highly collinear variables have large standard errors causing reduced power in hypothesis tests and increased width of confidence intervals for these regression coefficients.

• Prediction of Y may be highly inaccurate in a new sample when a different combination of values for the strongly collinear explanatory variables is not representative of the collinearity found in the sample used to fit the model.

• Unexpected results can occur, e.g., the estimated regression coefficient of a strongly collinear explanatory variable could have an opposite sign than expected on the basis of prior substantive knowledge. The following example is illustrative.

Consider a multiple regression analysis where a memory test score is the response variable and age and minutes spent in daily moderate exercise are the explanatory variables. Researchers were surprised that the regression coefficient for the exercise variable had a negative sign, which indicated that the more minutes the individuals spent in daily

---

[2]A technical explanation for this is that regression coefficients for perfectly collinear explanatory variables cannot be uniquely estimated in a multiple linear regression analysis by the method of ordinary least squares.

moderate exercise the worse the memory test score was. This result is inconsistent with results from previous studies which suggest that daily moderate exercise boosts rather degrades memory test scores. The surprising finding in this study could be related to the fact that the explanatory variables age and exercise were strongly correlated in the researchers' sample. In their sample of individuals aged 65–100, the amount of time spent in moderate exercise was negatively correlated with an individual's age, the older the individual the less time spent in moderate exercise. This strong collinearity between the explanatory variables exercise and age occurring in the researchers' sample resulted in unreliable least squares estimation of the regression coefficients for these explanatory variables.

The effect of exercise on the response variable, memory test score, could not be reliably adjusted for the effect of age on the response variable because these two explanatory variables, exercise and age, were so strongly correlated with each other in the sample. Similarly the effect of age on the response variable could not be reliably adjusted for the effect of exercise on the response variable. This makes intuitive sense when we recall in multiple linear regression the regression coefficient for a particular explanatory variable, say X_1, is the expected change in the response variable per unit change in X_1 while the other explanatory variables in the model are held constant. However, if $X1$ is strongly correlated with other explanatory variables, then when the value of X_1 changes the values of the other explanatory variables correlated with X_1 also change which makes it difficult to disentangle the separate effects of the collinear explanatory variables on the response variable.

Recommendations
The possibility of obtaining misleading results underscores the importance of evaluating collinearity. Before interpreting a multiple regression analysis, we recommend applying methods that assess collinearity such as those described in Chapter 6. When strong (but not perfect) collinearity detected in the sample is not representative of what is occurring in the target population, remedial measures described in Section 14.11 should be considered if the model is being used:

- to understand the underlying relationship between the outcome and explanatory variables in the target population

- to predict the outcome variable in a new sample which has a combination of data values for the collinear variables that is slightly different from the sample used in estimating the model

---

5.13 Why the Estimation of Variance of Y Given the X Variables is Critical in Multiple Linear Regression

Some key facts to remember:

- In a multiple regression framework, the variance of Y given the X variables is the variance of Y within the population subgroups which have the same combination of values for the X (explanatory variables) in the multiple regression model.

- An estimate of the variance of Y given the X variables is required in multiple regression for many model checking procedures, hypothesis tests, confidence intervals, and prediction intervals.

- In multiple linear regression, the Mean Square Error (Section 2.13.1) is used as a measure of the extent that factors **other** than the X variables affect the response variable, Y. As the Mean Square Error is a model-dependent estimator of the variance of Y given the X variables, its validity depends on whether the model's ideal inference conditions are reasonable for the sampled population.

- If the model's ideal inference conditions are not reasonable for the sampled population, then misleading conclusions can be made from model checking methods, hypothesis tests, confidence intervals, and prediction intervals which are based on the Mean Square Error.

5.14   Chapter Summary

This chapter:

- describes the characteristic features of a multiple linear regression model and lists reasons why you would want to use this model

- discusses how you can estimate the values for the parameters of a multiple linear regression model from sample data via the method of least squares

- describes how these least squares estimated values of the model parameters (the Y intercept and regression coefficients for the explanatory variables) are used in predicting the mean of Y of a sampled population subgroup with a specific combination of values for the explanatory variables and also in predicting the value of Y for an individual from a sampled population subgroup with a specific combination of values for the explanatory variables

- illustrates how easy it is to obtain these least squares parameter estimates using the REG procedure

- describes ideal inference conditions for hypothesis testing and confidence and prediction interval estimation in multiple linear regression and discusses potential difficulties when these ideal inference conditions are not reasonable for the sampled population

- gives reasons why researchers should worry about explanatory variables being strongly correlated among themselves in a multiple linear regression analysis

- discusses why it is critical to get a valid estimate of the variance of Y given the X variables

- provides an example of a SAS program which applies a multiple linear regression analysis

Appendix

5.A Program 5.1

The following SAS program reads in the data for Example 5.1 and stores it in the temporary SAS data set arbitrarily named HTDTA. The PRINT procedure is used to list the data and the REG procedure is used to apply a multiple linear regression analysis to these data and obtain a predicted and residual value for each case.

```
data htdta;
input Record ht18 birthl mht18 fht18;
datalines;
1 67.2 19.7 60.5 70.3
2 69.1 19.6 64.9 70.4
3 67.0 19.4 65.4 65.8
4 72.4 19.4 63.4 71.9
5 63.6 19.7 65.1 65.1
6 72.7 19.6 65.2 71.1
7 68.5 19.8 64.3 67.9
8 69.7 19.7 65.3 68.8
9 68.4 19.7 64.5 68.7
10 70.4 19.9 63.4 70.3
11 67.5 18.9 63.3 70.4
12 73.3 20.8 66.2 70.2
13 70.0 20.3 64.9 68.8
14 69.8 19.7 63.5 70.3
15 63.6 19.9 62.0 65.5
16 64.3 19.6 63.5 65.2
17 68.5 21.3 66.1 65.4
18 70.5 20.1 64.8 70.2
19 68.1 20.2 62.6 68.8
20 66.1 19.2 62.2 67.3
;
proc print data=htdta;
   title Example 5.1;
run;
proc reg data=htdta plots=none;
   model ht18=birthl mht18 fht18/p;
   id record;
quit;
```

5.B   Example 5.1 Data Values

The following are the data used in Example 5.1:

Example 5.1

Obs	Record	ht18	birthl	mht18	fht18
1	1	67.2	19.7	60.5	70.3
2	2	69.1	19.6	64.9	70.4
3	3	67.0	19.4	65.4	65.8
4	4	72.4	19.4	63.4	71.9
5	5	63.6	19.7	65.1	65.1
6	6	72.7	19.6	65.2	71.1
7	7	68.5	19.8	64.3	67.9
8	8	69.7	19.7	65.3	68.8
9	9	68.4	19.7	64.5	68.7
10	10	70.4	19.9	63.4	70.3
11	11	67.5	18.9	63.3	70.4
12	12	73.3	20.8	66.2	70.2
13	13	70.0	20.3	64.9	68.8
14	14	69.8	19.7	63.5	70.3
15	15	63.6	19.9	62.0	65.5
16	16	64.3	19.6	63.5	65.2
17	17	68.5	21.3	66.1	65.4
18	18	70.5	20.1	64.8	70.2
19	19	68.1	20.2	62.6	68.8
20	20	66.1	19.2	62.2	67.3

6

Before Interpreting a Multiple Linear Regression Analysis. . .

6.1   Introduction

Before interpreting results from a multiple linear regression analysis, it is important to assess how much confidence you can have in the results by evaluating whether:

- there are **any data input or recording errors**

- **the ideal inference conditions** for a multiple regression analysis (Section 5.10) are reasonable

- the estimated linear multiple regression model largely depends on only one or a few **influential cases** in the sample data

- there is collinearity among the explanatory variables

6.2   Evaluating Collinearity

To centre or not to centre

There has been controversy as to whether data on the explanatory variables should be mean-centred or not when investigating the presence of collinearity. Data are mean-centred when the sample mean of a variable is subtracted from each of its original data values. Some authors refer to mean-centred data simply as centred data.

Belsley (1991) and Draper and Smith (1998) have warned that collinearity involving the intercept term cannot be detected when the data are mean-centred. Others have suggested that it does not matter that this type of collinearity involving the intercept cannot be detected if the intercept term is of no substantive interest to the researcher. However, even if the intercept term is of no substantive interest in itself, detecting collinearity involving the intercept term can be useful as it implies that at least one of the explanatory variables or some linear combination of the explanatory variables has very low variance, which can result in a large variance of the regression coefficients for these variables. Simon and Lesage (1988) suggested that it may be important to detect collinearity involving the intercept term even if the intercept is of no substantive interest as their simulations demonstrated that collinearity involving the intercept not only degrades the numerical accuracy of the intercept estimate but also the numerical accuracy of regression coefficients estimates for the explanatory variables that are collinear with the intercept.

However, Belsley (1991) did not entirely rule out mean-centring a predictor when assessing collinearity. He stated (p. 185–187) that collinearity diagnostics are most meaningful

DOI: 10.1201/9780429107368-6

when applied to variables that are "structurally interpretable". He defined structural interpretable variables as those "whose numeric values and (relative) differences derive meaning and interpretation from the context they measure". He advised that investigators should choose an appropriate origin of measurement for an explanatory variable to attain structural interpretability. Mean-centring is one such approach for attaining structural interpretability. We will discuss this approach in detail in Chapter 9, where we consider an example of mean-centring and its role in aiding interpretability in a multiple regression model.

6.2.1   A Method for Diagnosing Collinearity

Belsley et al. (1980) described a method for diagnosing collinearity. This method has the capability of:

- detecting more than one linear relationship that is occurring among the predictors in the sample

- indicating the strength of each collinear relationship occurring in the sample and identifying which predictors are strongly involved in the relationship

This method suggested by Belsley et al. (1980) involves performing a principal components analysis on the explanatory variables. The technical details of this multivariate procedure are beyond the scope of this book but Section 5.2 in Belsley (1991) provides a detailed summary in matrix algebra terms for those who are interested.

The following quantities are employed as collinearity diagnostics in this method:

(1) a **scaled condition index**, which indicates the strength of a particular linear relationship among the predictors in the sample

(2) **variance-decomposition proportions**, which indicate the extent of involvement of each predictor in a particular linear relationship in the sample

These collinearity diagnostics can be obtained requesting the COLLIN option in the MODEL statement of the REG procedure. The COLLIN option does not by default centre the data.

6.2.2   Interpreting Collinearity Diagnostics from the Method Proposed by Belsley (1991)

Belsley (1991) suggested the following approximate guidelines to assess whether a collinear relationship occurring among predictors in the sample might **potentially** cause problems in interpreting the least squares estimates of the regression coefficients:

- If the largest scaled condition index associated with a linear relationship among the predictors in the sample is between 5 and 10, there is little cause for concern about collinearity.

- If a scaled condition index is in the range of 30–100 indicating a moderate linear relationship among the predictors in the sample **and** if the "weights" (the variance-decomposition proportions) of two or more predictors associated with that linear relationship are 0.5 or greater, the collinearity occurring in the sample data **may** be causing problems such as those discussed in Section 5.12.

- If a scaled condition index is in the range of 1000–3000 indicating a strong linear relationship in the sample **and** if the "weights" (the variance-decomposition proportions) of two or more predictors associated with that linear relationship are 0.5 or greater, collinearity in the sample data **may** be causing severe problems in interpreting the least squares estimates of the regression coefficients for the explanatory variables involved in that relationship.

- When a scaled condition index is enormous (in the hundreds of thousands) indicating **perfect collinearity**, i.e., an exact linear dependence among two or more predictors in the sample, the least squares estimation process breaks down and cannot provide unique estimates of regression coefficients for the collinear variables. The REG procedure issues a warning message when this occurs: An illustrative example is provided in Littell et al. (2002).

As Belsley (1991) pointed out, these guidelines merely are indications that collinearity in the sample may have a harmful impact on the study's objectives. Even when there is moderate to strong collinearity there is no magic number for any collinearity diagnostic that informs researchers that their objectives have been harmed by a collinearity relationship in their sample. When there is moderate to strong collinearity and an estimated regression coefficient has an opposite sign than expected given prior research (Section 5.12), there would be legitimate concern that the collinearity is causing harm if the primary objective is understanding the relationship between the response and explanatory variables. Collinearity in the sample is also harmful if it causes the values of the estimated regression coefficients to be highly dependent on the particular data values found in the research sample such that another sample with slightly different values might produce quite different estimated values for the regression coefficients.

6.2.3   Variance Inflation Factor: Another Measure of Collinearity

The variance inflation factor (VIF) of a regression coefficient for an explanatory variable, say X_1, is a measure of how much the variance of the regression coefficient for X_1 is inflated as a result of X_1's being linearly correlated with one or more of the other explanatory variables as compared to X_1's being completely independent of the other explanatory variables. VIF values can be obtained by requesting the VIF option in the MODEL statement of the REG procedure.

Equation for the Variation Inflation Factor

The equation for the VIF for the regression coefficient for explanatory variable X_j is

$$\text{VIF}_j = \frac{1}{(1 - R_j^2)} \tag{6.1}$$

where R_j^2 is the coefficient of multiple determination for the linear regression where X_j is regressed against all the other explanatory variables in the original multiple regression model (e.g., R^2 for birth length in Example 5.1, Section 5.3) is the coefficient of multiple determination for the linear regression where birth length is the dependent variable and mother's height at age 18 and father's height at age 18 are the explanatory variables.

It can be shown that

$$\text{VIF}_j = \frac{\widehat{\text{SE}}(\hat{\beta}_j)^2 \sum_{i=1}^{n} (X_{ij} - \bar{X}_j)^2}{\text{Mean Square Error}} \tag{6.2}$$

where
$\widehat{\text{SE}}(\hat{\beta}_j)$ is the estimated standard error of the estimated regression coefficient for variable j and can be obtained from the REG procedure

$\bar{X}_j$ is the sample mean of X_j

Mean Square Error is the model dependent estimate of the variance of Y given the X variables (Section 5.13)

Interpreting VIF Values for Multiple Regression Coefficients

Before interpreting VIF values for the regression coefficients of the explanatory variables in a multiple regression model, it is important to initially evaluate whether the model's ideal inference conditions are reasonable for the sampled population. If these conditions are not reasonable, the Mean Square Error may not be a valid estimate of the variance of Y given the X variables which in turn can result in misleading VIF values.

Various cutoff values for VIF, ranging from 4 to 30, have been suggested as an indication that collinearity in the sample is a cause for concern. It has been commonly suggested that a VIF greater than 10 would cause concern. However, these numerical cutoffs are not based on any underlying theory and are essentially arbitrary. O'Brien (2007) lamented that often too much emphasis is given to these arbitrary cutoff values for a VIF and that exclusive focus on the VIF values can lead to unnecessary action such as implementing procedures purported to remediate collinearity – often considered more harmful than the presence of collinearity itself – or rejection of a paper for publication because a VIF was greater than one of these arbitrary cutoff values. He concured with Belsley (1991) that a high VIF value does not indicate that collinearity in the sample has been harmful in a study when a confidence interval for a regression coefficient is narrow enough for a researcher's study objectives. Myers (1990) also advised caution in interpreting threshold values for a collinearity diagnostic and wrote "reliable rules of thumb are not always reliable".

6.2.4   Further Comments Regarding Collinearity Diagnosis

In a total diagnosis of collinearity among predictors in a model, it is important to consider the following quantities:

- the scaled condition indexes from the principal components analysis of the explanatory variables, which indicate the severity of a particular collinear relationship

- the variance decomposition proportions from this principal components analysis which inform us which predictors are involved and their extent of involvement in a particular collinear relationship

- VIF values for the predictors which help determine if there has been harmful impact of any collinearity in the sample based on inflation of the standard errors for the predictors' regression coefficients

For those who are interested in a technical discussion of scaled condition indexes, variance decomposition proportions and VIF values based on matrix algebra, we recommend Myers (1990).

6.3   A Ten-Step Approach for Checking Ideal Inference Conditions in Multiple Linear Regression

In this section we propose a ten-step approach for checking whether the ideal inference conditions for a multiple linear regression analysis are reasonable.

Step (1) Screen for any data input errors – the first essential step of any data analysis. See Section 1.14.

Step (2) Screen the data for any recording errors by identifying and investigating all cases that are Y-outliers, X-outliers, or contextual outliers to find out if these unusual cases are the result of recording errors. Outlying observations for the Y and X variables can be identified using the UNIVARIATE procedure to provide a listing of the unusual (extreme) values for these variables along with their ID number. Contextual outliers, i.e., cases with unusual combinations of values for some or all of the model variables in the context of the sampled population (e.g., a five-year-old who weighs 200 pounds) can be identified with various scatter plots of these variables. When there are many explanatory variables in the model the process of looking for contextual outliers can be laborious but nevertheless an important part of getting acquainted with the data.

If any outlying value(s) have resulted from an error in recording, the error(s) should be corrected at this step before proceeding to the subsequent model checking steps. This would seem an obvious recommendation but it is not always carried out in practice. Step (2) is sometimes delayed until a much later stage in the multiple regression analysis, which can result in a lot of wasted time and effort if a recording error is found at a later stage. The discussion in Section 3.12 regarding what to do about outlying values for Y and X variables detected in Step 2 is relevant here as well.

Step (3) Confirm that the response variable is continuous.

If the response variable is continuous, proceed to Step (4).

If the response variable is not continuous, see recommendations given in Step (3) for simple linear regression (Section 3.13), which are also applicable for multiple linear regression.

Step (4). Review the details of the sampling design for your study to see if there are any obvious aspects of the design that would suggest that a more complicated model is required. Examples of sampling designs which require a more complicated model are given in Chapter 14. One such example requiring a more complicated model is where there are repeated measures on the same individual at yearly intervals. Violation of independence of errors would occur if a multiple regression model were applied to these data. When there is an obvious aspect of the sampling design that would make your multiple linear regression model inappropriate, then stop at this step and consider a different model with a more complicated error structure to reflect how you obtained your sample data. If, on the other hand, there is nothing immediately obvious about your sampling design that would make your multiple linear regression model inappropriate, then proceed to Step (5).

Step (5) Obtain regression diagnostics for evaluating ideal inference conditions by applying a multiple regression to the data using the REG procedure.

Step (6) Informally evaluate the ideal inference conditions of linearity and equality of variances in the sampled population. Examine a plot of internally studentized residuals against their corresponding predicted values for violations of the ideal inference conditions of linearity and equality of variances. Linearity is indicated in this plot by a linear band of points centred about the line $Y = 0$. It is the pattern of a linear band that suggests linearity rather than the pattern of the residuals being centred about the line $Y = 0$ because in least squares regression, the mean of the residuals is always zero or close to zero whether or not ideal inference conditions hold. Equality of variances is indicated in this plot if the internally studentized residuals do not systematically increase or decrease in size as the size of the predicted values increases. If the ideal inferences conditions of linearity, and equality of variances do not look reasonable, see Chapter 14 for other approaches. If these ideal inferences conditions look reasonable, go to Step (7).

Note Regarding Omitting Step (7): You can omit Step (7), which further investigates independence of errors using internally studentized residuals and proceed directly to Step (8) if your investigation was a well designed, controlled randomized trial (experiment).

Step (7) Use graphical displays of internally studentized residuals to informally evaluate whether the ideal inference condition of independence of errors has been violated for the multiple linear regression model due to the errors being systematically related to one or more of the explanatory variables in the model or to any other variable not included in the model. This violation of independence of errors can occur when:

- the sampling units (e.g., individual mice) for the study have not been randomly sampled from the same population

- all aspects related to the measurement of the response variable have not been held constant in an observational study or in a quasi-experiment (i.e., a non-randomized experiment)

You can informally evaluate independence of errors in these situations by examining separate plots of internally studentized residuals against each explanatory variable in your model as well as against any time, space, or other key variable related to the details of the sampling and measurement of the cases. If the internally studentized residuals appear to be exhibiting a systematic pattern in any of these plots (subsection 3.9.1), then this suggests that the model errors are not independent of the variable in the plot and you should stop checking the appropriateness of your multiple linear regression model at this step and consider other approaches such as adding additional explanatory variables to your multiple linear regression model.

Step (8)   Evaluate the ideal inference condition of normality. Examine a Q-Q plot of internally studentized residuals (available in the UNIVARIATE procedure, e.g., Program 6.1B, Section 6.B). When the ideal inference condition of normailty is reasonable, the residuals will lie approximately along the reference line as in Figure 3.8. The discussion of other plots in subsection 3.9.3 illustrating violations of normality are also relevant for this step.

Step (9) Identify and investigate regression outliers using studentized deleted residuals. Studentized deleted residuals can be appropriately used for regression outlier detection because previous steps have suggested that the ideal inference conditions of independence of errors, linearity, equality of variances, and normality are reasonable. Investigate any cases with an absolute value of a studentized deleted residual (RSTUDENT) greater than 2.

A listing of the values of RSTUDENT and the corresponding ID for the cases can be readily obtained by specifying the INFLUENCE option in the MODEL statement of PROC REG. However, instead of examining the entire listing of RSTUDENT values to find those cases with absolute values that exceed a recommended cutoff, it is easier if you get SAS to check if there are any such cases and if so print out their ID number and their RSTUDENT values as well as any other values of interest (e.g., Program 6.1B, STEP 9).

If investigation of a regression outlier reveals a recording error, correct the error and then go back to Step (5) and apply the multiple linear regression analysis on the revised data set. As discussed in Section 3.2, if specific facts, not merely speculation, can be forwarded as evidence that a regression outlier has resulted from a recording error, which is impossible to correct at that stage in the research, then the regression outlier may be deleted from the data set. If this is the situation, go back to Step (5) and apply the multiple linear regression analysis on this revised data set where any cases with recording errors have been removed. However, whenever one or more cases are removed from an analysis, it is important to report the removal and the justification for the removal in the description of the study's methods. Another course of action that might be possible when a recording error has been documented is to consider such an error as a missing value and apply an imputation method (beyond the scope of this book) to impute a value to replace the error and then run the multiple linear regression analysis on this revised data set.

Step (10). Evaluate whether harmful collinearity is present See Sections 5.12 and 6.2.

Note: We have not included absence of measurement errors in the explanatory variables in our ten-step approach to evaluate ideal inference conditions because the issue of measurement errors in a study is typically related to the specific subject matter being investigated and should be initially addressed at the planning phase of a study.

6.4   Example – Evaluating Ideal Inference Conditions

To illustrate our ten-step approach, we consider checking ideal inference conditions for the multiple linear regression model applied to Example 5.1. The results and conclusions reached from applying this ten-step approach are as follows:

Step (1) Screen for any data input errors. No data input errors were found in the listing of the data produced by the PRINT procedure in Program 6.1A.

Step (2) Screen the data for possible recording errors by identifying and investigating outliers. The listings of extreme values in the data for the Y and X variables by the UNIVARIATE procedure in Program 6.1A did not reveal any outliers for these variables. Scatter plots from the SGSCATTER procedure in Program 6.1A did not reveal any contextual outliers that would suggest recording errors.

Step (3) Confirm that the response variable is continuous. The response variable, height at age 18 is continuous.

Step (4). Review the details of the sampling design of the study to see if there are any obvious aspects of the design that would suggest that a different model with a more complicated error structure than a multiple linear regression model is required. The following details[1] suggest a more complicated model is not required to reflect the sampling design of the study:

---

[1] Although details (ii) to (v) were not described by Graybill and Iyer (1994), we surmise that these details were part of the study design as they would be desirable if the researchers had planned to apply a multiple linear regression analysis to their data.

(i)   The individuals in the study were a random sample without replacement from the same village.

(ii)   None of the individuals had the same biological mother and/or father. Therefore the error of one individual in the sample was not dependent on the error of any other individual in the sample due to their sharing the same biological parent. If a number of individuals in the sample had come from the same biological parent, a model with a more complicated error structure than that of multiple linear regression would be required. If this were the situation, the researchers would be advised to consult a professional statistician for help.

(iii) The Y values, the individuals' height at age 18, were all measured by the same research assistant using a strict protocol for height measurement.

(vi)   Birth lengths were obtained from records from the same hospital.

(v)   Measurement error of birth length, mother's and father's height at age 18 was investigated at the planning stage of the study and found to be negligible.

As these study design details suggest that, at least at this stage of the evaluation process, a more complicated model is not required and that we should proceed to Step (5).

Step (5). Obtain regression diagnostics for evaluating ideal inference conditions by applying a multiple regression to the data. This was carried out using the REG procedure (Program 6.1B, STEP 5).

Step (6) Informally evaluate the ideal inference conditions of linearity and equality of variances in the sampled population. The plot of the internally studentized residuals against the predicted values (Figure 6.1) does not suggest any serious departures from these ideal inference conditions. There is no evidence of curvature in the band of points and the size of the internally studentized residuals does not appear to be systematically related to the size of the predicted values.

Step (7) Use graphical displays of internally studentized residuals to informally evaluate whether the ideal inference condition of independence of errors has been violated for the multiple linear regression model due to the errors being systematically related to one or more of the explanatory variables in the model or to any other variable not included in the model. Plots of internally studentized residuals against each of the explanatory variables (Figure 6.2) do not reveal a systematic pattern that would suggest the errors were highly dependent on the values of these variables. A key variable related to the recording of the data measurements in the study was kept constant (i.e., only one research assistant did the recording). Other aspects of the study design that might have varied from subject to subject that might have caused a violation of independence of errors (for example, time and date of recording) were not included in the data set and could not be investigated.

Step (8) Evaluate the ideal inference condition of normality. The Q-Q plot of the internally studentized residuals from the model (Figure 6.3) does not suggest a major departure from normality.

Step (9) Identify and investigate regression outliers using studentized deleted residuals. Two of the cases, Records 2 and 5, have a studentized deleted residual slightly greater than 2. The values of the response and explanatory variables do not suggest anything unusual about these cases. In Program 6.1C we investigate whether these cases are influential in the regression analysis.

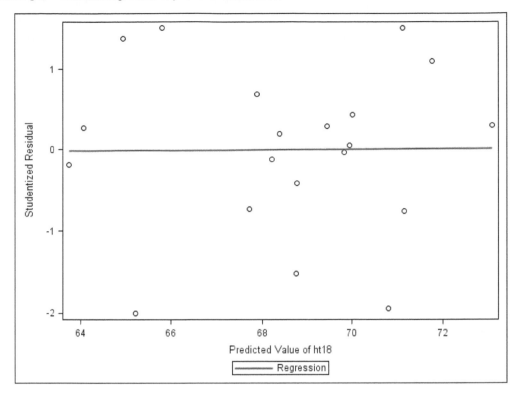

FIGURE 6.1
Internally studentized residuals vs. predicted values for Example 5.1.

Step (10) Evaluate whether harmful collinearity is present. Collinearity diagnostics (subsection 6.2.1), proposed by Belsley et al. (1980), were obtained from the COLLIN option in the REG procedure of Program 6.1B. Using Belsley's (1991) guidelines (subsection 6.2.2), the following collinearity diagnostics identify birth length and mother's height as moderately strong collinear predictors because the scaled condition index associated with Number 3^2 is 109.22926 and the "weights" (labelled as Proportion of Variance) associated with Number 3 are 0.76667 and 0.60573 respectively for birth weight and mother's height. No other collinear relationships are identified as potentially causing harm using Belsey's (1991) guidelines.

```
                         Collinearity Diagnostics
               Number      Eigenvalue      Condition Index

                  1          3.99840          1.00000
                  2          0.00108         60.80869
                  3        0.00033513        109.22926
                  4        0.00018307        147.78472



               -----------------Proportion of Variation---------------
      Number       Intercept          birthl           mht18            fht18
```

---

[2]The third principal component, for those familiar with principal component analysis.

1	0.00001543	0.00003444	0.00002569	0.00005856
2	9.427416E-8	0.12629	0.03672	0.56785
3	0.01246	0.76667	0.60573	0.06297
4	0.98752	0.10701	0.35752	0.36912

Belsley (1991) suggested that moderate collinearity among the explanatory variables in a sample might be harmful as it could result in large standard errors for the estimated regression coefficients of the collinear variables. However, the following results from Program 6.1B show that the VIF values for birth length and mother's height at age 18 are quite low (respectively 1.25 and 1.22). These low VIF values suggest that the moderate collinearity between birth length and mother's height at age 18 occurring in the sample is not a cause for concern in terms of inflating the standard error of the estimated regression coefficients for these collinear variables.

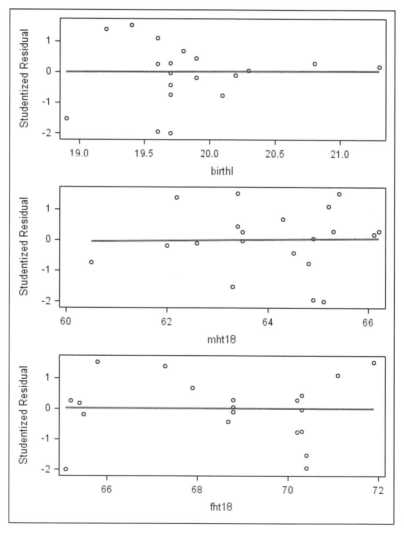

FIGURE 6.2
Internally studentized residuals vs. predictor variables: Example 5.1.

Parameter Estimates

Variable	DF	Parameter Estimate	Standard Error	t Value	Pr > |t|	Variance Inflation
Intercept	1	-78.38040	13.25051	-5.92	<.0001	0
birthl	1	1.31994	0.44712	2.95	0.0094	1.25048
mht18	1	0.70482	0.16023	4.40	0.0004	1.22276
fht18	1	1.10172	0.09903	11.12	<.0001	1.03284

The objective of this study was solely prediction of the height of an individual at age 18 in the sampled population. Fortunately, the objective was not to elucidate the relationship between the response variable, height at age 18 and the explanatory variables, birth length, mother's height at age 18, and father's height at age 18. If the latter were the objective, there would have been concern that the separate effect of birth length on the response variable would be somewhat difficult to disentangle from the effect of mother's height on the response variable because of the correlation between these two explanatory variables.

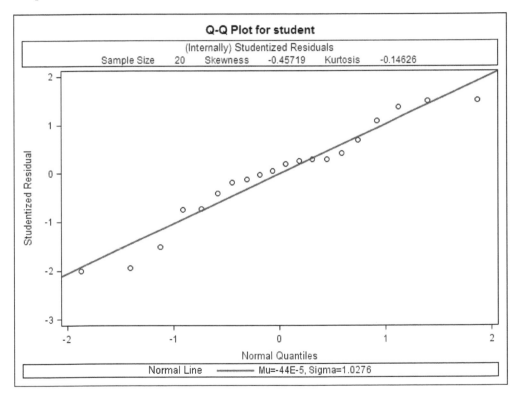

FIGURE 6.3
QQ Plot of internally studentized residuals from Example 5.1.

6.5 Identifying Influential Cases

The description and discussion of influence diagnostics for simple linear regression (Section 3.15) are applicable to multiple linear regression as well. We will continue to use Example

5.1 as illustration. Evaluation of the ideal inference conditions of independence of errors, linearity, and equality of variance appeared to be reasonable for Example 5.1 (Section 6.4), so we can expect the influence diagnostics, DFFITS, DFBETAS, and Cook's D to be helpful in identifying influential cases in this example.

We used Program 6.1C (subsection 6.C) to identify cases that may merit investigation because they have values of DFFITS, DFBETAS, and Cook's D which exceed suggested cutoff values (discussed in subsections 3.16.3, 3.17.2).

6.6 Summary of Results Regarding Potentially Influential Cases based on Cook's D, DFBETAS, and DFFITS Values Output from Program 6.1C

Cook's D

None of the cases has a Cook's D greater than 0.5, the cutoff value suggested by Cook (1999) for identifying cases which might merit investigation because of their influence on all the predicted values in the sample (Figure 6.4). The following partial output from the log file for Program 6.1C also provides this information:

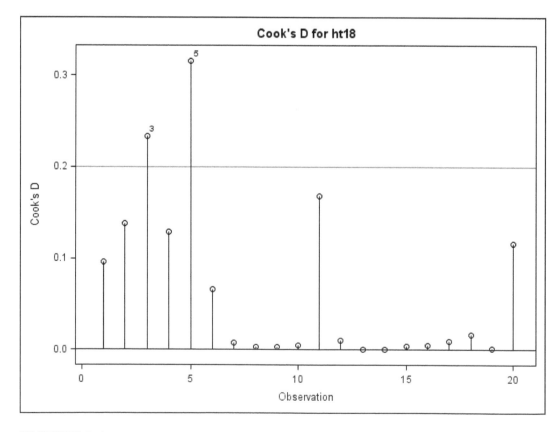

FIGURE 6.4

Plot of Cook's D values for Example 5.1.

```
NOTE: No observations were selected from data set WORK.INFLUENCE.
NOTE: There were 0 observations read from the data set WORK.INFLUENCE.
      WHERE cookd>0.5;
```

6.6.1   DFBETAS

None of the absolute values of DFBETAS displayed in the panel plots of Figure 6.5 exceeds the value of 1.00, the cutoff value suggested by Kutner et al. (2005) for DFBETAS in small and medium samples. However, if you choose to use $2/\sqrt{n}=0.45$, the size-adjusted cutoff value proposed by Belsley et al. (1980) as a benchmark, it may be noted from Figure 6.5 that: case 20 has an absolute value of DFBETA that exceeds this cutoff value for the intercept; cases 3, 5, and 11 have an absolute value of DFBETA that exceeds this cutoff value for the DFBETA for birthl; cases 1, 2, 3, and 5 have an absolute value of DFBETA that exceeds this cutoff value for the DFBETA for mht18; and cases 3, 4, and 5 have an absolute value of DFBETA that exceeds this cutoff value for the DFBETA for fht18.

6.6.2   DFFITS

The following results from Program 6.1C lists cases which may merit investigation for their potential influence according to a suggested cutoff absolute value used for DFFITS: i.e., the cutoff absolute value of 1 for small to medium samples (Kutner et al., 2005); the

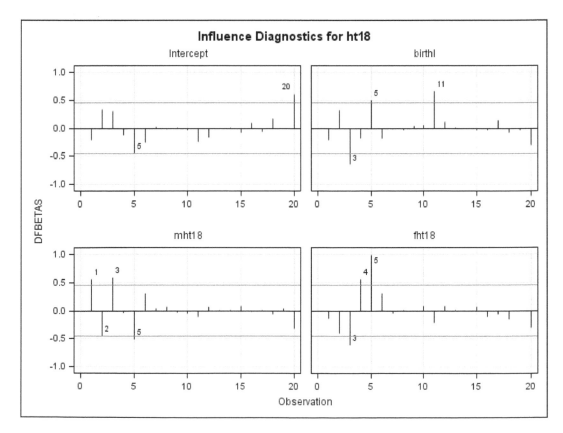

FIGURE 6.5
Plot of DFBETAS values for Example 5.1.

size-adjusted cutoff absolute value of $2\sqrt{p/n}$=0.89 (Belsley et al., 1980); or the size-adjusted cutoff absolute value of $1.5\sqrt{p/n}$=0.67 (Staudte and Sheather, 1990). We have printed the input data values for these cases in each list along with the absolute value of dffits and rstudent (externally studentized residuals) to allow one to see at a glance if anything unusual about these cases stands out that might explain why they might be considered influential. Nothing unusual stands out.

Cases with DFFITS Absolute Value > 1.00

Obs	record	absdffits	ht18	birthl	mht18	fht18	predicted
3	3	1.01000	67.0	19.4	65.4	65.8	65.8152
5	5	1.25763	63.6	19.7	65.1	65.1	65.2285

Obs	rawresid	student	rstudent	leverage
3	1.18485	1.51102	1.58011	0.29006
5	-1.62848	-2.00578	-2.24469	0.23891

Cases with DFFITS Absolute Value > Belsley et al. (1980) Cutoff Value

Obs	record	absdffits	ht18	birthl	mht18	fht18	predicted
3	3	1.01000	67.0	19.4	65.4	65.8	65.8152
5	5	1.25763	63.6	19.7	65.1	65.1	65.2285

Obs	rawresid	student	rstudent	leverage
3	1.18485	1.51102	1.58011	0.29006
5	-1.62848	-2.00578	-2.24469	0.23891

Cases with DFFITS Absolute Value > Staudte and Sheather (1990) Cutoff Value

Obs	record	absdffits	ht18	birthl	mht18	fht18	predicted
2	2	0.82416	69.1	19.6	64.9	70.4	70.7947
3	3	1.01000	67.0	19.4	65.4	65.8	65.8152
4	4	0.75183	72.4	19.4	63.4	71.9	71.1260
5	5	1.25763	63.6	19.7	65.1	65.1	65.2285
11	11	0.85689	67.5	18.9	63.3	70.4	68.7430
20	20	0.70462	66.1	19.2	62.2	67.3	64.9483

Obs	rawresid	student	rstudent	leverage
2	-1.69465	-1.94891	-2.16085	0.12700
3	1.18485	1.51102	1.58011	0.29006
4	1.27399	1.51511	1.58511	0.18365
5	-1.62848	-2.00578	-2.24469	0.23891
11	-1.24298	-1.51762	-1.58817	0.22547
20	1.15168	1.38066	1.42435	0.19661

Figure 6.6 displays a plot of DFFITS values for Example 5.1.

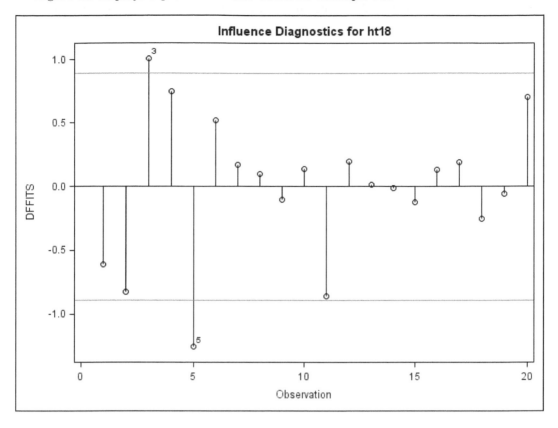

FIGURE 6.6
Plot of DFFITS values for Example 5.1.

6.6.3 Potentially Influential Cases

Based on these preceding results, cases 3 and 5 may be potentially influential in terms of the researchers' objective. A simple way to check this out is to compare the conclusion from the analysis with all cases included to the conclusion from the analysis with case 3 excluded. Similarly investigate case 5's influence on the research conclusion. We will carry out this approach in the next chapter where inference procedures for a multiple linear regression analysis are discussed.

6.7 Chapter Summary

This chapter:

- emphasizes that before interpreting results from a multiple linear regression analysis, it is important to assess: whether there are data input or recording errors; whether the ideal inference conditions are reasonable; and whether there are any regression outliers or influential cases

- points out that the diagnostic procedures for model checking a simple linear regression model described in Chapter 3 can also be used for checking a multiple linear regression model

- discusses the issue of centring or not centring the data when evaluating collinearity

- describes a method for detecting collinearity proposed by Belsley et al. (1980)

- explains how quantities known as variance inflation factors can estimate harm caused by collinearity

- proposes a ten-step approach for checking ideal inference conditions for multiple linear regression and provides an illustrative example along with SAS programs that implement this approach

- discusses detection of potentially influential cases in multiple linear regression and gives a SAS program which facilitates identification of such cases in an illustrative example

Appendix

6.A Program 6.1A

Program 6.1A applies Steps (1) and (2) of Section 6.3 to the data of Example 5.1 stored in the temporary SAS data set HTDTA created by Program 5.1 (Section 5.A). Program 6.1A lists the data so the analyst can screen for data input errors. This program also facilitates the search for outliers by generating scatter plots of the response variable vs. each of the explanatory variables in the multiple regression model and by reporting the identification numbers of cases with extreme values for these variables.

Please note:
If you attempt to execute Program 6.1A after having logged out from the SAS session in which you executed Program 5.1, you will get error messages in most operating environments. One way to avoid this is by executing Program 5.1 in your current SAS session prior to running Program 6.1A or alternatively by copying the DATA step statements and data from Program 5.1 and pasting them at the beginning of Program 6.1A.

```
*Program 6.1A;
*STEP 1;
proc print data=htdta;
run;
*STEP 2;
ods select ExtremeValues Frequencies;
ods exclude TestsForLocation;

proc univariate data=htdta;
   var ht18 birthl ht18 mht18 fht18;
   id record;
run;

/*The following plots can be examined to check for contextual,
predictor and response outliers*/;
ods listing style=journal2 image_dpi=200;
proc sgscatter data=htdta;
   plot ht18 *(birthl mht18 fht18)/columns=1 rows=3 ;
run;
*End of STEP 2;
```

6.B   Program 6.1B

Program 6.1B implements Steps 5–10 of the ten-step approach (Section 6.4) for checking ideal inference conditions for the multiple linear regression model applied to the **temporary** SAS data set HTDTA created by Program 5.1.

Please note:
If you attempt to execute Program 6.1B after having logged out from the SAS session in which you executed Program 5.1, you will get error messages in most operating environments. One way to avoid this is by executing Program 5.1 in your current SAS session prior to running Program 6.1B or alternatively by copying the DATA step statements and data from Program 5.1 and pasting them at the beginning of Program 6.1B.

```
*Program 6.1B;
*STEPs 5 and 10;
/* Apply a multiple linear regression analysis to obtain
 regression diagnostics and a plot of rstudent by H (Leverage) values
 to be used later in STEP 9 if STEPS 7 and 8 suggest that
inference conditions appear reasonable and hence
values of rstudent are appropriate to detect regression outliers*/
/* Collinearity diagnostics are obtained for STEP 10 by specifying the COLLIN
option on the MODEL statement for the REG procedure*/
proc reg data=htdta plots(only label)=  RStudentByLeverage;
    model ht18=birthl mht18 fht18/ vif r collin p;
    output out=diag rstudent=rstudent student=student p=p;
run;
*STEPS 6 and 7;
/*Obtain plots of internally studentized residuals vs. predicted Y,
and  studentized residuals vs. predictor variables*/
proc sgscatter data=diag;
    plot (student)*(p birthl mht18 fht18)/columns=2 rows=2 reg;
run;
*STEP 8 EVALUATE NORMALITY;
* Obtain a Q-Q normal plot for internally studentised residuals;
proc univariate data =diag;
    var student ;
    qqplot student  /normal (mu=est sigma=est);
    inset n='Sample Size' skewness kurtosis
    /pos=tm header='(Internally) Studentized Residuals';
run;
 /*STEP 9 Get a listing of variables of interest for cases identified as
    regression outliers based on having an absolute value of
    the studentized deleted residual (rstudent) greater than 2 */
proc print data=diag;  where abs(rstudent) gt 2.0;
run;
```

6.C   Program 6.1C

Program 6.1C facilitates identification of potentially influential cases in the multiple linear regression analysis applied to the SAS temporary data set HTDTA, created in Program 5.1. Program 6.1C computes and generates graphical displays of Cook's D, DFFITS, and DFBETAS values and calculates recommended cutoff values for DFBETAS and DFFITS that are appropriate for the sample size (n = 20) and the number of model parameters (p = 4) for Example 5.1.

Please note:
If you attempt to execute Program 6.1C after having logged out from the SAS session in which you executed Program 5.1, you will get error messages in most operating environments. One way to avoid this is by executing Program 5.1 in your current SAS session prior to running Program 6.1C or alternatively by copying the DATA step statements and data from Program 5.1 and pasting them at the beginning of Program 6.1C.

```
* Program 6.1C;
 /*Apply a multiple linear regression analysis to generate plots
of Cook's D, DFFITS, DFBETAS*/
proc reg data=htdta plots (only label)= (cooksd dffits dfbetas );
   id record;
   model ht18=birthl mht18 fht18/ vif r collin p;
   output out=diag r=rawresid student=student rstudent=rstudent
   p=predicted cookd=cookd dffits=dffits h=leverage;
run;
/*CALCULATE SIZE-ADJUSTED CUTOFF VALUES
for DFBETAS: Belsley et al. 1980
for DFFITS Belsley et al. 1980,Staudte and Sheather 1990*/
data cutvalues;
   *parm=number of parameters in model and n=sample size;
   /*dfbetacut is cutoff value for DFBETA recommended by Belsley 1980*/
   /*dffitscutbels is cutoff value for DFFITS recommended by Belsley 1980*/
   /*dffitscutss is cutoff value for DFFITS recommended
   by Staudte and Sheather 1990*/
   parm=4;
   n=20;
   nminusp=n-parm;
   dfbetacut=2/(sqrt(n));
   dffitscutbels=2*(sqrt(parm/n));
   dffitscutss=1.5*(sqrt(parm/n));
   drop nminusp;
run;
proc print data=cutvalues ;
   title Size-Adjusted Cutoff Values for DFBETAS and DFFITS Example 5.1;
run;
data influence;
   if _n_=1 then set cutvalues;
   set diag;
   absdffits=abs(dffits);
run;
```

```
proc print data=influence;
   where cookd gt 0.5 ;
   title Cases with COOKD Value > 0.5 ;
   var record cookd ht18 birthl mht18 fht18 predicted
   rawresid  student rstudent leverage ;
run;
proc print data=influence;
   where absdffits gt 1.00;
   title Cases with DFFITS Absolute Value > 1.00;
   var record absdffits ht18 birthl mht18 fht18 predicted
   rawresid  student rstudent leverage ;
run;
proc print data=influence;
   where absdffits gt dffitscutbels;
   title Cases with DFFITS Absolute Value > Belsley et al Cutoff Value;
   var record absdffits ht18 birthl mht18 fht18 predicted
   rawresid  student rstudent leverage ;
run;
proc print data=influence;
   where absdffits gt dffitscutss;
   title Cases with DFFITS Absolute Value > Staudte and Sheather Cutoff;
   var record absdffits ht18 birthl mht18 fht18 predicted
   rawresid  student rstudent leverage ;
run;
```

7

Interpreting an Additive Multiple Linear Regression Model

7.1 Introduction

In this chapter our focus is on interpreting results from an additive multiple linear regression model. An additive model assumes that, in the sampled population, no interactions occur among any of the model's predictor (explanatory) variables. Absence of interaction between model predictors occurs when the effect of one predictor on the response variable remains the same no matter what the values of the other model predictors are, provided these values are in the observed range of values for the predictors. We will defer a detailed discussion of multiple regression interaction models until Chapters 8 and 9.

When the ideal inference conditions for a multiple regression model are not reasonable for the sampled population, results from the inferential methods presented in this chapter may be spurious. We, therefore, have used the height at age 18 study (Example 5.1) as an illustrative example in this chapter as no serious departures from ideal inference conditions were detected for this example in Section 6.4. Moreover, the researchers' primary objective in this example was prediction (estimation). When prediction is the primary objective, researchers typically want a parsimonious model (a model involving as few terms as possible) that can predict the response in the sampled population accurately enough for their research purposes. For this reason, when prediction is a study's primary goal, adding interaction terms to the model is not necessarily required. If the predictive accuracy of the model without any interaction term(s) is sufficient for the researchers' objectives, then a parsimonious model without interaction effects is preferred over a more complicated model with interaction terms. If, however, the primary objective of a study is understanding the process underlying the relationship between the response and predictor variables, evaluation of interaction effects is an important step which we will explore in Chapters 8 and 9.

7.2 Height at Age 18 Example

The objective of the investigation in Example 5.1, Section 5.3 was prediction of height at age 18 for males in a particular village and the predictor variables in the model were a male's birth length, his mother's height at age 18, and his father's height at age 18.

DOI: 10.1201/9780429107368-7

7.3   The Model F-Test

The first question a researcher usually asks in multiple linear regression is: "Can one or more of the predictors in the model provide any information at all about the response variable in the sampled population?" This question can be evaluated by the Model F-test, sometimes referred to as the global test of the model. The null hypothesis for the Model F-test when there are k predictor variables, $X_1, \ldots, X_k$, in the model is that all of the regression coefficients corresponding to each of the predictor variables in the model are equal to zero, i.e., $\beta_1 = \beta_2 = \cdots = \beta_k = 0$. The two-sided alternative hypothesis is that at least one of the regression coefficients in the model is not equal to zero, i.e., there is one or more predictors when considered together in the model that can provide information about the response variable in the sampled population. We have already encountered a Model F-test in simple linear regression where the null hypothesis is that β_1, the regression coefficient for the single predictor variable in the model, is equal to zero. The equation for the test statistic for the Model F-test has been given previously (Equation 4.3.1).

The Model F-test is obtained from the REG procedure by default. The following SAS statements generate the Model F-test and the other output given in subsection 7.3.1 for the additive multiple linear regression model applied to the data for the height at age 18 example:

```
proc reg data=htdta;
model ht18=birthl mht18 fht18;
run;
```

7.3.1   Model F-Test and Some Other Results Generated by the Model Statement of the REG Procedure for the Height at Age 18 Example

The REG Procedure
Model: MODEL1
Dependent Variable: ht18

Number of Observations Read	20
Number of Observations Used	20

Analysis of Variance

Source	DF	Sum of Squares	Mean Square	F Value	Pr > F
Model	3	131.88802	43.96267	50.76	<.0001
Error	16	13.85748	0.86609		
Corrected Total	19	145.74550			

Root MSE	0.93064	R-Square	0.9049
Dependent Mean	68.53500	Adj R-Sq	0.8871
Coeff Var	1.35791		

The calculated F-value of 50.76 in this example has an extremely small p-value (less than 0.0001). We can assume that this p-value is reasonably accurate as no serious departures from ideal inference conditions for a multiple linear regression model were detected for these data. The next subsection discusses the interpretation of this particular Model F-test result as well as the general interpretation of any Model F-test for an additive multiple linear regression model with a small p-value.

7.3.2 Interpretation When the Model F-Test Statistic for a Multiple Linear Regression Model Has a Small p-Value

When the Model F-test statistic has a small p-value for a multiple linear regression model, you can conclude the alternative hypothesis, that one or more predictor variables when considered together in the model can give information about the response variable, Y, in the sampled population, **provided**:

- The sample used to estimate the model parameters is representative of the sampled population.

- The ideal inference condition of independence of errors for the multiple linear regression model is not violated.

- The ideal inference conditions of linearity, equality of variances, and normality for the multiple linear regression model are reasonable for the sampled population.

- The model is used to predict Y within the region jointly defined by predictor values observed in the sample. The model may or may not be able to predict Y with any reasonable accuracy outside of this region. Extrapolation in multiple regression will be discussed in Section 7.15.

In the height at age 18 example, where the Model F-test has a very small p-value (< 0.0001), we reject the null hypothesis that the regression coefficients for birth length, mother's height at age 18, father's height at age 18 are all equal to zero and conclude that one or more of these predictor variables, when considered together in an additive multiple linear regression model, can provide information about the height of males at age 18 in the particular population sampled. The preceding provisos for interpreting a Model F-test with a low p-value were not a concern for this example, as the sample was obtained at random from the population of interest using a formal random sampling procedure, the ideal inference conditions for the model were reasonable, and the researchers were careful to avoid extrapolation when they used the estimated model based on their sample data to predict the Y variable.

What a Small p-Value Associated with a Model F-Test Statistic for an Additive Multiple Linear Regression Model Does Not Imply

A small p-value associated with a Model F-test statistic for an additive multiple linear regression model does not imply that:

- the response variable is caused by one or more of the predictor variables in the model. As noted previously, the design of the study, not the statistical analysis, can enable researchers to conclude cause and effect (Section 1.12).

- the additive multiple linear regression model is the best way to describe the relationship between the response and predictor variables in the sampled population. For example, a nonlinear model could be a better way to describe this relationship.

- this model is useful for a researcher's objective.

- all of the predictor variables in the model are required in order to provide information about Y in the sampled population. Even when a Model F-test has a very low p-value, it does not indicate that all the predictor variables are required in the model to accurately predict the response.

7.3.3   Possible Interpretations When the Model F-Test Statistic for an Additive Multiple Linear Regression Model Has a Large p-Value

A large p-value associated with a Model F-test statistic for an additive multiple linear regression model merely informs us we have insufficient information to say whether the model can provide any information about Y in the sampled population. The model predictors may actually provide information about Y in the sampled population but the Model F-test statistic for an additive multiple linear regression model has a large p-value for one or more of the following reasons:

- non-random sampling

- insufficient sample size

- in the sample, the variability of Y within subgroups having particular values of the predictor variables is large, resulting in a large Mean Square Error and thus a small calculated F-value with a large p-value

- there is violation of one or more of the ideal inference conditions for multiple linear regression, viz., independence of errors, linearity, and equality of variances, to such an extent that the Mean Square Error (the model dependent estimate of the variance of Y given the X variables) is unduly large resulting in an unduly small F-value and low power for the test

- there is a serious departure from the ideal inference condition of normality and the sample size is not sufficiently large to compensate for this departure

- additional predictor variables are required in the additive multiple linear regression model

- an interaction effect is present among two or more of the model predictor variables, and an interaction regression model, not an additive model, is required, as will be discussed in Chapters 8 and 9

7.4 R^2 in Multiple Linear Regression

R^2 in multiple linear regression is called the coefficient of multiple determination and is defined as the proportion of the total variation of the Y values **in the sample** that is explained by the predictor variables when they are considered together in a multiple linear regression model.

7.4.1   Equation

$$R^2 = \frac{\text{Model Sum of Squares}}{\text{Total Sum of Squares}} \tag{7.1}$$

where

Model Sum of Squares is a measure of the variation of the observed Y values that is explained by the predictor variables when considered together in the model

Total Sum of Squares is a measure of the variation of the observed Y values about the overall sample mean of Y

7.4.2 Issues Regarding R^2

The issues regarding R^2 discussed in subsection 4.9.2 carry over to the multiple regression model. We do not recommend that evaluation of a model's usefulness for prediction be solely based on the value of R^2.

In multiple linear regression it is important to note that the value of R^2 varies according to the number of predictors in the model. As the number of model predictors increases, the value of R^2 will increase, or at least not decrease, regardless of whether the predictors added are improving or degrading the capability of the model to predict Y. In fact, the value of R^2 can be increased to 1, simply by adding additional predictor variables to the model, even though these additional variables may degrade the capability of the model to provide information about Y. Therefore, it is not recommended to compare R^2 values of models with an unequal number of predictors.

7.4.3 Example

The R^2 for the height at age 18 example (given in subsection 7.3.1) is 0.90, which indicates that 90% of the total variation of the values of Y, height at age 18, in the sample is explained by the additive multiple linear regression model with an intercept and birth length, mother's height at age 18, and father's height at age 18 as predictor variables. However as R^2 is an optimistically biased estimate of the population multiple correlation coefficient, this interpretation may overestimate the extent of the linear relationship between a male's height at age 18 and the model predictors in this example.

7.5 Adjusted R^2

The quantity known as adjusted R^2 adjusts the value of R^2 so that R^2 does not necessarily increase when additional terms are added to the model regardless of whether they are needed.

7.5.1 Equation for Adjusted R^2

The following form of the equation for adjusted R^2 shows how this quantity takes the number of model parameters and the sample size into account when computing R^2

$$\text{Adjusted R}^2 = 1 - \left(\frac{n-1}{n-p}\right)(1 - R^2) \tag{7.2}$$

where

p is the number of parameters in the model

n is the number of cases in the sample

7.5.2   Issues Regarding Adjusted R^2

The issues associated with interpreting R^2 (subsections 7.4.2, 4.9.2) also apply to adjusted R^2 with the exceptions that:

- adjusted R^2 does not automatically increase when additional terms are added to the model

- adjusted R^2 is a less biased estimate of the square of the population multiple correlation coefficient, $\rho_{Y|X_1,\ldots,X_k}$ than R^2 and therefore adjusted R^2 is less likely to overestimate the extent of the linear relationship between Y and the model predictors (Mickey et al., 2004).

7.5.3   Example

The adjusted R^2 for the height at age 18 example (given in subsection 7.3.1) is 0.89, which indicates that 89% of the total variation of the values of Y, height at age 18, in the sample is explained by the additive multiple linear regression model with an intercept and birth length, mother's height at age 18, and father's height at age 18 as predictor variables. As previously noted, a large adjusted R^2 does not necessarily imply that there is a linear relationship between the outcome and predictor variables. However, in this example, evaluation of the linearity ideal inference condition in Step (6), Section 6.4, does not reveal there is a serious departure from linearity.

7.6   Root Mean Square

The root mean square can be interpreted as an average difference in the sample between the observed Y value and the Y value predicted by the model. In the height at age 18 example, we interpret the root mean square, 0.93 of an inch (Root MSE in subsection 7.3.1), as the average difference in the sample between the observed height at age 18 and the height at age 18 predicted by the model. As the ideal inference conditions of this model seem reasonable, we infer that this multiple regression model would have similarly good prediction capability in the sampled population, provided we do not extrapolate beyond the combination of predictor values observed in the sample.

7.7   Coefficient of Variation for the Model

Recall from Section 4.11, the coefficient of variation is expressed as a percentage and is a unit-less measure of a model's fit to the sample data. The smaller the value of the coefficient of variation the better the fit of the model to the sample data, provided all the Y values are positive. As one would expect, the Y values, height at age 18 in Example 5.1, are all positive. Thus the coefficient of variation for this example, 1.36%, indicates a close fit in the sample between the observed values and the values predicted by the model for height at age 18.

7.8 Partial *t*-Tests in a Multiple Linear Regression Model

A partial *t*-test can be helpful in deciding whether a predictor variable in a multiple regression model provides information about the response variable in the sampled population. As previously noted, a Model *F*-test with a small p-value does not necessarily imply that all the predictor variables in the model are needed. The null and alternative hypotheses of a two-sided partial *t*-test are

H$_0$: The predictor variable, X_j provides no additional improvement in predicting Y over and above that contributed by the other predictor variables in the model, i.e., $\beta_j = 0$ where β_j denotes the regression coefficient for X_j in the multiple linear regression model.

H$_A$: X_j does provide additional improvement in predicting Y over and above that contributed by the other predictor variables in the model, i.e., $\beta_j \neq 0$.

7.8.1 Equation of the Test Statistic for a Partial *t*-Test of a Regression Coefficient

The equation for a partial *t*-test of an individual regression coefficient being equal to zero in an additive multiple linear regression model is

$$t = \frac{\hat{\beta}_j - 0}{\text{SE}(\hat{\beta}_j)} \tag{7.3}$$

where
  $\hat{\beta}_j$ is the estimated regression coefficient of X_j adjusted for all the other predictor variables in the model as well as the intercept term
  $\text{SE}(\hat{\beta}_j)$ is the standard error of $\hat{\beta}_j$ in an additive multiple linear regression model and is is typically estimated by

$$\widehat{\text{SE}}(\hat{\beta}_j) = \sqrt{\frac{(\text{Mean Square Error})\text{VIF}_j}{\sum_{i=1}^{n}(X_{ij} - \bar{X}_j)^2}} \tag{7.4}$$

where
  VIF$_j$ is the variance inflation factor for the predictor variable, X_j, in the model (Equation 6.1)
  It can be seen from Equation 7.4 that the standard error of $\hat{\beta}_j$ in an additive multiple linear regression model depends on:

(i) the Mean Square Error, i.e., the variability of Y unexplained by the model

(ii) $1/\sum(X_{ij} - \bar{X}_j)^2$, which is the reciprocal of a measure of the amount of variation in the values of the predictor, X_j

(iii) VIF$_j$, which is a measure of X_j's collinearity with one or more of the other predictors in the model

Thus, a *t*-value of the partial *t*-test is small with a large p-value when the denominator of this statistic, i.e., the standard error of $\hat{\beta}_j$, is large which occurs when:

(i) the variability of Y unexplained by the model is large

(ii) the variation in the sample values of the predictor variable, X_j about the sample mean of X_j is small

(iii) X_j is highly collinear with one or more of the other predictors in the model

7.8.2   Partial *t*-Test Examples

The REG procedure computes the following results for partial *t*-tests for the regression coefficients of the predictor variables, birth length of the subject at birth (birthl), mother's height at age 18 (mht18) and father's height at age 18 (fht18) in the multiple regression model in Example 5.1 (the height prediction example):

Parameter Estimates

Variable	DF	Parameter Estimate	Standard Error	t Value	Pr > |t|	Variance Inflation
Intercept	1	-78.38040	13.25051	-5.92	<.0001	0
birthl	1	1.31994	0.44712	2.95	0.0094	1.25048
mht18	1	0.70482	0.16023	4.40	0.0004	1.22276
fht18	1	1.10172	0.09903	11.12	<.0001	1.03284

By default, the REG procedure provides a p-value for each two-sided partial *t*-test. A two-sided partial *t*-test is appropriate for this example because the researchers' alternative hypothesis was that the regression coefficient was not equal to zero. The degrees of freedom associated with the *t*-value for each of these partial *t*-tests are $n-p$, where p is the number of estimated parameters in the multiple regression model. There are four parameters estimated in this example (estimates for the intercept and the three predictor variables) and the sample size, n, is 20, therefore 16 $(20 - 4)$ degrees of freedom are associated with the preceding *t*-values for these partial *t*-tests.

Each p-value reported for a partial *t*-test by the REG procedure by default is the **per comparison error rate**, i.e., the probability of Type I error, for the single test under consideration and does not take into account whether multiple tests were carried out in the analysis.

From the preceding "Parameter Estimates" results generated by the REG procedure, we note the *t*-value, 2.95, for birth length (birthl) is the calculated *t*-value from the partial *t*-test of the null hypothesis that birth length provides no additional improvement in predicting height at age 18 over and above that contributed by mother's height at age 18 and father's height at age 18. The probability (p-value) associated with a *t*-value of 2.95 in the distribution of *t*-values for 16 degrees of freedom where the null hypothesis is true would be less than or equal to 0.0094. As the ideal inference conditions for the model appear to be reasonable for the sampled population in this example, it can be assumed that this observed p-value, 0.0094 for this **single** partial *t*-test, is reasonably accurate. On the basis of this small observed p-value < 0.0094 associated with this individual partial *t*-test for birth length, the researchers rejected the null hypothesis of this test and concluded that birth length does provide additional improvement in predicting height at age 18 over and above that contributed by mother's height at age 18 (mht18) and father's height at age 18 (fht18).

Similarly the observed p-values from the partial *t*-tests of the regression coefficient being equal to zero for mother's height at age 18 (p < 0.0004) and for father's height at age 18 (p<0.0001) are both very small. Therefore, the researchers concluded that each predictor provides additional improvement in predicting height at age 18 over and above that contributed by the other predictors in the model.

Typically the partial *t*-test for the intercept is not of substantive interest as the null hypothesis of this partial *t*-test is that the intercept is equal to zero, when all other predictors in the model are equal to zero. In this example zero values for the predictors (birth length, mother's height, father's height) are meaningless.

7.9 Partial *t*-Tests and the Issue of Multiple Testing

It is important to note that when multiple hypothesis tests are carried out, such as partial *t*-tests of regression coefficients for all the model predictors, the issue of multiple testing arises. The multiple testing issue (sometimes referred to as the multiplicity or multiple comparison issue) simply stated is that the greater the number of hypothesis tests one carries out, the greater the probability of making an incorrect decision. An incorrect decision occurs in a hypothesis test either when one rejects the null hypothesis but the null hypothesis is actually true in the sampled population (Type I error) or when one does not reject the null hypothesis although the null hypothesis is actually false in the sampled population (Type II error). In Chapter 12 we describe methods that address the multiple testing issue and discuss situations where these methods are generally advised.

7.10 The Model *F*-Test, Partial *t*-Tests, and Collinearity

Harmful collinearity can provide an explanation of the seemingly paradoxical situation where a Model *F*-test has a small p-value yet all of the partial *t*-tests of the regression coefficients for the predictors in the model do not have small p-values. When a partial *t*-test for a regression coefficient of a predictor has a large p-value, this indicates that the test is not able to demonstrate that this predictor provides any additional information about Y over and above that provided by the other predictors in the multiple regression model. However, as discussed in Section 5.12, when predictor variables are highly collinear, it is difficult to disentangle the separate effects of the collinear variables on the Y variable. As a consequence, the estimated regression coefficients of highly collinear variables can have large standard errors resulting in large p-values for the partial *t*-tests of the regression coefficients of the collinear variables although these collinear variables, when considered together in the model, can provide information about Y resulting in a low p-value for the Model *F*-test.

In the height at age 18 example, the Model *F*-test as well as the partial *t*-tests for the regression coefficients of the model predictors have very small p-values and the signs of the estimated regression coefficients make substantive sense (Section 5.12) suggesting harmful collinearity is not present in the sample. These results are consistent with the assessment of collinearity for this example (Section 6.4) where the moderately strong collinearity detected between birth length and mother's height using the method of Belsley et al. (1980) was not considered to be harmful as the VIF for these predictor variables were observed to be quite low.

7.11 Why Estimate a Confidence Interval for β_j?

Rejecting the null hypothesis in a partial *t*-test of $\beta_j = 0$ tells us little about how useful X_j is as a predictor of Y in the model – it tells us only that X_j is not completely useless as a predictor. Knowing the value of the sample estimate for β_j, for example 1.31994 for birth length (subsection 7.8.2), provides some information about the importance of X_j as a linear predictor of Y in the model. However, a sample estimate of β_j, although unbiased, may not be very accurate. To get an idea of the accuracy of the sample estimate, we can estimate the lower and upper limits of a confidence interval for the true value of β_j in the sampled population. The CLB option in the MODEL statement of the REG procedure estimates a conventional (i.e., not adjusted for multiplicity) 95% confidence interval for β_j. In the context of confidence intervals for the β_j, multiplicity is the issue that the greater the number of confidence intervals considered, the greater the coverage error probability, i.e., the greater the probability that one or more of these estimated confidence intervals for the β_j will not include the true value of the parameter being considered. We will discuss applications of multiplicity-adjusted confidence intervals in Section 12.11.

7.11.1   Example

The following SAS statements requests that lower and upper limits for 95% conventional intervals be estimated for the regression coefficients for the predictor variables in the multiple regression model for the height at age 18 example.

proc reg data=htdta;
model ht18=birthl mht18 fht18/clb;
run;

The conventional confidence limits generated by the preceding SAS statements are as follows:

Variable	DF	95% Confidence Limits	
Intercept	1	-106.47022	-50.29058
birthl	1	0.37210	2.26779
mht18	1	0.36515	1.04449
fht18	1	0.89179	1.31166

From these results we see that the estimated 95% conventional confidence interval for the regression coefficient for birth length, (0.37, 2.27), excludes zero, which suggests, as the ideal inference conditions seem to be reasonable, that birth length provides additional information about height at age 18 over and above that provided by the other predictor variables in the model. This is the same suggestion indicated by the partial t-test for the regression coefficient for birth length. However, the 95% estimated confidence interval provides us with more information than the partial t-test. The estimated confidence interval not only suggests that zero may be an implausible value for the population regression coefficient, but also gives us a range of plausible values for the true value of the population regression coefficient for birth length in the multiple regression model considered. The 95% estimated conventional confidence limits for the regression coefficients for mother's height at age 18 and father's height at age 18 also do not include zero. Thus it is suggested by these results that each of these predictors provides additional information about height at age 18 over and above that provided by the other predictor variables in the model. The intercept is not of substantive interest in this example. However, in later chapters we will consider examples where the intercept is meaningful in the context of the research study.

7.11.2   Equation for a Conventional Confidence Interval Estimate for β_j

A $100(1 - \alpha)\%$ conventional confidence interval estimate for a population regression coefficient, say β_j, in a multiple regression model with an intercept is given by

$$\hat{\beta}_j \pm t(1 - \frac{\alpha}{2}; n - p) \, \widehat{\text{SE}}(\hat{\beta}_j) \tag{7.5}$$

where

n is the sample size

p is the number of estimated parameters in the model

$t(1 - \frac{\alpha}{2}; n - p)$ is the appropriate critical value from the Student's t-distribution with $(n - p)$ degrees of freedom for a $1 - \alpha$ confidence interval

$\widehat{SE}(\hat{\beta}_j)$ is the estimated standard error of $\hat{\beta}_j$ and is given by Equation 7.4

7.12 Estimating a Conventional Confidence Interval for the Mean Y Value in a Population Subgroup with a Particular Combination of the Predictor Variables

In Section 5.8 we discussed how the predicted value from an estimated regression model equation can be used to estimate the mean Y value of a sampled population subgroup which has a specified combination of values for the predictor variables. Provided ideal inference conditions are reasonable for the model, a confidence interval can give us an idea about how accurately such a population subgroup mean is estimated by the predicted value from the model. Typically a 95% confidence interval is considered and if this confidence interval is wide, the population subgroup mean of Y has not been very accurately estimated by the model. A conventional[1] 95% confidence interval estimate for the subgroup mean of Y can be obtained from the REG procedure by specifying the CLM option in the MODEL statement. Conventional confidence interval estimates other than 95% can also be requested as illustrated in Section 4.A.

7.12.1 Example

The researchers would like to get an idea of how accurately the estimated regression model estimates the true mean height at age 18 of various subgroups defined by the values of the predictor variables in the sample. The CLM option in the MODEL statement of the REG procedure gives a 95% conventional confidence interval estimate for the mean of the subgroup to which each case belongs as illustrated for the height at age 18 example.

```
proc reg data=htdta plots=none;
   model ht18=birthl mht18 fht18/ clm;
   id record;
run;
```

You can get a listing of the predictor values for each individual in the sample to find out each individual's subgroup by using the PRINT procedure, for example,

```
proc print data=htdta;
title Height at Age 18 Example;
run;
```

The first line generated by the PRINT procedure for the height at age 18 example reveals that the individual identified as Record = 1 belongs to the subgroup with birth length equal to 19.7 inches, mother's height at age 18 equal to 60.5 inches and father's height at age 18 equal to 70.3 inches.

---

[1] Unadjusted for multiplicity.

Height at Age 18 Example

Obs	Record	ht18	birthl	mht18	fht18
1	1	67.2	19.7	60.5	70.3
2	2	69.1	19.6	64.9	70.4
3	3	67.0	19.4	65.4	65.8
.					
.					
.					
19	19	68.1	20.2	62.6	68.8
20	20	66.1	19.2	62.2	67.3

As shown below and reported in the "Output Statistics" of the REG procedure, the lower and upper limits of the 95% conventional confidence interval estimated by the CLM option for first case (Record = 1) are respectively 66.44 and 69.00. Provided we consider only this confidence interval (i.e., not multiple confidence intervals) this 95% conventional confidence interval suggests that (66.44, 69.00) is a range of plausible values for the true value of the mean height in inches at age 18 in the sampled population for the subgroup that has the same predictor values as this first case, i.e., the subgroup whose birth length is 19.7 inches, whose mother's height at age 18 is 60.5 inches, and whose father's height at age 18 is equal to 70.3 inches.

The REG Procedure
Model: MODEL1
Dependent Variable: ht18

Output Statistics

Obs	Record	Dependent Variable	Predicted Value	Std Error Mean Predict	95% CL Mean		Residual
1	1	67.2	67.7153	0.6039	66.4350	68.9955	-0.5153
2	2	69.1	70.7947	0.3316	70.0916	71.4977	-1.6947
3	3	67.0	65.8152	0.5012	64.7526	66.8777	1.1848
.							
.							
.							
19	19	68.1	68.2028	0.3993	67.3563	69.0493	-0.1028
20	20	66.1	64.9483	0.4126	64.0735	65.8231	1.1517

7.13 Estimating a Prediction Interval for an Individual Y Value Given Specified Values for the Predictor Variables

A prediction interval can provide information about how accurately a multiple regression model can predict a Y value for a single individual having a certain combination of values for the model predictors. The width of this interval can give an idea of the accuracy of the predicted Y for such an individual. We can specify the CLI option of the MODEL statement in the REG procedure to generate a conventional prediction interval estimate.

Recall that the accuracy of a prediction interval can be degraded by a departure from the ideal inference condition of normality, even if the sample size is large (van Belle et al., 2004). Moreover, the prediction intervals generated by the CLI option for the individuals in the input data set are not adjusted for multiplicity: Therefore, the greater the number of conventional prediction intervals considered, the greater the probability that one or more of these intervals will not include the true value. Thus caution is advised in interpreting the multiple prediction intervals generated by the CLI option.

With these caveats in mind, we consider the 95% conventional prediction interval estimated for the height of a male at age 18 for the first observation (Record = 1) in the data set reported in the "Output Statistics" of the REG procedure:

Output Statistics

Obs	Record	Dependent Variable	Predicted Value	Std Error Mean Predict	95% CL Predict		Residual
1	1	67.2	67.7153	0.6039	65.3634	70.0671	-0.5153

Assuming independence of errors holds and the other ideal inference conditions of this multiple linear regression model especially normality are reasonable, and assuming we are considering only this prediction interval (not multiple prediction intervals), we interpret the prediction interval estimate for the case with Record=1 (Obs 1) as follows: The interval (65.36, 70.07) contains a range of plausible values for the height at age 18 in inches of an individual with birth length, mother's height at age 18, and father's height at age 18 equal to 19.7, 60.5, and 70.3 inches, respectively.

A Y value of an individual given that individual's values of the model predictor variables is always estimated by the model with less accuracy than the population mean Y value of the subgroup characterized by those predictor values. This is illustrated by the first observation Record = 1 with birth length, mother's height at age 18, and father's height at age 18, respectively, equal to 19.7, 60.5, and 70.3 inches. The width of the 95% prediction interval estimate (70.07, 65.36) for the Y value of an individual with the same predictor values as that of Record = 1 is 4.7 inches. The width of the 95% confidence interval estimate (69.00, 66.44) for the population mean of the subgroup characterized by the predictor values of RECORD = 1 is only 2.6 inches (subsection 7.12.1).

7.14 Further Evaluation of Influence in the Height at Age 18 Example

In the previous chapter (Section 6.6), Cases 3 and 5 from the height at age 18 example were flagged as potentially influential cases. In this section the influence of these cases is evaluated by comparing substantive conclusions from this chapter's inferential methods applied to the full data set of this example to conclusions reached when either one of these cases was excluded from the multiple regression analysis. That is, we investigated whether Case 3 influenced the study's substantive conclusions by repeating the multiple regression analysis with Case 3 excluded (see Program 7.1 in Section 7.A of this chapter's appendix). Similarly, we carried out a separate analysis with Case 5 excluded.

The general conclusions from the multiple regression analyses with either Case 3 or Case 5 excluded are consistent with conclusions from the original analysis which included all cases. The Model F-test and the partial t-tests of the predictors' regression coefficients still have small p-values with Case 3 or 5 excluded. There are some differences in the values of the estimated regression coefficients when Case 3 or Case 5 was excluded as compared to those obtained when the full data set was analysed. However, the estimated regression coefficients from the analyses with either Case 3 or 5 excluded are all well within the 95% confidence intervals for the population regression coefficients estimated from the full data set. Hence, we conclude that neither Case 3 nor Case 5 is a cause of concern because of its influence on the study's general substantive conclusions. We do not suspect that these cases are masking each other's influential effect because these cases do not lie close together in the multidimensional predictor space. Although the values of birth length and mother's height are somewhat similar for the two cases, the value of father's height at age 18 for Case 3 is not close to that for Case 5.

7.15   Extrapolation in Multiple Regression

In simple linear regression avoiding extrapolation is straightforward as it simply involves using the estimated regression model to predict Y only for a case that has a predictor value which is within the range of maximum and minimum values observed in the sample. For example, if age is the predictor variable in the simple linear regression model, and the minimum and maximum ages observed in the sample are 5 and 65 respectively, then extrapolation is avoided by not using the model to predict Y for ages that are greater than 65 or less than five. In multiple linear regression, however, the problem of "hidden extrapolation" may arise when we use the model to predict Y for a "new" case not found in the original sample. Hidden extrapolation refers to the situation where we predict Y for a new case and that case is outside the region jointly defined by the predictor values of the cases in the sample, although that case's value for each model predictor lies within the observed range for that predictor in the sample. Thus, simply checking whether the values of the model predictors for a new case falls within the ranges observed for each of those predictors in the sample will not necessarily inform us as to whether we would be extrapolating if we were to predict Y for the new case. We illustrate a scenario where hidden extrapolation occurs in subsection 7.15.1.

Cook and Weisberg (1999) have suggested the following approach as an approximate guide when deciding whether a new case is outside the region jointly defined by the observed predictor values in the sample:

- Compute a hat value for each case in the sample. A hat value in multiple linear regression is a measure of the distance a case is from the centre of the region jointly defined by the predictor values of the cases in the sample.

- Compute a hat value for the new case, based on that case's predictor values.

- If the hat value for the new case exceeds the maximum hat value observed in the sample, then this new case is outside the region of the cases in the sample. Predicting Y for this new case would therefore involve extrapolation beyond the region jointly defined by the predictor values of the observed data.

7.15.1 Delivery Time Data Example

A soft drink company was interested in predicting delivery time involved in stocking vending machines. Two predictor variables were identified as being important: the number of boxes of product stocked (BOXES) and the distance walked by the driver from the truck when delivering the stock (DISTANCE). In Program 7.2, Section 7.B, we used a subset of data given by Montgomery et al. (2012) and applied an additive multiple linear regression where the Y variable, denoted as TIME, was delivery time and the predictor variables were BOXES and DISTANCE. We assume that the ideal inference conditions for this model hold in the population sampled. Suppose we wanted to use the model to predict delivery time for the following two new cases (scenarios): (i) the new case, denoted as RECORD 21, where the number of boxes is 8 and the distance the driver had to walk from the truck to deliver the stock is 275 feet and (ii) the new case, denoted as RECORD 22, where the number of boxes is 20 and the distance the driver had to walk is 350 feet.

In Program 7.2 we generated hat values via the REG procedure to decide whether either RECORD 21 or RECORD 22 is outside the region jointly defined by the predictor values in the original sample. The diagonal hat values for the new cases, RECORD 21 and RECORD 22 are equal to 0.0786 and 0.888 respectively. Examination of the hat values for all the cases in the original sample (RECORDS 1–20) given in subsection 7.15.2 reveals that 0.653 is the largest hat value. Thus, using Cook's and Weisberg's suggestion (1999), prediction of delivery time for RECORD 21 would not involve extrapolation as its hat value of 0.078 is less than 0.653. However, if we were to use this model to predict delivery time for RECORD 22, we would be extrapolating outside the region jointly defined by the predictor values observed in the sample, as the hat value of RECORD 22, viz., 0.888, is greater than the maximum hat value observed in the original sample. It can be noted in Figure 7.1 that RECORD 22's value for BOXES, i.e., 20, falls within the observed range for this predictor variable in the original sample, i.e., 2–30 boxes, and that RECORD 22's value for DISTANCE, i.e., 350 feet, also falls within the observed range for DISTANCE in the original sample, i.e., 36–1460 feet. This illustrates why we cannot rely on using the ranges of the individual predictor variables observed in the original sample to identify whether a new case is an extrapolation point. As illustrated by Figure 7.1, when there are just two predictors in the model, the region jointly defined by the values of the model predictors can be easily determined visually by examining a scatter plot of observed values for these predictors in the sample. However, when there are more than two predictors in the model, visually assessing extrapolation with graphics is difficult and the approach of using hat values is helpful.

7.15.2 Cases Sorted by Hat Values from Program 7.2

Delivery Time Data Sorted in Descending Order by Hat Value (h)

Obs	RECORD	TIME	BOXES	DISTANCE	h
1	22	.	20	350	0.88884
2	9	79.24	30	1460	0.65317
3	10	21.50	5	605	0.28491
4	16	29.00	10	776	0.21100
5	12	21.00	10	215	0.19609

6	20	35.10	17	770	0.15073
7	18	19.00	7	132	0.14354
8	11	40.33	16	688	0.13739
9	3	12.03	3	340	0.13625
10	1	16.68	7	560	0.13303
11	19	9.50	3	36	0.11687
12	4	14.88	4	80	0.10613
13	5	13.75	6	150	0.10223
14	14	19.75	6	462	0.10087
15	7	8.00	2	110	0.09665
16	8	17.83	7	210	0.08994
17	2	11.50	3	220	0.08680
18	21	.	8	275	0.07857
19	17	15.35	6	200	0.07721
20	13	13.50	4	255	0.07367
21	6	18.11	7	330	0.05246
22	15	24.00	9	448	0.05105

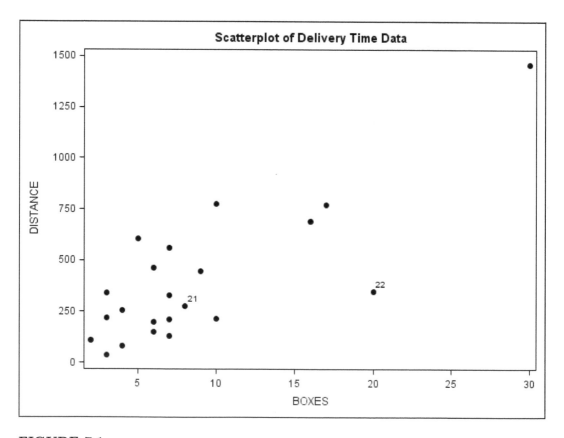

FIGURE 7.1

Scatterplot from delivery time example where DISTANCE (the distance walked by the driver from the truck) and BOXES (the number of boxes delivered by the driver) are the predictor variables in the multiple regression model to predict delivery time. New cases are labelled by RECORD NUMBERS (21 and 22).

7.16 Chapter Summary

This chapter:

- describes a multiple linear regression additive model

- discusses interpretation of the Model F-test for this model

- discusses R^2, adjusted R^2, root mean square, and the coefficient of variation as measures of fit of a multiple linear regression model to the data

- describes partial t-tests

- discusses the role of collinearity when results from a Model F-test and partial t-tests are conflicting

- discusses why a confidence interval estimate for a regression coefficient of a model predictor provides more information than a partial t-test for the regression coefficient for that predictor

- introduces the multiplicity issue associated with partial t-tests of model predictors and with confidence and prediction intervals in multiple linear regression

- investigates the possible influence of two cases on the general substantive conclusions of the height at age 18 example

- illustrates detection of hidden extrapolation in multiple linear regression

Appendix

7.A   Program 7.1

Using the **temporary** SAS data set HTDTA created by Program 5.1, Program 7.1 applies a multiple regression analysis to the data of Example 5.1 where RECORD 3 is excluded.

Please note:
If you attempt to execute Program 7.1 after having logged out from the SAS session in which you executed Program 5.1, you will get error messages in most operating environments. One way to avoid this is by executing Program 5.1 in your current SAS session prior to running Program 7.1 or alternatively by copying the DATA step statements and data from Program 5.1 and pasting them at the beginning of Program 7.1.

```
*Program 7.1;
title Excluding Case 3 from the Height at Age 18 Example;
proc reg data=htdta plots=none;
   where record ne 3;
   model ht18=birth1 mht18 fht18/p clb;
   id record;
quit;
```

7.B   Program 7.2

Program 7.2 generates hat values to implement Cook and Weisberg's (1999) approach (Section 7.15) for deciding whether prediction of delivery time for two new cases would involve extrapolation beyond the original data.

```
title Delivery Time Data;
/* subset of data on page 100 Montgomery et al 4th edn*/
data one;
input RECORD TIME BOXES DISTANCE;
datalines;
1   16.68 7 560
2 11.50  3 220
3   12.03 3 340
4 14.88 4 80
5 13.75 6 150
6 18.11 7 330
7 8.00 2 110
8 17.83 7 210
```

```
9 79.24 30 1460
10 21.5 5 605
11 40.33 16 688
12 21 10 215
13 13.5 4 255
14 19.75 6 462
15 24.00 9 448
16 29.00 10 776
17 15.35 6 200
18 19.00 7 132
19 9.5 3 36
20 35.1 17 770
21 . 8 275
22 . 20 350
;
proc reg data=one plots=none;
   model time=boxes distance/ influence;
   id record;
   output out=hatvalues h=h;
run;
proc sort data=hatvalues;
   by descending h;
run;
proc print data=hatvalues;
title Delivery Time Data Sorted in Descending Order by Hat Value (h);
run;
```

7.B.1   Explanation of Program 7.2

Cook's and Weisberg's (1999) approach is easily implemented using SAS. Hat values for cases not belonging to the original sample can be readily obtained in the REG procedure. We simply input a period, "." to indicate that the observed Y value is missing for the new case in a data step along with the values for the predictor variables for the new case of interest, as is shown for RECORDS 21 and 22 in the first data step of Program 7.2. This data input results in a hat value being computed for each of these new cases by the REG procedure.

Hat values are obtained by specifying the INFLUENCE option in the MODEL statement of the REG procedure. To facilitate identification of the maximum hat value for the original cases we listed the hat values for the data in descending order using the SORT and PRINT procedures. As we have already explained these procedures in previous programs, we will not repeat an explanation here.

8

Modelling a Two-Way Interaction Between Continuous Predictors in Multiple Linear Regression

8.1 Introduction to a Two-Way Interaction

A two-way interaction, the simplest type of interaction in multiple linear regression, occurs between two explanatory variables (predictors), say X_1 and X_2, when the effect of one explanatory variable, X_1 on the outcome variable, Y, depends on the values of the other explanatory variable, X_2. Moreover, as a two-way interaction is symmetric, the effect of X_2 on Y depends on the values of X_1. A two-way interaction is sometimes referred to as a first-order interaction because the effect of an explanatory variable on the outcome depends on the values of only one other explanatory variable. When we speak of an explanatory variable's effect on the outcome variable, Y, in an observational study, we are referring to its effect associated with Y, not its effect causing Y as would be true in a well-designed randomized experimental study.

Our focus in this chapter is on evaluating an interaction effect between two continuous predictor variables. Consider a hypothetical study where researchers wanted to investigate if a teacher's effectiveness score based on classroom evaluation (the response variable) could be predicted by two variables, which were scores on two different standardized tests written by the teacher (denoted as TEST1 and TEST2). Figure 8.1 illustrates the scenario when there is no interaction between the two predictor variables, TEST1 and TEST2. In this graph we show simple linear regression lines for $Y1$ and TEST1 for three subgroups of teachers with selected values of TEST2 scores:

  (i)   teachers with a low TEST2 score of 20 (solid line)
  (ii) teachers with an average TEST2 score of 50 (broken line)
  (iii) teachers with the highest TEST2 score of 100 (dash dot line)

There is no interaction between TEST1 and TEST2 on the response variable, $Y1$, in this example because the linear relationship between $Y1$ vs. TEST1 remains identical regardless of the value of a teacher's score on TEST2. This is immediately obvious in Figure 8.1 because the simple regression lines for each value of TEST2 score displayed in the graph are parallel, i.e., the three simple regression lines have the same slope.

In contrast, Figure 8.2 shows an example where an interaction occurs between the predictor variables, TEST1 and TEST2, as the linear relationship between the response variable, labelled as $Y2$, in the figure is different for the high, average, and low values of TEST2. The **presence of a two-way interaction** between the predictor variables is obvious as the simple regression lines of $Y2$ vs. TEST1 for the selected values of TEST2 are **not parallel** which means the simple regression lines have **different slopes**. In other words the prediction of a teacher's effectiveness score ($Y2$) by his/her score on TEST1 depends on

DOI: 10.1201/9780429107368-8

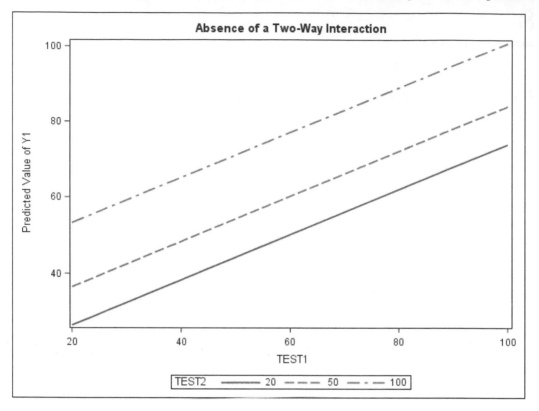

FIGURE 8.1
Comparison of estimated simple linear regressions of teacher effectiveness score ($Y1$) vs. TEST1 score at selected values of scores for TEST2 where no interaction occurs between TEST1 and TEST2.

the value of his/her score on TEST2. If a teacher has an abysmally low score on TEST2, e.g., a score of 20, then TEST1 score does not predict a teacher's effectiveness score very well. The simple linear regression line for $Y2$ vs. TEST1 for TEST2 scores of 20 is almost horizontal. This is not true for teachers who have average or high scores on TEST2.

Figures 8.1 and 8.2 are examples of a conditional effect plot, which can be a helpful visual tool when evaluating a two-way interaction in multiple regression. A conditional effect plot shows the regression of the response variable, Y, on one of the predictors, say $X1$, for selected values of the other predictor variable, say $X2$, in a multiple regression model.

8.2   A Two-Way Interaction Model in Multiple Linear Regression

Nonparallel regression lines in a conditional effect plot based on sample data inform us that an interaction is present in the sample. However, the interaction effect observed in a conditional effect plot based on sample data may not actually occur in the population from which the sample was drawn. It could be that the different slopes of the nonparallel regression lines observed in the sample data may be due solely to sampling variation and in the sampled population these slopes are in fact identical. Therefore, we need a statistical inferential

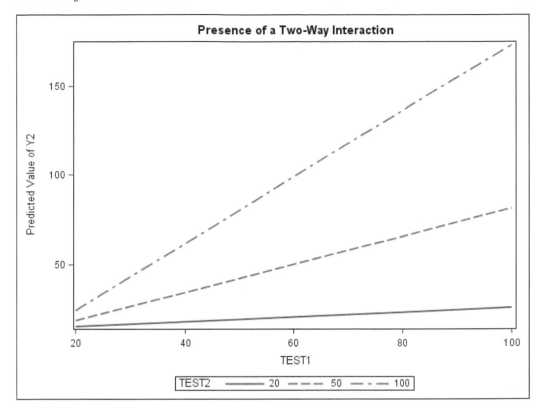

FIGURE 8.2
Comparison of estimated simple linear regressions of teacher effectiveness score ($Y2$) vs. TEST1 score at selected values of scores for TEST2 where an interaction occurs between TEST1 and TEST2.

method to assess whether the slopes are actually different, i.e., whether an interaction is occurring between the predictors in the sampled population.

A model that is typically used to assess the presence and effect size of an interaction between two continuous predictors, say X_1, X_2 in the population sampled, is given by the following equation.

$$Y = \beta_0 + \beta_1 X_1 + \beta_2 X_2 + \beta_3 X_1 X_2 + \epsilon \tag{8.1}$$

where

β_0 is the Y intercept

β_1 is the regression coefficient for the predictor X_1 when the predictor X_2 is equal to zero

β_2 is the regression coefficient for the predictor X_2 when the predictor X_1 is equal to zero

β_3 is the regression coefficient for the interaction term

$X_1 X_2$ is the interaction term which is the simple product of the predictors X_1 and X_2

The hierarchy principle is applied in forming this model by including in the model not only the higher order interaction term, $X_1 X_2$, but also the predictors, X_1 and X_2. Applying the hierarchy principle prevents confounding of the interaction effect with that of its constituent predictors, thereby making it possible to disentangle the separate effects of X_1, X_2, and $X_1 X_2$ on the Y variable.

In the model given by Equation 8.1 it may be noted that:

- The effect of the predictor X_1 on Y depends on what the value of X_2 is. Specifically, this model implies that if X_1 increases by one unit while X_2 remains constant, Y changes by the amount equal to $(\beta_1 + \beta_3 X_2)$ In other words, the effect of X_1 on Y depends on the value of X_2.

- The effect of the predictor X_2 on Y depends on what the value of X_1 is. Specifically this model implies that if X_2 increases by one unit while X_1 remains constant, Y changes by the amount equal to $(\beta_2 + \beta_3 X_1)$. In other words, the effect of X_2 on Y depends on the value of X_1.

- β_1 is the regression coefficient for the predictor X_1 when the predictor X_2 is equal to zero. This statement is easily confirmed by setting X_2 equal to zero in Equation 8.1 which simplifies to

$$Y = \beta_0 + \beta_1 X_1 + \beta_2 0 + \beta_3 X_1 0 + \epsilon$$
$$= \beta_0 + \beta_1 X_1 + \epsilon$$

  Thus, β_1, the regression coefficient for the predictor X_1 is not meaningful if the predictor X_2 does not have a meaningful zero value as we will later illustrate with an example.

- β_2 has the interpretation as the regression coefficient for the predictor X_2 when the predictor X_1 is equal to zero which is confirmed by setting X_1 equal to zero in Equation 8.1. Thus, β_2, the regression coefficient for the predictor X_2 is not meaningful, if the predictor X_1 does not have a meaningful zero value.

  In Section 8.5 we will discuss how to facilitate the interpretation of a regression coefficient for a continuous predictor involved in a two-way interaction when the other predictor does not have a meaningful zero value.

- The interaction term described in this model is most commonly used. More complex interaction terms, such as the product $log(X_1)X_2^2$ (Cook and Weisberg, 1999), are also possible but seldom used.

8.3   Investigating a Two-Way Interaction Using Sample Data

A two-way interaction in a sampled population can be investigated using sample data by:

- applying the method of ordinary least squares to the sample data to get parameter estimates for the model given in Equation 8.1

- assessing whether ideal inference conditions for this model are reasonable for the sampled population

- estimating a confidence interval for β_3, the regression coefficient for the interaction term, provided the ideal inference conditions for the model are reasonable

  Often a 95 % confidence interval is constructed. If the confidence interval estimate for β_3 does not include zero, then the interpretation is that the data provide evidence to suggest an interaction effect between X_1 and X_2 occurs in the population sampled. If the confidence interval estimate for β_3 includes zero, the interpretation is that there may or may not be an interaction effect in the sampled population. We cannot rule out the possibility of an interaction effect as the estimated upper and lower limits of the constructed confidence interval for β_3 suggest a range of other plausible values which are not equal to zero.

- applying a partial t-test of the null hypothesis that β_3 is equal to zero, provided the ideal inference conditions for the model are reasonable. Applying this partial t test is not necessary if a confidence interval, e.g., a 95% confidence interval, has been constructed.

A small p-value obtained from this partial t-test suggests that an interaction may exist between X_1 and X_2 in the population sampled. A large p-value indicates that, on the basis of applying this test to the sample data, there is insufficient evidence to show that the interaction effect between X_1 and X_2, as modelled in Equation 8.1, occurs in the sampled population.

Details regarding implementing these methods for investigating a two-way interaction will be illustrated with an example in Section 8.6.

8.4 Example 8.1: Physical Endurance as a Linear Function of Age, Exercise, and their Interaction

Researchers wanted to investigate whether physical endurance of an adult (as measured by the duration in minutes of sustained jogging on a treadmill) could be explained by age at the time of the treadmill test and the number of years spent in vigorous physical exercise prior to the test. Common knowledge would suggest that the effect of number of years spent in vigorous physical exercise on physical endurance (the Y variable) would be different depending on one's age at the time of the test, i.e., there would be a significant interaction effect between the explanatory variables, age at time of test and previous vigorous physical exercise. Or likewise the effect of age on physical endurance would depend on years spent in vigorous physical exercise. We describe in subsequent sections how a multiple linear regression model can be applied to evaluate an interaction effect between age and previous vigorous exercise using a subset ($n = 50$) of data from this study reported in Cohen et al (2003). The subset of data we used is listed in Program 8.1 (Section 8.C) and is displayed in Figure 8.3.

8.5 Facilitating Interpretation in Two-Way Interaction Regression via Centring

We have shown in Section 8.2 how the interpretation of a regression coefficient of a predictor involved in a two-way interaction is not meaningful if the other predictor does not have a meaningful zero value. One way to circumvent this difficulty involves mean-centring the original data values.

Recall from Section 6.2, mean-centring of a predictor can be achieved by subtracting the sample mean of a predictor from each value of the predictor. Measures other than the mean can be used in centring a continuous predictor involved in a two-way interaction to facilitate interpretation. If the continuous predictor does not have a symmetrical distribution (e.g.,

Observed Data Points for Example 8.1

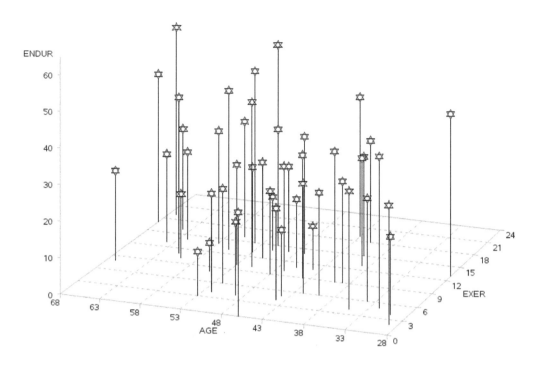

FIGURE 8.3

Observed data for Example 8.1 where ENDUR denotes duration in minutes of sustained jogging during a treadmill test, AGE denotes age at the time of the test, and EXER denotes number of years spent in vigorous physical exercise prior to the test.

a right skewed distribution), the sample mean may not be the centre of location, i.e., the sample mean may not be representative of an average value in the sample. In this case, some other measure, such as the median, that is more typical of the average sample value could be used, i.e., the continuous predictor could be centred by subtracting the median from each raw data value for the predictor. Alternatively, a value based on expert option in the particular research may be used instead of the mean or median estimated from the sample data.

Up to this point we have discussed the merit of centring a continuous predictor involved in a two-way interaction when the zero value for that predictor is not meaningful. However, centring a continuous predictor is also recommended to avoid extrapolation beyond the range of the data observed in the sample, in those situations where zero values for a predictor, although meaningful, do not occur in the researcher's sample.

Centring the predictors involved in a two-way interaction does not change the shape of the overall regression surface between the outcome and the predictors. Results from the Model F-test, R-square, and the estimate of the regression coefficient for the interaction term and its standard error are identical regardless of whether or not predictors are mean-centred in the analysis (Aiken and West, 1991).

It is not necessary to centre the Y variable if you are centring predictors involved in a two-way interaction. In fact, it is usually recommended to leave the Y variable in its original form because it simplifies prediction when the predicted values from the model are in the same metric as the original data for Y (Cohen et al., 2003).

8.5.1 Example of Centring to Facilitate Interpretation

Suppose the researchers in the physical endurance example described in Section 8.4 applied the following model to the original data values:

$$Y = \beta_0 + \beta_1 \text{AGE} + \beta_2 \text{EXER} + \beta_3 \text{ AGEEXER} + \epsilon \tag{8.2}$$

where

Y is minutes a subject endured on the treadmill test

AGE is the age of subject (rounded to the nearest year) when the subject took the treadmill test

EXER is number of years (rounded to the nearest year) spent in vigorous physical exercise prior to the treadmill test

AGEEXER is the interaction term formed by multiplying a subject's age at the time of the test by the subject's years spent in previous vigorous physical exercise

β_1 is the regression coefficient for AGE when EXER is equal to zero

β_2 is the regression coefficient for EXER when AGE is equal to zero

β_3 is the regression coefficient for the interaction term

ϵ is variation in Y not explained by AGE, EXER, and AGEEXER in the model

The interpretation of β_2, the regression coefficient for EXER, the previous vigorous exercise predictor is not meaningful when it is estimated using the original (raw) data values for AGE (age at time of treadmill test) where value of zero is meaningless, i.e., a subject who has an age of zero at the time of the treadmill test is nonsense. One way to avoid this problem is by centring the original data values for AGE and applying the regression analysis to the centred data.

We decided to use the sample mean age, 48 years (rounded to the nearest year) for centring. We used the mean because results from the UNIVARIATE procedure for AGE in Program 8.1 suggest that the mean is a reasonable measure of the centre of the distribution for AGE in the sample. These univariate results revealed that AGE does not have a markedly skewed distribution in this sample and that the sample mean and median AGE are the same when rounded to the nearest year.

Mean-centring AGE is achieved by subtracting the sample mean, rounded to the nearest year, i.e., 48, from each subject's age. If a subject is 48 years old, his mean-centred value for age at the time of the treadmill test is zero $(48 - 48)$. A zero value is meaningful for mean-centred data of AGE, as it indicates that a subject was at the average age of the sample at the time of the treadmill test. Thus, β_2, the regression coefficient for EXER can have a meaningful interpretation when the original values of AGE are mean-centred in the analysis.

8.6 A Suggested Approach for Applying a Multiple Linear Regression Two-Way Interaction Model with Two Continuous Predictors

In this section we suggest an approach for applying a model which includes an interaction term for two continuous predictors, using the model given in Equation 8.1 as a prototype. In subsequent sections we illustrate this approach with a specific example. The steps involved in this suggested approach are:

Step A: Screen for and correct any data input errors.

Step B: Screen for and correct if possible any data recording errors.

Step C: If zero is not a meaningful value for a continuous predictor involved in a two-way interaction or if zero, although meaningful for that predictor, does not occur in the sample, centre the original data values for that predictor (Section 8.5) and use the centred data values for that predictor in the subsequent steps.

Step D: Compute a value for the interaction term for each case in the sample by multiplying each case's data values for X_1 and X_2. For example, if case i has a value of 3 for X_1 and a value of 2 for X_2, then case i will have a value of 3x2=6 for the interaction term, $X_1 X_2$.

Step E: Apply the multiple linear regression interaction model with X_1, X_2, and the interaction term, $X_1 X_2$, as predictors to obtain regression diagnostics.

Step F: Evaluate ideal inference conditions. In evaluating ideal inference conditions outlined in Section 5.10, confirm that the response variable is continuous and examine details of study design as well as regression diagnostics and graphics obtained from Step E. Step F encompasses Steps (3) to (10) previously described in Chapter 6, Section 6.3.

If Step F suggests ideal inference conditions for the model are reasonable, proceed to Step G. Otherwise consider other approaches described in Chapter 14.

Step G: Check whether any cases merit investigation due to their potential disproportionate influence in this regression using methods described in Chapter 3.

Step H: Interpret the results from the interaction regression analysis. Typically you would first check if the Model F-test has a small p-value. If it has, this provides some protection against cumulative Type I error that arises from multiple testing in this analysis (e.g., multiple partial t-tests for regression coefficients of the model predictors). If the Model F-test has a large p-value, you are advised to interpret with caution the results from these multiple partial t-tests that are unadjusted for multiple testing.

You would then evaluate the two-way interaction effect by considering the confidence interval estimate for β_3 and/or the partial t-test of $\beta_3 = 0$. If this evaluation suggests that the hypothesized interaction regression model rather than an additive regression model (no interaction term) is appropriate, you can proceed to interpret the rest of the results based on the interaction model. If however, this evaluation of the two-way interaction effect does not indicate an interaction model is required, researchers often do not proceed with interpreting results from the interaction model, but rather drop the interaction term and apply an additive model to the data. There are situations, however, where prior substantive knowledge about a specific interaction effect might induce a researcher to use an interaction regression model even if the confidence interval estimate for β_3 and the partial t-test of $\beta_3 = 0$ suggest the interaction effect could not be detected in the researcher's own study.

8.7 An Example Using the Suggested Approach for Applying an Interaction Model with Two Continuous Predictors

This section illustrates our suggested approach for applying a multiple linear regression interaction model using the physical endurance example (described in Section 8.4).

Steps A and B: Screen for data input and recording errors.
No data input errors nor data recording errors were found when output from Program 8.1 (Section 8.C in this chapter's appendix) was examined.

Step C: If zero is not a meaningful value for a continuous predictor involved in a two-way interaction or if zero, although meaningful for that predictor, does not occur in the sample, centre the original data values for that predictor (Section 8.5) and use the centred data values for that predictor in the subsequent steps.

The predictor AGE (age of subject at the time of the treadmill test) does not have a meaningful zero value, so we centred the original data values for AGE in the data step in Program 8.2 (Section 8.D). We decided to use the sample mean age, 48 years (rounded to the nearest year) for centring. We used the mean because results from the UNIVARIATE procedure for AGE in Program 8.1 suggest that the mean is a reasonable measure of the centre of the distribution for AGE in the sample. These univariate results revealed that AGE does not have a markedly skewed distribution in this sample and that the sample mean and median AGE are the same when rounded to the nearest year. The following statement in the DATA STEP of Program 8.2 was used to compute each case's centred value for age which we arbitrarily named AGEC:

```
AGEC= AGE-48;
```

We did not centre the values for EXER (years of previous vigorous exercise) because a zero value is meaningful for this predictor in the context of the study and also because we saw that a zero value for EXER occurs in this sample, so zero is not beyond the range of the observed data values. (Results from the UNIVARIATE procedure in Program 8.1 showed zero as the minimum value in the data for EXER.)

Step D: Compute a value for the interaction term for each case in the sample.
This was computed in the data step in Program 8.2 using the centred AGE values and the original values for EXER. The following statement in the data step was used to compute each case's value for the interaction term which we arbitrarily named AGECEXER:

```
AGECEXER=AGEC*EXER;
```

Step E: Apply the multiple linear regression interaction model to obtain regression diagnostics and graphics. The model for this example was specified in the REG procedure in Program 8.2 by the following model statement:

```
model endur=agec exer agecexer
```

Program 8.2 generated regression diagnostics and graphics for this model.

Step F: Evaluate ideal inference conditions As summarized in Section 8.A in this chapter's appendix, Step F suggests ideal inference conditions are reasonable for this interaction model in the sampled population. We can therefore proceed to Step G.

Step G: Check whether any cases merit investigation due to their disproportionate influence in this regression using methods described in Chapter 3

From influence diagnostics output by Program 8.2 it may be concluded that no case in the sample merited investigation for disproportionate influence on the regression, as summarized in Section 8.B of this chapter's appendix.

Step H: Interpret the results from the interaction regression analysis. Results for this analysis are reported and interpreted in Sections 8.8 and 8.9.

8.8 Results from Application of a Multiple Regression Interaction Analysis with Centred AGE and Uncentered EXER in the Physical Endurance Example

Analysis of Variance

Source	DF	Sum of Squares	Mean Square	F Value	Pr > F
Model	3	1796.75892	598.91964	7.50	0.0003
Error	46	3675.16108	79.89481		
Corrected Total	49	5471.92000			

Root MSE	8.93839	R-Square	0.3284	
Dependent Mean	29.96000	Adj R-Sq	0.2846	
Coeff Var	29.83441			

Variable	DF	Parameter Estimate	Standard Error	t Value	Pr > |t|	Variance Inflation
Intercept	1	18.78299	3.18192	5.90	<.0001	0
AGEC	1	-1.05811	0.27732	-3.82	0.0004	5.08746
EXER	1	0.88704	0.27744	3.20	0.0025	1.27868
AGECEXER	1	0.08305	0.02339	3.55	0.0009	4.77806

Parameter Estimates

Variable	DF	95% Confidence Limits	
Intercept	1	12.37813	25.18786
AGEC	1	-1.61633	-0.49989
EXER	1	0.32858	1.44549
AGECEXER	1	0.03597	0.13014

The p-value of 0.00030 associated with the Model F-test suggests the interaction model fitted to the data in Example 8.1 can offer some explanation of the outcome variable,

ENDUR (duration in minutes of sustained jogging on a treadmill). The adjusted R-square of approximately 28% suggests that explanation of ENDUR by this model is not huge. The root mean square, i.e., the average difference in the sample between the observed duration in minutes and the duration predicted by the model is 8.9 minutes.

The results reported for the parameter estimates inform us that the estimated equation for this interaction model is

$$\widehat{\text{ENDUR}} = 18.78299 - 1.05811\text{AGEC} + 0.88704\text{EXER} + 0.08305\text{AGECEXER} \qquad (8.3)$$

The estimated 95% conventional confidence limits for the regression coefficient of the interaction term, AGECEXER, (0.03597, 0.13014), suggest that a plausible value for this regression coefficient does not include zero and lies in the range from 0.036 to 0.130. Moreover the test statistic of the partial t-test of the regression coefficient for the interaction term, AGECEXER, being equal to zero in the sampled population has a very small p-value (t=3.55, P > $|t|$=0.0009). These results suggest that, in the sampled population, there is an interaction effect between one's age at the time of the treadmill test and the years one spent in previous vigorous exercise.

The estimated 95% conventional confidence limits and the p-values associated with the partial t-tests for the regression coefficients of AGEC and EXER suggest that the true value of these regression coefficients is not equal to zero. The interpretation of these estimated regression coefficients is given in the next section.

8.9 Interpretation of Regression Coefficient Estimates for the Interaction Model Fitted to Example 8.1

AGEC

The estimated regression coefficient of -1.05811 for AGEC can be interpreted as follows: It is estimated that for every yearly increase in age, endurance on the treadmill test decreases an average of 1.06 minutes for those individuals who have had zero years of vigorous exercise prior to the test. We interpret that endurance on the treadmill test **decreases** as a linear function of age of subject because the regression coefficient for AGEC has a **negative** sign. We interpret that this negative effect of age applies to those individuals who have had zero years of vigorous exercise prior to the treadmill test because the original values for EXER have been used in the analysis.

EXER

As the values for age have been mean-centred in this analysis, the estimated regression coefficient of 0.89 for EXER can be interpreted as follows: It is estimated that the endurance on the treadmill test increases an average of 0.89 minutes for each year spent previously in vigorous exercise by 48-year-olds, i.e., those individuals who are at the average age observed in the sample. We interpret that endurance on the treadmill test **increases** as a linear function of previous vigorous exercise because the regression coefficient for EXER has a **positive** sign. We interpret that this positive effect of previous vigorous exercise applies to those individuals who are at the average age observed in the sample because the predictor age has been mean-centred (Section 8.5).

AGECEXER

We interpret the estimated regression coefficient of 0.08305 for AGECEXER as follows:

- It is estimated that with every unit increase in years spent in previous vigorous exercise, the regression coefficient for AGEC changes by the positive amount of 0.08305. This suggests that the negative effect of age (or some unmeasured correlate(s) of age) on endurance is mitigated by the effect of years spent in previous vigorous exercise or some correlate(s) of previous vigorous exercise. This interpretation applies only to the combination of values of age and previous vigorous exercise that are within the range of values observed in the sample on which this analysis was based. For example, as there were no centenarians in the sample, it would not be wise to extrapolate and conclude that this mitigative effect of previous vigorous exercise would necessarily apply to centenarians.

- With every unit (i.e., yearly) increase in age, it is estimated that the regression coefficient for EXER changes by the positive amount of 0.08305. A possible interpretation of this result is that the effect of previous vigorous exercise or the effect of some unmeasured correlate(s) of previous vigorous exercise is greater for older aged adults or some correlate of being an older aged adult. Again it would not be good statistical practice to extend this interpretation beyond the combination of the range of values of age and previous vigorous exercise observed in the sample on which this analysis is based.

8.10   Visualization of the Interaction Effect Observed in Example 8.1

Figure 8.4, a conditional effect plot based on predicted values from the fitted interaction model, illustrates the estimated interaction effect between previous vigorous exercise and age. The estimated regression lines depicting predicted values of ENDUR vs. EXER for selected values of AGE in this figure are obviously not parallel. It may be noted that the regression lines shown for each AGE do not involve extrapolation as these lines do not extend beyond the range of the observed exercise values for each age. We have used the original rather than the mean-centred values of AGE in this figure to facilitate interpretation. As previously noted in Section 8.5, a regression coefficient for an interaction term is identical regardless of whether the predictors involved are centred or not: Thus, an identical pattern would be displayed for Figure 8.4 regardless of whether the original or centred values of AGE are used in fitting the model.

    Figure 8.4 suggests the following insight into the nature of the observed interaction effect: Previous years of vigorous exercise or some factor(s) correlated with previous years of vigorous exercise are increasingly beneficial for older men in this observational study. This insight is suggested because the slopes of the estimated regression lines for ENDUR vs. EXER become increasingly greater for older aged men. Figure 8.5, a three-dimensional visualization of predicted values for new cases, also provides this insight into the nature of the interaction effect. It may be noted that none of the predicted values of the cases displayed in this figure involves extrapolation beyond the scope of the model, as all of the displayed cases have a combination of values for AGE and EXER that lie within the region of the observed data.

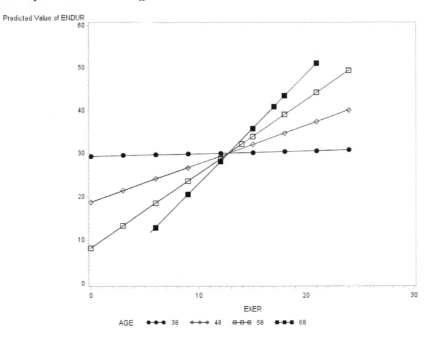

FIGURE 8.4
Conditional effect plot comparing predicted values from the interaction model vs. years of previous exercise (EXER) for selected ages.

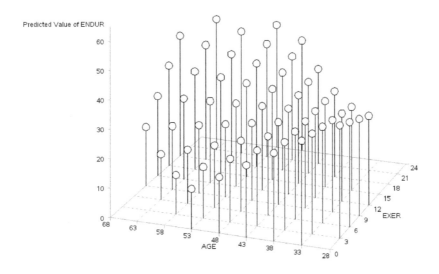

FIGURE 8.5
Predicted values of treadmill endurance in minutes (ENDUR) for new cases generated by the estimated interaction model in Example 8.1.

8.11   Predicting Endurance on a Treadmill

This section illustrates how Equation 8.3 can be used to predict treadmill endurance in the sampled population, given one's age (mean-centred) and previous vigorous exercise experience. For example, suppose one wants to predict the average treadmill time in minutes for a subgroup of 60-year-olds from the sampled population who have had 10 years of previous vigorous exercise experience. The predicted value from the model for the average treadmill time for this subgroup can be obtained by substituting the centred value for AGEC in years $(60 - 48 = 12)$, 10 years for EXER, and 120 as the value for the interaction term AGECEXER (10x12) into Equation 8.3 as follows:

$$\widehat{ENDUR} = 18.78299 + -1.05811(12) + 0.88704(10) + 0.08305(120)$$
$$= 24.92$$

Recall the predicted value of a model also predicts the Y value for an individual from the sampled population given that individual's values for the model predictors. Thus 24.92 minutes is also the endurance time estimated for an individual from the sampled population who is 60 years old and has had 10 years of previous vigorous exercise experience.

As Program 8.2 illustrates, the P option of the MODEL statement in the REG procedure generates a predicted value and its standard error given the values of the predictor variables for each individual in the sample. The CLM option estimates a conventional[1] confidence interval for the **mean** Y value in a population subgroup with a particular combination of the values of the predictor variables. The CLI option estimates a conventional prediction interval for the Y value of an **individual** from the sampled population given that individual's values of the predictor variables. These statistics are given in "Observation-wise Statistics" under the subheading "Output Statistics" by the REG procedure. We report these results for the first observation (RECORD=1) for illustration.

```
                        Output Statistics


                                         Std
                                        Error
             Dependent   Predicted       Mean
Obs RECORD   Variable        Value    Predict          95% CL Mean
  1     1          18       24.9222    2.2021     20.4896      29.3549
  .
  .
  .

Obs RECORD      95% CL Predict
  1     1       6.3922    43.4522
```

As evaluation of the ideal inference conditions for this interaction model suggests that these conditions appear to be reasonable for the physical endurance example (Section 8.A), we interpret (20.4896, 29.3549), the estimated limits for the 95% conventional confidence interval (denoted as 95% CL Mean in the output from the REG procedure), as a range of plausible values for the average endurance time in minutes for the subgroup of 60-year-olds from the sampled population who have had 10 previous years of vigorous exercise. We note, however, that this conventional confidence interval estimate is unadjusted for multiplicity.

---

[1] Unadjusted for multiplicity.

Actual Data with Corresponding Predicted Values for Example 8.1.
Stars are Actual Data and Balloons are Corresponding Predicted Values

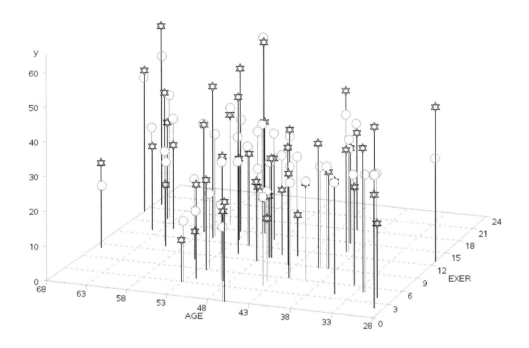

FIGURE 8.6
Actual values (stars) with their corresponding predicted values (balloons) for Y, minutes of endurance on the treadmill test, for observed cases in Example 8.1.

Caution is advised interpreting a prediction interval estimate (denoted as 95% CL Predict in the output of the REG procedure) because as previously noted the accuracy of a prediction interval can be degraded by a departure from the ideal inference condition of normality, even if the sample size is large (van Belle et al., 2004). Moreover, caution is also advised if we consider multiple conventional prediction intervals simultaneously, as the multiplicity issue applies. With these caveats in mind, we interpret the conventional 95% prediction interval estimate (6.3922, 43.4522) as a range of plausible values for the endurance time in minutes for a 60-year-old individual who has had 10 years of prior vigorous exercise history.

Figure 8.6, a three dimensional plot of actual and corresponding predicted values for the observed cases in Example 8.1, provides an informal visualization of the prediction accuracy of the model.

8.12   Chapter Summary

This chapter:

- describes the general situation when a two-way interaction effect between predictor variables occurs in a multiple linear regression model

- describes a multiple linear regression model with a two-way interaction term where the predictor variables involved in the interaction are continuous variables and outlines why you would want to use such a model

- discusses how centring can facilitate interpretation of regression coefficients of predictor variables involved in a two-way interaction

- suggests a step-by-step approach for applying a multiple linear regression model with a two-way interaction between continuous predictor variables and illustrates this approach with an example

- interprets the illustrative example's estimated regression coefficients from the two-way interaction model and provides three-dimensional graphics and a conditional effect plot to facilitate interpretation

- gives SAS programs which implement the step-by-step approach we recommend for applying a two-way interaction model with continuous predictors

Appendix

8.A Summary of Evaluation of Ideal Inference Conditions for the Physical Endurance Example with a Two-Way Interaction Model

Step F suggests ideal inference conditions are reasonable for this interaction model in the sampled population. As previously noted, Step F encompasses Steps (3) and (4) and Steps (6) to Step (10) previously described in Chapter 6, Section 6.3. The conclusions drawn from these steps are the following:

- The response variable, duration in minutes of sustained jogging on a treadmill, is continuous (Step 3).

- Although we do not know the details of the study design, we assume that the model given by Equation 8.2 reflects the sampling design of the study and that a model with a more complicated error structure is not required to satisfy ideal inference conditions (Step 4).

- The plot of the internally studentized residuals vs. the predicted values from this model (Figure 8.7) does not suggest any serious departure from linearity or equality of variances (Step 6).

- Plots of the internally studentized residuals against centred age values (AGEC), original exercise values (EXER), and the interaction term (AGECEXER) displayed in Figure 8.7 do not reveal a systematic pattern indicating that the model errors are dependent on the values of these variables. Other aspects of the study design that might have varied among subjects that would have caused violation of independence of errors were not described or included in the data set and therefore could not be investigated (Step 7).

- The normal Q-Q plot of the internally studentized residuals (Figure 8.8) does not reveal a major departure from normality (Step 8).

- Although three cases (Records 36, 42, and 26) have an absolute studentized deleted residual from the model greater than two, it may be noted from the "Observation-wise" "Output Statistics" that the absolute studentized deleted residuals of these cases were quite close to two (respectively 2.25 2.009, 2.068). Moreover, it can be seen in Figure 8.9 that these cases do not have a high leverage (hat) value (Step 9).

- The variance inflation factor (VIF) values for AGEC, EXER, amd AGECEXER reported in Section 8.8 were all much less than 10, indicating harmful collinearity is not present among the explanatory variables for this model. The following results generated by Program 8.2 corroborate this conclusion. Recall from Section 6.2.2, Belsley (1991) suggested there is little cause for concern about collinearity when the largest scaled condition index is between 5–10.

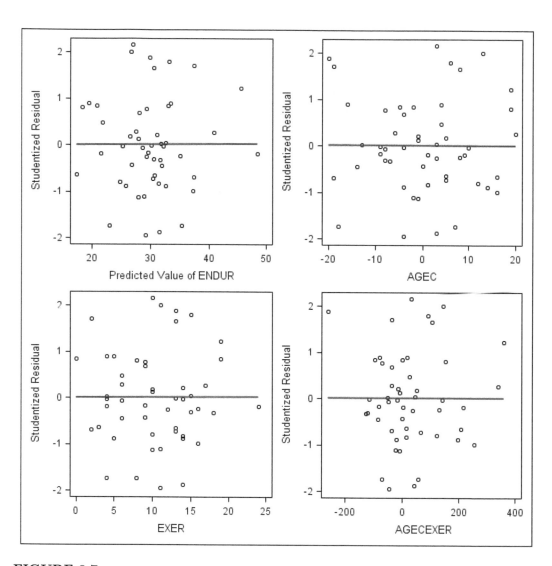

FIGURE 8.7

Plots of internally studentized residuals from the interaction model for the physical endurance example vs. the predicted values and vs. the explanatory variables, centred AGE (AGEC), exercise (EXER) and the interaction (AGECEXER).

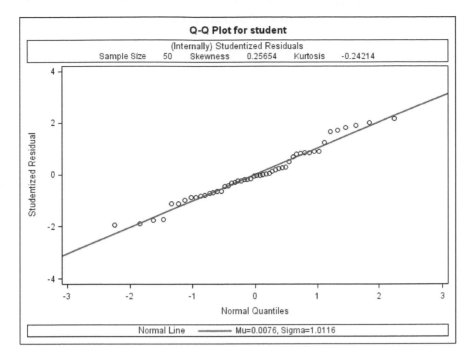

FIGURE 8.8
Normal Q-Q plot of internally studentized residuals from the interaction model for the physical endurance example.

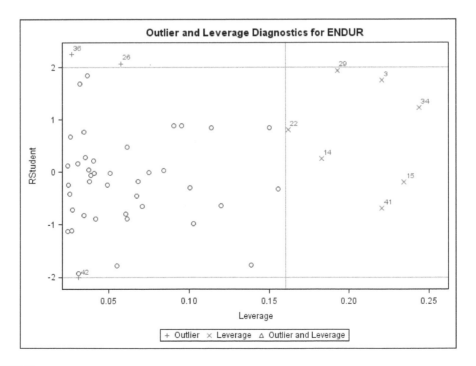

FIGURE 8.9
A scatter plot of externally studentized residuals (Rstudent) by leverage from the interaction model for the endurance example.

Physical Endurance Example

The REG Procedure
Model: MODEL1
Dependent Variable: ENDUR

Collinearity Diagnostics

Number	Eigenvalue	Condition Index
1	2.32514	1.00000
2	1.47914	1.25378
3	0.12217	4.36252
4	0.07355	5.62239

Collinearity Diagnostics

Number	Intercept	AGEC	EXER	AGECEXER
		----------------Proportion of Variation----------------		
1	0.01472	0.01725	0.01983	0.02415
2	0.03298	0.04269	0.01873	0.02571
3	0.09810	0.49477	0.22653	0.71326
4	0.85420	0.44529	0.73491	0.23688

8.B Summary of Influence Diagnostics for the Physical Endurance Example with a Two-Way Interaction Model

It was concluded that:

- No case in the sample merited investigation for its influence on **all** predicted values in the sample, as no case had a Cook's D value greater than 0.5, the cutoff value for Cook's D recommended by Cook and Weisberg (1999).

- No case in the sample merited investigation for its influence on its **own** predicted value, as no case had an absolute value of DFFITS value greater than 1, the cutoff value for the absolute value of DFFITS in small to medium-sized samples recommended by Kutner et al. (2005).

- No case in the sample merited investigation for its influence on the value of the estimated regression coefficients, as no case had an absolute value of DFBETAS value greater than 1, the cutoff value for the absolute value for DFBETAS in small to medium-sized samples recommended by Kutner et al. (2005).

8.C   Program 8.1

Program 8.1 screens for data input and recording errors in the physical endurance example.
The statements in Program 8.1 have been explained in previous programs.

```
*Program 8.1;
data one;
    title Physical Endurance Example with Uncentered Predictors;
    *subset of data (n=50) page 275 Cohen et al. 2003;
    input RECORD AGE EXER ENDUR;
datalines;
    1     60     10     18
    2     40      9     36
    3     29      2     51
    4     47     10     18
    5     48      9     23
    6     42      6     30
    7     55      8      8
    8     43     19     40
    9     39      9     28
   10     51     14     15
   11     54     15     49
   12     52      4     27
   13     53      3     12
   14     68     17     43
   15     57     24     47
   16     30      4     21
   17     35      4     32
   18     56     16     33
   19     62     14     25
   20     39     13     30
   21     32      5     41
   22     67      8     25
   23     56     13     45
   24     47     14     33
   25     47     10     29
   26     61     11     44
   27     40     15     28
   28     49      4     20
   29     28     13     45
   30     40      6     28
   31     44      5     18
   32     41     18     29
   33     53     13     24
   34     67     19     55
   35     52      6     26
   36     51     10     46
   37     46     11     19
   38     44      4     25
   39     64     16     29
```

```
40    58    14    32
41    29     2    32
42    44    11    12
43    51    12    27
44    51    15    33
45    53    10    28
46    44     9    34
47    46     0    28
48    49    14    24
49    34     6    28
50    64    13    25
   ;
run;
*STEP A;
proc print;
run;
*STEP B;
proc univariate data=one;
   ods select ExtremeValues Frequencies;
   ods exclude TestsForLocation;
   id record;
   run;

/*The following plots can be examined for contextual,
predictor and response outliers that merit investigation
for recording errors*/
ods listing style=journal2 image_dpi=200;
proc sgscatter data=one;
   plot endur *(age exer)/columns=1 rows=2;
run;
proc sgplot;
   scatter x=age y=exer;
run;
```

8.D   Program 8.2

Using the temporary SAS dataset ONE, created in Program 8.1, Program 8.2 evaluates an interaction model for the physical endurance example where AGE is centred and EXER (previous exercise) is not centred.

Please note:
If you attempt to execute Program 8.2 after having logged out from the SAS session in which you executed Program 8.1, you will get error messages. One way to avoid this is by executing Program 8.1 in your current SAS session.

```
*Program 8.2;
data two;
   title Physical Endurance Example;
```

```
    *subset of data (n=50) page 275 Cohen et al. 2003;
    set one;
    /*Center AGE and call centered age value AGEC*/
    AGEC=AGE-48;
    /*Compute value of interaction term
    using centered AGE values and original EXER values*/
    AGECEXER=AGEC*EXER;
proc print;
run;
    /*Apply a linear regression analysis to obtain
    regression diagnostics and measures of influence*/

proc reg data=two plots (only label)=(RStudentByLeverage cooksd
dffits dfbetas);
        id record;
      model endur= agec exer agecexer/ vif r collin p cli clm clb influence;
      output out=diag r=r rstudent=rstudent
      student=student p=p cookd=cookd dffits=dffits;
run;


/*Obtain plots of internally studentized residuals vs. predicted Y,
and vs. centered predictor variables */

proc sgscatter data=diag;
    title;
    plot (student)*(p agec exer agecexer)/columns=2 rows=4 reg;
run;
 *EVALUATE NORMALITY;
 * Obtain a Q-Q normal plot for internally studentised residuals;
proc univariate data =diag;
    var student ;
    qqplot student  /normal (mu=est sigma=est);
    inset n='Sample Size' skewness kurtosis/
    pos=tm header='(Internally) Studentized Residuals';
run;


/*CHECK IF THERE ARE ANY REGRESSION OUTLIERS BASED ON RSTUDENT VALUES*/
/*Get a listing of variables of interest for cases identified as
    regression outliers based on a specified cutoff value for
    the studentized deleted residual (rstudent)*/

proc print data=diag;
    where abs(rstudent) gt 2;
run;


/*CALCULATE SIZE-ADJUSTED CUT-OFF VALUES
for DFBETAS: Belsley et al. 1980
for DFFITS Belsley et al. 1980,Staudte and Sheather 1990*/
*parm=no of parameters in model and n=sample size;
*dffitscutbels is cutoff value recommended by Belsley et al. 1980;
*dffitscutss is cutoff value recommended by Staudte and Sheather 1990;
```

```
data cutvalues;
   parm=4;
   n=50;
   nminusp=n-parm;
   dfbetacut=2/(sqrt(n));
   dffitscutbels=2*(sqrt(parm/n));
   dffitscutss=1.5*(sqrt(parm/n));
run;
 proc print data=cutvalues;
    title Size-Adjusted Cutoff Values for DFBETAS and DFFITS;
run;
 data influence;
   if _n_=1 then set cutvalues;
   set diag;
   absdffits=abs(dffits);
run;

proc print data=influence;
   where cookd gt 0.5 ;
   title Cases with COOKD Value > 0.5 ;
  var record age exer endur p r student rstudent cookd dffits;
run;
proc print data=influence;
   where absdffits gt 1.00;
    title Cases with DFFITS Absolute Value > 1.00;
   var record age exer endur p r student rstudent cookd dffits;
run;
proc print data=influence;
   where absdffits gt dffitscutbels;
   title Cases with Abs. DFFITS Value > Belsley et al. 1980 Cutoff Value;
   var record age exer endur p r student rstudent cookd dffits;
run;
proc print data=influence;
   where absdffits gt dffitscutss;
   title Cases with Abs. DFFITS Value > Staudte and Sheather 1990
   Cutoff Value;
   var record age exer endur p r student rstudent cookd dffits;
run;
```

8.D.1   Explanation of SAS Statements in Program 8.2

The SAS statements in Program 8.2 have been explained previously in earlier programs and in Section 8.7, Steps C and D. The exception is the SET statement in the first DATA step of this program. The SET statement in SAS requests that all variables and observations be read from the specified input data set. Thus in Program 8.2 the statement SET ONE requests that all variables and observations be read from the SAS **temporary** data set ONE created in Program 8.1.

9

Evaluating a Two-Way Interaction Between a Qualitative and a Continuous Predictor in Multiple Linear Regression

9.1 Introduction

In this chapter we first describe how information on a qualitative variable can be incorporated in a multiple linear regression model. We then discuss a simple example which illustrates evaluation of a two-way interaction between a qualitative and a continuous predictor.

9.2 How to Include a Qualitative Variable in a Multiple Linear Regression Model

A qualitative variable can be incorporated as a predictor in a multiple regression model by:

1. creating one or more **indicator variables**, also known as **dummy variables**, which capture the information on the categories of the qualitative variable

2. including the indicator variable(s) to represent the qualitative variable in the multiple regression model

The term indicator or dummy refers to the fact that the values assigned to an indicator (or dummy) variable do not reflect any sort of ranking but merely indicate whether or not an individual belongs to a particular category of the qualitative variable. There are various prescribed coding schemes (set of rules) that can be used to represent a qualitative variable as a predictor in multiple regression. In this chapter we will discuss one of these prescribed coding schemes for qualitative variables known as **reference cell coding**. It is important to realize that all prescribed coding schemes yield identical predicted Y values, identical residuals, an identical R^2, an identical Mean Square Error value, as well as identical results for the Model F-test for the multiple linear regression model. However, the regression coefficient(s) associated with the indicator variable(s) will have different interpretations (and sometimes even different values) depending on the particular coding scheme used. It is essential therefore to provide details of the coding scheme used when reporting on a multiple regression analysis which includes a qualitative predictor.

9.3   Full Rank Reference Cell Coding

Characteristics of full rank reference cell coding, usually referred to as reference cell coding, typically are:

- The intercept, β_0, is included in the multiple regression model.

- The number of indicator variables created to represent a qualitative variable in the regression model with an intercept is always one less than the number of categories for that qualitative variable. In other words, **k-1** indicator variables are created to represent a qualitative variable that has **k** categories. For example, we would only create **one indicator variable** to represent the **two categories of the qualitative variable**, sex of sea star, male and female. This rule avoids the problem of perfect collinearity which would result if k indicator variables are used to represent the k categories of a qualitative predictor in a multiple linear regression model with an intercept. Recall when there is perfect collinearity the least squares process does not yield unique estimates for the regression coefficients of variables that are perfectly collinear.

- An individual in the sample is assigned a value of 0 or 1 for the indicator variable to indicate to which category of the qualitative variable the individual belongs, with 0 designating the reference category for the analysis. We will illustrate this with the example that follows.

9.4   Example 9.1: Mussel Weight

Researchers were interested whether the growth rate in freshwater mussels was different in two locations. They investigated this question by checking whether the weight of mussels was changing at a different rate in the two locations. Previous studies suggest that weight can be linearly related to age in mussels. Therefore, the researchers evaluated whether the slopes of the population regression lines of weight against age were different in the two locations. They evaluated this question by fitting a multiple linear regression interaction model where the outcome variable was weight of mussel in grams and the predictor variables were: age of mussel (continuous); an indicator variable indicating the mussel's location (qualitative); and a two-way interaction term between age and the indicator variable for location. The researchers then investigated whether the model's ideal inference conditions were reasonable for their data. As no serious departures from the model's ideal inference conditions were detected, the researchers then evaluated whether there was a difference in the growth rates for the two mussel populations by:

- obtaining a confidence interval estimate for the regression coefficient of the model's interaction term

- applying a partial t-test of the null hypothesis that the regression coefficient for the interaction term is equal to zero vs. the alternative that it is not equal to zero

Details of the analysis are provided in the next section.

9.5 Example Using a Nine-Step Approach for Applying a Multiple Linear Regression with a Two-Way Interaction between a Continuous and Qualitative Predictor

The basic steps outlined in Section 8.6 are also applicable for implementing a multiple linear regression with a two-way interaction between a qualitative and a continuous predictor. However, when modelling a qualitative variable, the additional step of creating indicator variable(s) to capture the information on the qualitative variable is required.

This section uses the mussel example described in Section 9.4 to illustrate our approach for applying a linear regression with a two-way interaction between a continuous and a qualitative predictor.

Steps A and B: Screen for data input and recording errors. No data input errors nor data recording errors were found when output from Program 9.1 (Section 9.B) was examined.

Step C: Create one or more indicator variables, as required, to capture information on the categories of the qualitative variable for the regression analysis, if this step has not already been carried out at the data recording phase of the study. If you decide to use reference cell coding, create k-1 indicator variables to represent the k categories of the qualitative variable.

In this example, we used reference cell coding and created **a single** indicator variable, which we arbitrarily called ILOC in order to capture information on the **two** categories of the qualitative predictor, location, for the regression analysis. We created this indicator variable in the DATA STEP of Program 9.2 (Section 9.C) where we assigned the values for ILOC as follows:

ILOC = 0 if a mussel was from location A.
ILOC = 1 if a mussel was from location B.

In the absence of a substantive reason to do otherwise, we designated location A as the reference category for the analysis as the sample size for location A ($n = 21$) was considerably larger than that for location B ($n = 13$).

Step D: If zero is not a meaningful value for a continuous predictor involved in a two-way interaction or if zero, although meaningful for that predictor, does not occur in the sample, centre the original data values for that predictor (Section 8.5) and use the centred data values for that predictor in the subsequent steps.

In this sample, no zero values were observed for age rounded to the nearest year. We therefore centred this predictor variable to facilitate interpretation of the analysis. Results from the UNIVARIATE procedure (Program 9.1) suggest the values of 13 or 14 would seem a reasonable choice for centring age, as the age classes of 13 and 14 were found in both locations; the median age for the samples from locations A and B was respectively 13 and 14; and the mean age for location A and B samples was respectively 13.24 and 15.08. However, upon consultation, the investigators advised that 13 would be a more meaningful choice for centring as they were especially interested in the age cohort that was 13 years old at the time of sampling.

In Program 9.2 the value of 13 was used to centre age in the multiple regression analysis and the results reported in Section 9.8 and discussed in Section 9.9 are based on using 13 for centring age. It is important to note that the estimated regression coefficient for the interaction term and its interpretation are identical regardless of whether 13 or 14 is used for centring age. Thus the primary research question investigated in this multiple regression analysis via the interaction term (as discussed in Section 9.4) is not affected by using 13 or 14 to centre AGE. However, as will be discussed in later sections, interpretations of the estimated model parameter for the intercept (Section 9.9.6) and for the estimated regression coefficient for ILOC, the explanatory variable indicating location (Section 9.9.5) are affected by the particular value used for centring age.

Step E: Compute a value for the two-way interaction term for each case in the sample by multiplying a case's value for the indicator variable by that case's data value for the continuous predictor. We implemented Step E using Program 9.2 (Section 9.C).

Step F: Apply a two-way interaction regression model to the sample data to obtain regression diagnostics and graphics. We applied the following model to the sample data

$$\mu_{Y|\text{ILOC, AGEC, AGECILOC}} = \beta_0 + \beta_1 \text{ILOC} + \beta_2 \text{AGEC} + \beta_3 \text{AGECILOC} \qquad (9.1)$$

where

$\mu_{Y|\text{ILOC, AGEC, AGECILOC}}$ is the mean mussel weight given the values of the model predictors, ILOC, AGEC, AGECILOC, in the populations sampled
ILOC denotes the indicator variable which represents for the qualitative variable, location, in the analysis
AGEC denotes the centred predictor for AGE
AGECILOC denotes the interaction term obtained by the product of AGEC and ILOC
β_0 is the Y intercept
β_1 is the regression coefficient for the indicator variable, ILOC
β_2 is the regression coefficient for the predictor, AGEC
β_3 is the regression coefficient for the interaction term, AGECILOC

Step G: Evaluate ideal inference conditions for the model. As summarized in Section 9.A in this chapter's appendix, the ideal inference conditions for this interaction model seem reasonable for the sampled population.

Step H: Check whether any cases merit investigation due to their disproportionate influence on the regression. It was concluded that no case merited investigation due to its disproportionate influence on the regression (Sections 9.6, 9.7).

Step I: Interpret results from the two-way interaction regression analysis. Results generated by Program 9.2 for the two-way interaction regression analysis given in Section 9.8 are interpreted and discussed in Section 9.9.

9.6   Evaluating Influence in Multiple Regression Models which Involve Qualitative Variables

Cook's D (Cook, 1977), DFFITS (Belsley et al., 1980) and DFBETAS (Belsley et al., 1980) were developed as measures of influence for continuous predictors and the commonly reported guidelines for cutoff values for these measures have been based on multiple regression models which involve only continuous predictors. This is not surprising as Cook's D, DF-FITS, and DFBETAS are functions of leverage (Equations 3.12, 3.9, 3.10 of this book and page 13 of Belsley et al. (1980) who show DFBETAS as a function of leverage based on matrix algebra). We have learned that a leverage (hat) value for the ith case, in a multivariable setting, is a measure of geographic distance of the point for the ith case, $(X_{i1}, X_{i2}, \ldots, X_{ip})$, from the centre point of the predictor space, $(\bar{X}_{i1}, \bar{X}_{i2}, \ldots, \bar{X}_{ip})$, as defined by the means of the predictor values for all n cases in the analysis. However, the X values of an indicator (dummy) variable representing a qualitative predictor have no ranking or no meaning in terms of geometric distance, as these X values merely indicate the category of the qualitative variable to which a case belongs. Therefore, we cannot expect that Cook's D, DFBETAS, and DFFITS to be meaningful measures of a case's influence on a multiple regression model which involves qualitative variables. Cook's D, DFFITS, and DFBETAS can be used, however, to measure the influence of a case on the simple linear regression within the category of the qualitative variable to which the case belongs as illustrated in Program 9.3 (Section 9.D).

A simple way to evaluate influence in a multiple linear regression model which involves qualitative variables is by examining any case with a large studentized deleted residual which indicates that the case is a regression outlier and comparing the substantive conclusions from the multiple regression analyses with and without this case. If there is a change in the substantive conclusions, then you can conclude that this case is influential. It is always good practice to investigate cases that have been found to be influential to see if you can find out why these cases have outlying predicted response values. Often valuable information about a research problem is revealed when influential regression outliers are fully investigated. Of course, if there is a masking problem, which can occur when there are two or more outlying data points that lie close together (subsection 3.8.4), a regression outlier and hence a potentially influential case may not be detected by the size of its studentized deleted residual value.

Evaluating influence in linear models has been an ongoing area of research. For those who wish to pursue this topic, a literature review of deletion diagnostics is provided in Ganguli et al. (2016).

9.7 Summary of Influence Evaluation in the Multiple Regression Interaction Analysis for the Mussel Example

It was concluded that no case merited investigation based on influence diagnostics obtained when a simple linear regression model of WEIGHT vs. AGE was fitted within each location (Program 9.3, Section 9.D). Results from Program 9.3 reveal that only one case, Record 22, has an absolute value of a studentized deleted residual greater than two: However its studentized deleted absolute value, 2.0292, just barely exceeds two and this case has a low leverage value equal to 0.1830. Therefore Record 22 would not be expected to be an influential case. When the interaction multiple regression model was fitted to the entire data set (cases from both locations) none of the cases had an absolute studentized deleted residual larger than two. If one wanted to be extra cautious, the multiple regression interaction analysis could be rerun with Record 22 excluded from the original data set. When Record 22 is excluded as compared to being included from the analysis, the same primary substantive conclusion is reached that it is unlikely that the growth rate of mussels is the same in the sampled populations from locations A and B.

9.8   Results from Program 9.2: Two-Way Interaction Regression Analysis between a Qualitative Predictor (Reference Cell Coding) and a Centred Continuous Predictor

Mussel Data

The REG Procedure
Model: MODEL1
Dependent Variable: WEIGHT

Number of Observations Read	34
Number of Observations Used	34

Analysis of Variance

Source	DF	Sum of Squares	Mean Square	F Value	Pr > F
Model	3	239.71227	79.90409	70.83	<.0001
Error	30	33.84128	1.12804		
Corrected Total	33	273.55355			

Root MSE	1.06209	R-Square	0.8763
Dependent Mean	6.10118	Adj R-Sq	0.8639
Coeff Var	17.40801		

Parameter Estimates

Variable	DF	Parameter Estimate	Standard Error	t Value	Pr > |t|	Variance Inflation
Intercept	1	4.41367	0.23280	18.96	<.0001	0
ILOC	1	3.08139	0.40124	7.68	<.0001	1.14598
AGEC	1	0.25860	0.09182	2.82	0.0085	3.01834
AGECILOC	1	0.33488	0.11434	2.93	0.0064	3.21686

Parameter Estimates

Variable	DF	95% Confidence Limits	
Intercept	1	3.93823	4.88910
ILOC	1	2.26194	3.90085
AGEC	1	0.07109	0.44612
AGECILOC	1	0.10137	0.56840

9.9 Interpreting Results of the Model F-Test and Parameter Estimates Reported in Section 9.8

9.9.1 Introduction

As ideal inference conditions for the model given by Equation 9.1 seem reasonable (Section 9.A), we assume that the conclusions based on interpreting the Model F-test and parameter (regression coefficient) estimates reported in Section 9.8 are not misleading. For statistical enthusiasts, explanations underlying the interpretation of the regression coefficients are also provided in this section.

9.9.2 Interpretation of Model F-Test

The result of the Model F-test (F-value = 70.83, $\Pr > F < 0.0001$) reveals that the hypothesized model for the mussel example estimated by Equation 9.2 can provide information about the outcome variable, weight of mussel.

$$\hat{Y} = 4.41 + 3.08\text{ILOC} + 0.26\text{AGEC} + 0.33\text{AGECILOC} \tag{9.2}$$

9.9.3 Interpretation of the Estimated Regression Coefficient for AGECILOC

The regression coefficient (parameter) estimate for the interaction term, AGECILOC, 0.33, represents the difference between the estimated slopes of the regression lines, WEIGHT vs. AGE, for location A and location B, where this difference is obtained by subtracting the estimated slope for location A (the reference category) from the estimated slope for location B.

The sign of the regression coefficient for this interaction is positive, which indicates that the slope of the line, WEIGHT vs. AGE, is greater for location B than for location A. We therefore conclude that the mussels from location B are gaining on average an estimated 0.33 grams more weight each year as they age than the mussels from location A. The estimated interaction effect (i.e., difference between estimated slopes for location A and B) is shown in Figure 9.1.

Figure 9.1 shows that centring AGE has no impact on the regression coefficient of the interaction effect as the difference between estimated slopes of the regression WEIGHT vs. AGE where age values are not centred in the left hand panel is identical to that in the right hand panel where age values have been centred (AGEC) for the interaction regression analysis. However, although both conditional effect plots show an identical interaction effect, the plot which has AGE, with the original age values, as the horizontal variable is more readily interpretable and hence would be the more desirable plot to include in a report of the study.

As reported in Section 9.8, the p-value of the partial t-test of the population regression coefficient for AGECILOC being equal to zero is small ($t = 2.93$, $\mathrm{P} > |t| = 0.0064$), which suggests that there is a difference between the slopes of the population regression lines, WEIGHT vs. AGE, for the two locations. In other words, the linear change in weight in grams per unit (i.e., yearly) increase in age is different in the sampled populations of the two locations. This conclusion is corroborated by the 95% confidence interval estimate for the regression coefficient of the interaction term, AGECILOC. This estimate (0.10137, 0.56840), excludes zero and suggests that the difference between the two locations' slopes

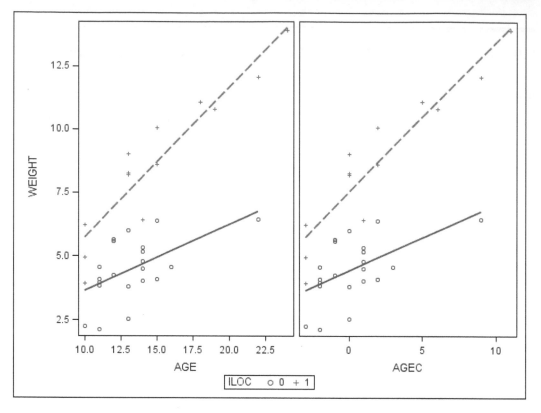

FIGURE 9.1
Comparison of estimated simple linear regressions of mussel weight with uncentered age in locations A and B (left panel) with estimated simple linear regressions of mussel weight with centred age of mussels (AGEC) in locations A and B (right panel).

of the population regression lines, WEIGHT vs. AGE, could lie between 0.10 and 0.57.

Explanation
To understand why the regression coefficient for the interaction term, AGEILOC, is interpreted as the difference between the slopes of the estimated simple regression lines of WEIGHT vs. AGE in locations A and B, we first represent location B by substituting ILOC = 1 into our estimated multiple regression interaction model equation as follows:

$$
\begin{aligned}
\hat{Y}_{\text{ILOC}=1} &= \hat{\beta}_0 + \hat{\beta}_1(1) + \hat{\beta}_2\text{AGEC} + \hat{\beta}_3((\text{AGEC})(1)) \\
&= \hat{\beta}_0 + \hat{\beta}_1 + \hat{\beta}_2\text{AGEC} + \hat{\beta}_3\text{AGEC} \\
&= 4.41 + 3.08 + (\hat{\beta}_2 + \hat{\beta}_3)\text{AGEC} \\
&= 7.49 + (\hat{\beta}_2 + \hat{\beta}_3)\text{AGEC}
\end{aligned}
\tag{9.3}
$$

Therefore, as shown in Equation 9.3, when we substitute ILOC = 1 into the multiple regression interaction model equation, it simplifies to the equation for the simple linear regression for location B in which WEIGHT is regressed against AGEC and the estimated regression coefficient of AGEC for location B in this simple linear regression (i.e., the slope of the regression line of WEIGHT vs. AGEC for location B) is $\hat{\beta}_2 + \hat{\beta}_3$ from the multiple regression interaction model.

To represent location A we substitute ILOC = 0 into our estimated interaction model equation and obtain the following estimate of the simple linear regression line for WEIGHT vs. AGEC for location A:

$$\hat{Y}_{\text{ILOC}=0} = \hat{\beta}_0 + \hat{\beta}_2\text{AGEC}$$

From this previous equation, we see that the estimated regression coefficient of AGEC (slope) for location A in the simple linear regression of WEIGHT vs. AGEC is $\hat{\beta}_2$.

Thus $\hat{\beta}_3$ is the difference obtained when the regression coefficient of AGEC in the simple linear regression of WEIGHT vs. AGEC for location A is subtracted from its counterpart in the simple linear regression of WEIGHT vs. AGEC for location B.

This difference can be verified numerically by examining the results for the simple linear regression of WEIGHT vs. AGEC for each location generated by REG procedure of Program 9.3 (Section 9.D).

The simple linear regression line estimated for location A (ILOC = 0) by Program 9.3 is

$$\hat{Y} = 4.41367 + 0.25860\text{AGEC} \tag{9.4}$$

The simple linear regression line estimated for location B (ILOC = 1) by Program 9.3 is

$$\hat{Y} = 7.49506 + 0.59349\text{AGEC} \tag{9.5}$$

Subtracting the slope of the regression line for location A from its counterpart for location B we obtain the value of 0.3349 ($0.59349 - 0.25860 = 0.3349$), which is identical to the value rounded to the fourth decimal place reported in Section 9.8 for $\hat{\beta}_3$, the parameter estimate for the interaction term AGECILOC.

9.9.4 Interpretation of the Estimated Regression Coefficient for AGEC

The estimated regression coefficient for the indicator variable AGEC, 0.26 can be interpreted as the estimated average change of mussel weight in grams per unit (i.e., yearly) increase in AGEC for mussels from location A. Thus, it is estimated that for each year of age the mussel population from location A gained on average 0.26 grams.

Explanation
We can see how we make this interpretation for AGEC, by substituting ILOC = 0 for location A into Equation 9.1, i.e.,

$$\begin{aligned}
\hat{Y}_{\text{ILOC}=0} &= \hat{\beta}_0 + \hat{\beta}_1\text{ILOC} + \hat{\beta}_2\text{AGEC} + \hat{\beta}_3\text{AGECILOC} \\
&= \hat{\beta}_0 + \hat{\beta}_1(0) + \hat{\beta}_2\text{AGEC} + \hat{\beta}_3(\text{a value of AGEC})(0) \\
&= \hat{\beta}_0 + \hat{\beta}_2\text{AGEC}
\end{aligned} \tag{9.6}$$

9.9.5 Interpretation of the Estimated Regression Coefficient for ILOC

The regression coefficient for the indicator variable ILOC, 3.08, can be interpreted as the difference obtained when the estimated mean weight of 13-year-old mussels in location A is subtracted from the estimated mean weight of 13-year-old mussels in location B.

Explanation
To show why we interpret the regression coefficient for ILOC in this way, we first represent the mean mussel weight of 13-year-olds from location B by substituting ILOC = 1 and AGEC = 0 in our interaction model. AGEC = 0 represents 13-year-old mussels because 13 was the value used for centring age in Program 9.2. ILOC = 1 represent mussels from

location B, because of the way we set up our reference cell coding scheme, i.e., assigning a value of 1 for the indicator variable, ILOC, if a mussel was from location B. Therefore we represent the mean mussel weight of 13-year-olds (AGEC = 0) from location B (ILOC = 1) in our two-way interaction model given by Equation 9.1 as follows

$$\hat{Y}_{\text{ILOC}=1, \text{ AGEC}=0} = \hat{\beta}_0 + \hat{\beta}_1(1) + \hat{\beta}_2(0) + \hat{\beta}_3(1)(0)$$

$$= \hat{\beta}_0 + \hat{\beta}_1 \qquad (9.7)$$

$$= 4.41 + 3.08$$

We then represent the mean mussel weight of 13-year-olds from location A in our model by substituting ILOC = 0 and AGEC = 0 in Equation 9.1. We use ILOC = 0 to represent mussels from location A in accordance with our reference cell coding scheme where we assigned a value of 0 to ILOC if a mussel was from location A. Therefore the mean weight of 13-year-old mussels from location A in our model is given by

$$\hat{Y}_{\text{ILOC}=0, \text{ AGEC}=0} = \hat{\beta}_0 + \hat{\beta}_1(0) + \hat{\beta}_2(0) + \hat{\beta}_3(0)(0)$$

$$= \hat{\beta}_0 \qquad (9.8)$$

$$= 4.41$$

Subtracting Equation 9.8 from Equation 9.7 we obtain $\hat{\beta}_1$, the regression coefficient for ILOC, equal to 3.08. Thus, the regression coefficient for ILOC can be interpreted as the difference obtained when the estimated mean weight of 13-year-old mussels from location A is subtracted from the estimated mean weight of 13-year-old mussels from location B.

9.9.6   Interpretation of the Sample Model Intercept

The sample intercept, 4.41 grams, for the interaction model (Equation 9.2) can be interpreted as the estimated mean weight (mean Y) in location A (the reference group) for mussels that were 13 years old at the time the sample was collected. The sample intercept in this mussel model can also be interpreted as the estimated weight of a 13-year-old mussel from the sampled population in location A.

Explanation
The sample intercept in the interaction model can be interpreted in this way for the following reasons:

- A predicted Y value ($\hat{Y}$) can be used to estimate a mean of Y given the values of the predictors.

- The definition of the sample intercept, $\hat{\beta}_0$, is the value of $\hat{Y}$ when the predictor variables in the model are equal to zero. Therefore, substituting zero for the predictor variables, ILOC, AGEC, and AGECILOC in the estimated mussel interaction model given by Equation 9.2, by definition gives the sample intercept

$$\hat{Y}_{\text{ILOC}=0, \text{ AGEC}=0, \text{ AGECILOC}=0} = \hat{\beta}_0 + \hat{\beta}_1(0) + \hat{\beta}_2(0) + \hat{\beta}_3(0)(0)$$

$$= \hat{\beta}_0 \qquad (9.9)$$

- When the model predictor ILOC is equal to zero, we are referring to cases in the reference category, i.e., to mussels in location A in this example because of the reference cell coding used to capture information on the qualitative variable, location.

- When the model predictor AGEC is equal to zero, we are referring to mussels that are 13 years old at the time of sampling.

Recall the predicted value in a regression model can also estimate the Y value of an individual given the values of the predictors when the individual is drawn from the sampled population. Therefore, the sample intercept in the mussel model of Equation 9.2 also can be interpreted as the estimated weight of a 13-year-old mussel from the sampled population in location A.

9.10   Chapter Summary

This chapter:

- explains how a qualitative variable can be represented as a predictor in a multiple linear regression model via indicator (dummy) variables

- describes the characteristics of a commonly used coding scheme for indicator variables known as **reference cell coding**

- emphasizes that researchers should be aware that there are other prescribed coding schemes for indicator variables that are also commonly used and the interpretation of regression coefficients for indicator variables depends on the particular coding scheme used. However, these prescribed different coding schemes for indicator variables yield identical predicted Y values, identical residuals, an identical R^2, an identical Mean Square Error, as well as identical results for the Model F-test for the multiple linear regression model

- provides an example of a multiple linear regression model with a two-way interaction between a qualitative and a continuous predictor, where the qualitative variable is represented by an indicator variable using reference cell coding and the continuous predictor has been centred

- suggests a step-by-step approach for evaluating a two-way interaction effect in the multiple linear regression model for this example

- interprets results of the Model F-test and parameter estimates for this example to illustrate interpretation of regression coefficients in a multiple linear regression analysis with a two-way interaction between a reference cell coded qualitative predictor and a centred continuous predictor

- explains why the model's parameter estimates are interpreted as described

- discusses evaluation of influence in models which include a qualitative variable

- provides an example of a SAS program which applies the suggested step-by-step approach to the illustrative example

Appendix

9.A   Evaluation of Ideal Inference Conditions for the Mussel Example

Evaluation of ideal inference conditions for the interaction model fitted to the mussel weight data (Equation 9.1) is summarized as follows:

- The response variable, mussel weight in grams is continuous.

- Although we do not have any details regarding the sampling design for this study we assume that the mussels were collected and measured from the two locations in such a way that independence of errors was not violated.

- In Figure 9.2 the plot of the internally studentized residuals from the model against the predicted values does not suggest a severe departure from linearity or the equal variances assumption. The plot of the model's internally studentized residuals against the indicator variable for location (ILOC) does not suggest that the ϵ_{ij} are related to location. In the plot of internally studentized residuals against AGE, insufficient data at the larger age values make it impossible to infer anything definite about age and the ϵ_{ij}.

- The normal Q-Q plot of the internally studentized residuals (Figure 9.3) does not reveal a major departure from normality.

- The variance influence factor (VIF) values for AGEC, ILOC and AGECILOC reported in Section 9.8 were all much smaller than 10, indicating harmful collinearity is not present among the explanatory variables for this model. The following collinearity diagnostics (generated by Program 9.2) proposed by Belsley (1991) for evaluating harmful collinearity corroborate this conclusion.

Collinearity Diagnostics

Number	Eigenvalue	Condition Index
1	2.01752	1.00000
2	1.42550	1.18967
3	0.38136	2.30007
4	0.17562	3.38939

Collinearity Diagnostics

Number	Intercept	------------------Proportion of Variation----------------		
		ILOC	AGEC	AGECILOC

1	0.03642	0.06152	0.04946	0.05855
2	0.15937	0.10464	0.04953	0.02963
3	0.60353	0.75737	0.01251	0.06766
4	0.20067	0.07647	0.88849	0.84415

Recall from Section 6.2.2, Belsley (1991) suggested there is little cause for concern about collinearity when the largest scaled condition index is between 5 and 10.

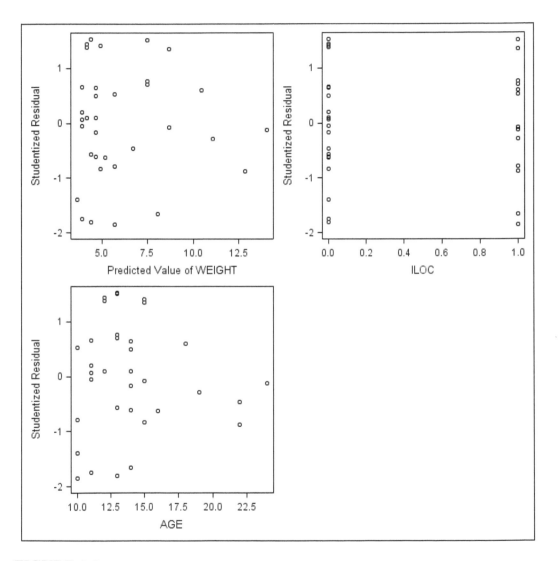

FIGURE 9.2
Internally studentized residuals vs. predicted values and internally studentized residuals vs. predictor variables from multiple regression of mussel data.

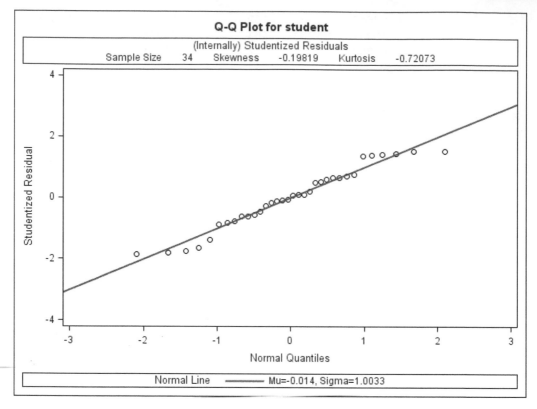

FIGURE 9.3
Normal Q-Q plot of internally studentized residuals from the mussel example.

9.B   Program 9.1: Getting to Know the Mussel Data

Program 9.1 reads in the data for Example 9.1 and stores it in the **temporary** SAS data set arbitrarily named MUSSEL. This program then generates output to facilitate screening for data input and recording errors (Steps A and B of Section 9.5).

```
*Program 9.1;
data mussel;
input RECORD LOC $ AGE WEIGHT @@;
 datalines;
1 A 10 2.26 2 A 11 4.10 3 A 11 4.56
4 A 11 2.12 5 A 11 3.96 6 A 11 3.84
7 A 12 5.58 8 A 12 5.64 9 A 12 4.26
10 A 13 6.00 11 A 13 2.54 12 A 13 3.82
13 A 14 4.50 14 A 14 5.18 15 A 14 4.78
16 A 14 5.34 17 A 14 4.04  18 A 15 6.38
19 A 15 4.08 20 A 16 4.56 21 A 22 6.44
22 B 10 3.94 23 B 10 6.22 24 B 10 4.96
25 B 13 9.02 26 B 13 8.20 27 B 13 8.26
```

```
28 B 14 6.40 29 B 15 10.06 30 B 15 8.60
31 B 18 11.06 32 B 19 10.78 33 B 22 12.04
34 B 24 13.92
;
proc sort data=mussel presorted out=sortmussel;
   by loc;
run;
proc print;
run;
proc univariate data=sortmussel mode;
   ods select ExtremeValues Frequencies;
   ods exclude TestsForLocation;
   id record;
   var age; by loc;
run;
/*The following plots can be informally examined for contextual,
predictor and response outliers that merit investigation for recording errors*/
proc sgplot data=sortmussel;
   scatter x=age y=weight;
   by loc;
run;
```

9.B.1   Explanation of Statements in Program 9.1

Program 9.1 is largely similar to Program 8.1, which was used to screen for data input and recording errors. An addition to Program 9.1 not found in Program 8.1 is the BY statement in the UNIVARIATE, SGPLOT, and FREQ procedures. A BY statement is specified so that the results from each of these procedures will be computed separately for each location (loc). Specifying the BY statement in a procedure requires that the data must be first sorted in the same manner using the SORT procedure as shown below:

```
proc sort data=mussel presorted out=sortmussel;
   by loc;
run;
proc print;
run;
proc univariate data=sortmussel mode;
   ods select ExtremeValues Frequencies;
   ods exclude TestsForLocation;
   id record;
   var age; by loc;
run;
```

New to this example is the PRESORTED option. You would use PRESORTED option if you suspect that the input data set has already been sorted according to the key variables in your BY statement of the SORT procedure. The PRESORTED option checks first whether the input data have been sorted in this way. We also have not previously encountered sorting a character variable as we do in this example where the input variable LOC is indicated as either A or B in the input data set as shown in Program 9.2. When SAS is run in UNIX and its derivatives, Windows, and OpenVMS operating environments, the SORT procedure uses by default the ASCII English-language sequence, i.e., consistent with following order starting with the smallest character first and the largest last:

ASCII Sort Order Example
blank ! " # $ % & ' ( ) * + , - . / 0 1 2 3 4 5 6 7 8 9 : ; < = > ? @
A B C D E F G H I J K L M N O P Q R S T U V W X Y Z [\ ] _
a b c d e f g h i j k l m n o p q r s t u v w x y z { } ~

9.C   Program 9.2: Two-Way Interaction Regression Analysis between a Qualitative Predictor (Reference Cell Coding) and a Centred Continuous Predictor

Program 9.2 implements Steps C-F and generates output to facilitate Steps H and I of the suggested approach described in Section 9.5. In its initial DATA step Program 9.2 creates a temporary SAS data set arbitrarily named SORTMUSSEL2 by using the SET statement to read the data stored in the **temporary** SAS data set SORTMUSSEL created by Program 9.1.

Please note:
If you try to execute Program 9.2 after having logged out from the SAS session in which you executed Program 9.1, you will get error messages. To avoid this, you could create the SAS data set SORTMUSSEL again by executing Program 9.1 in your current SAS session or alternatively by inserting the DATA STEP and the SORT procedure statements from Program 9.1 at the beginning of Program 9.2.

```
*Program 9.2;
data sortmussel2;
set sortmussel;
   *STEP C;
   /* Create an indicator variable ILOC using full rank
   reference cell coding*/
   if LOC = "A" then ILOC=0;
   else if LOC = "B" then ILOC=1;

   *STEP D;
   /*Center  AGE by subtracting 13 from each original age value
   and denote centered age value as AGEC*/
   AGEC=age-13;
   *STEP E;
   /* Compute the two-way interaction term*/
   AGECILOC=agec*iloc;

proc freq data=sortmussel2;
     table  iloc;
run;
   *STEP F;
   /* Apply the two-way interaction regression model to the data
   to obtain regression diagnostics and graphics.*/
proc reg data=sortmussel2 plots=none;
   id record;
```

```
     model weight=iloc agec ageciloc/vif r collin p clb ;
       output out=diag r=r rstudent=rstudent student=student p=p ;
run;
     *STEP G;
     *Evaluate Ideal Inference Conditions;
   /*Obtain plots of internally studentized residuals vs. predicted Y and
     vs. iloc agec age */

proc sgscatter data=diag;
   plot (student)*(p  iloc  agec age);
run;
 proc sgplot data=diag;
   reg x=AGE y=STUDENT /group=LOC;
run;
    * Evaluate normality;
    * Obtain a Q-Q normal plot for internally studentised residuals;

proc univariate data =diag;
    var student ;
    qqplot student  /normal (mu=est sigma=est);
    inset n='Sample Size' skewness kurtosis
    /pos=tm header='(Internally) Studentized Residuals';
run;
    /* Get a listing of variables of interest for cases identified
    as regression outliers based on a specified cutoff value for
    the studentized deleted residual (rstudent) */
proc print data=diag;
    where abs(rstudent) gt 2;
run;
/*Get a plot of rstudent vs. record ( case id)*/
proc sgscatter data=diag;
    plot (rstudent)* record;
run;
/* Get a sorted listing of rstudent  for the cases */
proc sort data =diag;
    by rstudent;
run;
proc print data=diag;
    var record rstudent;
run;
    /* Construct conditional effect plots to show interaction effect*/
    /*Create a panel of 2 conditional effect plots
    where the left side of the panel is the conditional effect plot
    of WEIGHT vs. AGE (uncentered)for location A and B and the right
    side of the panel is the conditional effect plot
    of WEIGHT vs. AGEC (centered) for location A and B.*/
proc sgscatter data=sortmussel2;
    compare y=weight x=(age agec)/
    reg spacing=4 group=iloc;
run;
```

9.C.1   Explanation of Selected SAS Statements in Program 9.2

This subsection discusses SAS statements in Program 9.2 that have not been encountered in previous programs.

The following IF THEN and ELSE IF statements create the indicator variable ILOC to represent the categories "A" and "B" of the qualitative variable, location.

```
*STEP C;
  /* Create an indicator variable ILOC using reference cell coding*/
 if LOC = "A" then ILOC=0;
else if LOC = "B" then ILOC=1;
```

The statement **if LOC = "A" then ILOC = 0;** checks if the current case has A input for the variable LOC, and if so a value of 0 is assigned to the indicator variable ILOC for that case and stored in the temporary SAS data set called SORTMUSSEL2.

The statement **else if LOC = "B" then ILOC = 1;** checks if the current case has B input for LOC, and if so a value of 1 is assigned to ILOC for that case and stored in the temporary SAS data set called SORTMUSSEL2.

If the current case has neither A nor B input for the variable LOC, then ILOC is labelled as a missing value for that case.

Although the reference cell coding of the qualitative variable location in the DATA STEP is straightforward, the coding may be checked by using the FREQ procedure to give the frequencies of the values for the indicator variable ILOC to confirm that these frequencies correspond to the actual sample sizes for location A and B.

```
/* Check coding of indicator varible iloc*/
 proc freq data=sortmussel2;
title Mussel Data;
table  iloc;
run;
```

The FREQ procedure can be especially helpful for checking more complicated coding than that implemented in this mussel example.

We have encountered the SGSCATTER procedure in a previous program but new to Program 9.2 is the COMPARE statement for this procedure with the REG, SPACING, and GROUP= options in the code that follows.

```
/* CONSTRUCT CONDITIONAL EFFECT PLOTS TO SHOW INTERACTION EFFECT*/
/*Create a panel of 2 conditional effect plots
where the left side of the panel is the conditional effect plot
of WEIGHT vs. AGE (uncentered) for location A and B and the right
side of the panel is the conditional effect plot
of WEIGHT vs. AGEC (centered) for location A and B.*/
proc sgscatter data=sortmussel2;
title;
compare y=weight
x=(age agec)/reg spacing=4 group=iloc;
run;
```

The COMPARE statement in Program 9.2 requests that a panel of two graphs be created where the outcome variable, WEIGHT, on the y axis is shared by both graphs and where the left graph in the panel has AGE on the x axis and the right graph in the panel has AGEC (centred age) on the x-axis. The REG and GROUP = options of the COMPARE statement specify that a simple linear regression line should be fitted for each level of ILOC in each panel. The SPACING = option requests spacing between the graphs in the panel.

9.D   Program 9.3

Program 9.3 inputs data from the **temporary** SAS data set SORTMUSSEL2 (created by Program 9.2) and applies a simple linear regression (WEIGHT vs. AGEC) to the observations within each location as a means of evaluating influence (Section 9.6).

Please note:
If you try to execute Program 9.3 after having logged out from the SAS session in which you executed Programs 9.1 and 9.2, you will get error messages. One way to avoid this is by executing Programs 9.1 and 9.2 in your current SAS session prior to running Program 9.3.

```
*Program 9.3;
proc reg data=sortmussel2 plots(only label)=(rstudentbyleverage cooksd);
    model weight=agec/influence r p;
    by iloc;
    output out=diag r=r rstudent=rstudent
    student=student p=p cookd=cookd dffits=dffits;
run;
```

10

Subset Selection of Predictor Variables in Multiple Linear Regression

10.1 Introduction

This chapter describes model selection procedures that can help researchers decide whether it is possible to exclude some of their predictor variables from a multiple linear regression model and still achieve their research objective.

If the primary research objective is to understand the underlying relationship between an outcome and its predictor variables, then relying solely on an automated statistical procedure to select regression model predictors is contentious (Zhang, 1992; Harrell, 2001; van Belle et al., 2004; Leeb and Pötscher, 2005; Wiegand, 2010; Vittinghoff et al., 2012; Fox, 2016). This is because the subset of predictors selected by an automated statistical selection procedure as the "best" often depends on the statistical procedure (Section 10.3) and the specific criterion (Section 10.4) used for selection.

If, however, the primary research objective is to reliably predict an outcome without necessarily understanding why one set of predictors appears to predict more reliably than another set, then a statistical selection procedure may work well. For example, Foster and Stine (2004) applied an automated statistical selection procedure to build a predictive model for personal bankruptcy from 67,160 candidate predictors using a database of 244,094 credit-card users. They showed that predictions from the model obtained from their statistical selection procedure compared favourably with those from models obtained from machine learning algorithms. Their study illustrates how statistical predictor selection can be an important component of predictive analytics, a research area which is experiencing rapid growth in this era of big data (Hastie et al., 2009; Kuhn and Johnson, 2013; Abbott, 2014).

Reasons why researchers would want to use an automated statistical procedure to select a subset of predictors from a specified set (referred to as the full set or the full model) include:

- They would like to use as simple a model as possible for reliable prediction (principle of parsimony).

- Their sample size is not large enough relative to the number of predictors in their full set to make accurate predictions possible.

- They want to know whether in the future they can omit collecting some predictor variables that are expensive or inconvenient to measure without sacrificing prediction accuracy.

In this chapter our focus is on a special case of model selection, as we consider predictor subset selection for a single model class, a multiple linear regression model. However, many of the procedures and issues we discuss can be related to a wider framework of model building (e.g., building nonlinear models).

DOI: 10.1201/9780429107368-10

10.2   Under-Fitting, Over-Fitting, and the Bias-Variance Trade-Off

Under-fitting occurs when relatively important predictors of the outcome variable are not included in a model. The consequences of an underfitted model (too few important predictors) is that the estimates of the regression coefficients are biased and the model may give inaccurate predicted values.

Over-fitting occurs when a model fits a researcher's sample so well that the model is not a good fit for other random samples drawn from the same population with slightly different data. Thus, an over-fitted model (too many unimportant predictors) may give inaccurate predicted values.

An issue to be considered when building linear regression models is the bias-variance trade-off. As the number of predictors is increased in a linear model, biased estimation of regression coefficients and predicted values is often reduced but potentially at the expense of increasing their variance. Thus, the art of model building in linear regression involves including a sufficient number of relatively important predictors in a model to reduce bias while avoiding having so many predictors that accuracy is a problem.

10.3   Overview of Traditional Model Selection Algorithms

Model selection algorithms that have been used traditionally and are still currently used in certain research disciplines include:

- forward selection (subsection 10.3.1)

- backward elimination (subsection 10.3.2)

- stepwise selection (Efromyson's algorithm) (subsection 10.3.3)

- all subsets selection algorithm (subsection 10.3.4)

Although these selection procedures typically identify only a single model as a final result for the "best" model, there may be a number of other models that might provide a similarly reasonable fit to the data. These other models may be substantively more plausible and/or may include predictors that are less costly or more convenient to measure. As we will illustrate in Section 10.7, the REG procedure provides an option that allows the data analyst to specify the number of the top "best" models to be included in the output for a given selection procedure and selection criterion. This facilitates inclusion of a researcher's expert opinion in the selection process and circumvents the disadvantage of focusing on only one model. However, other disadvantages associated with predictor selection are not so easily circumvented and are discussed in Section 10.5.

10.3.1   Forward Selection

A forward selection procedure begins with just the intercept and then finds among all the available candidate predictors, the single predictor that appears to best satisfy a researcher's selection criterion (specified before the analysis begins) and adds this predictor to the model. The procedure then searches among the remaining available predictors and selects, on the basis of the pre-specified selection criterion, a second predictor to include along with the intercept and the first predictor selected. After the first two predictors are determined, the forward selection procedure may select a third predictor to include along with the first two based on the selection criterion. This process can be continued until no additional predictors are assessed as useful in modelling the data based on the pre-specified criterion.

10.3.2 Backward Elimination

The backward elimination procedure begins by fitting a full model containing all of the candidate predictors specified by the researcher. The first step consists of deleting from the full model the predictor that is assessed as being the least useful, i.e., the predictor that is making the smallest contribution to the model based on the selection criterion pre-specified by the researcher. After removing the least useful predictor, the backward elimination procedure fits a new model containing all of the remaining predictors. The procedure then eliminates from this new model the predictor that is assessed as least useful based on the selection criterion and once again fits a new model with the remaining predictors. The procedure continues to remove predictors from the model one at a time at each step until all predictors remaining in the model are considered to be making a useful contribution to the model based on the researcher's pre-specified selection criterion.

10.3.3 Stepwise Selection

Stepwise selection, also known as Efromyson's algorithm (Efroymson, 1960), uses a combination of forward and backward selection procedures. It operates in a forward manner selecting one predictor at a time as in the forward selection method, but after a predictor is selected, it then checks to see if all of the predictors previously selected are still making a useful contribution to the model and if any predictors are not, then the procedure would remove the predictor that is least useful based on the prespecified selection criterion.

10.3.4 All Subsets Selection Algorithm

The all subsets selection algorithm involves selection from all possible subsets of given sizes that can be formed from the predictors in the full model. In other words, the algorithm generates all possible two-variable models that can be generated from the full model; all possible three-variable models that can be generated from the full model; all possible four-variable models that can be generated from the full model and so on until the data analyst tells the algorithm to stop. We will illustrate all subsets selection with an example in Section 10.7.

10.3.5 The Impact of Collinearity on Traditional Sequential Selection Algorithms

When applying a traditional sequential selection procedure, it is important to be aware that important predictors of the outcome variable, Y, may be omitted from the final selected model as a result of collinearity. This can occur in backward selection as follows: A predictor correlated with other predictors may be removed from the current model regardless of whether it is strongly related to Y because this predictor is redundant due to its correlation with predictors already in the model. In forward selection, an important predictor may not be entered into the current model because it is not needed due to its being correlated with predictors already in the current model. Similarly, important predictors may not be considered for the final model selected by stepwise selection which uses a combination of forward and backward sequential selection. In the all subsets algorithm, regardless of collinearity among any of the predictors, **all** possible subsets of predictors are formed from the predictors in the full model and are considered for selection for the final model. However, other potential problems associated with collinearity, as discussed in Section 5.12, apply to the all subsets algorithm as well as to forward, backward, and stepwise selection methods. Thus, using an automatic traditional selection algorithm to understand a relationship between predictors and an outcome in the presence of potentially harmful collinearity is contentious.

If the research objective of the subset selection is to reliably predict an outcome, severe collinearity among predictors in a model can cause harm by reducing prediction accuracy in a new sample from the same population with just slightly different data values from those in the original sample used in the subset selection method. Therefore, it is important to investigate collinearity at the outset of the predictor subset selection process. If severe collinearity is diagnosed, you might want to consider using an alternative method for predictor subset selection (Section 10.9) rather than a traditional sequential algorithm.

10.4 Fit Criteria Used in Model Selection Methods

Fit criteria (with corresponding SAS keywords in parentheses) that can be used in model selection methods include:

- adjusted R^2 (ADJRSQ), subsection 10.4.1

- Akaike's information criterion (AIC), subsection 10.4.2

- the corrected Akaike's information criterion (AICC), subsection 10.4.3

- average square error (ASE), subsection 10.4.4

- Mallows' C_p criterion (CP), subsection 10.4.5

- mean square error (MSE), subsection 10.4.6

- the PRESS statistic (PRESS), subsection 10.4.7

- the Schwarz (1978) Bayesian information criterion (SBC), subsection 10.4.8

- the significance level of the F-statistic (SL) used to assess a predictor's contribution to the fit when it is added or dropped from the model, subsection 10.4.9

10.4.1 Adjusted R^2

When two models are compared using the adjusted R^2 selection criterion as a measure of fit, the model with the highest adjusted R^2 is selected. Adjusted R^2 denoted by the keyword ADJRSQ in SAS, is given by

$$ADJRSQ = 1 - \frac{(n-1)(1-R^2)}{n-p} \tag{10.1}$$

where
$n =$ the sample size
$R^2 =$ the coefficient of determination (Equation 7.1)
$p =$ the number of model parameters estimated

R^2 is not a recommended criterion for comparing models with differing number of predictors. The value of R^2 increases or does not decrease as the number of predictors in a model increases regardless of whether the additional predictors are necessary. Adjusted R^2 addresses this problem by reducing the value of R^2 as a function of the sample size and p, the number of parameters in the model.

10.4.2 Akaike's Information Criterion

Based on a well known loss function in information theory (Kullback and Leibler, 1951), the Akaike information criterion (Akaike, 1973), commonly denoted as AIC, is a measure of information lost when an approximating model is used to approximate a full model. In model selection, the model with the minimum AIC value for a given data set is selected. The formula for AIC for a model, as given in SAS online documentation associated with SAS 9.2 and subsequent versions, is

$$\text{AIC} = n \log \left(\frac{\text{SSE}}{n} \right) + 2p + n + 2 \tag{10.2}$$

where
    n = the sample size
    SSE = sum of squares error obtained when a model is fitted to a given data set
    p = number of model parameters estimated

From Equation 10.2 it can be seen that AIC imposes a penalty for the number of predictors in a model as the value of AIC is increased by $2p$. Thus, the greater number of predictors in a model, the greater the model's value of AIC, which means the less likely the model will be chosen as the "best" model since the model with the minimum AIC is selected. Although there are various versions of formulas for AIC in the literature, all versions involve adding a penalty for the number of predictors in the model.

10.4.3 The Corrected Akaike's Information Criterion

As the number of predictors in a model increases relative to the sample size, AIC becomes negatively biased, i.e., it underestimates the information lost when an approximate model is substituted for the full model. This can lead to the selection of an overfitted model when the minimum value of AIC is used as the selection criterion for identifying the "best" model. To circumvent this difficulty, a corrected version of AIC, denoted as AIC_c, was derived (Sugiura, 1978; Hurvich and Tsai, 1989). The formula for AIC_c, as given in SAS online documentation associated with SAS 9.2 and subsequent versions, is

$$\text{AIC}_c = n \log \left(\frac{\text{SSE}}{n} \right) + \frac{n(n+p)}{n-p-2} \tag{10.3}$$

where
n = the sample size
    SSE = sum of squares error obtained when a model is fitted to a given data set
    p = number of model parameters estimated

Burnham and Anderson (2004) recommended that AIC_c should be used instead of AIC, unless n/p is greater than about 40 for the model with the largest p. Because AIC_c converges to AIC with increasing n, they suggested in practice, AIC_c should be used instead of AIC.

10.4.4 Average Square Error

The average square (squared) error for a model is given by

$$\text{ASE} = \text{SSE}/n \tag{10.4}$$

where
  SSE = the error sum of squares obtained when a model is fit to the data set
  n is the sample size.

10.4.5 Mallows' C_p Statistic

Mallows' C_p statistic (Mallows, 1973) can be used to assess candidate models in model selection. Mallows' C_p statistic for a model is defined as

$$C_p = \frac{\text{SSE}}{\hat{\sigma}^2} + 2p - n \tag{10.5}$$

where
  SSE = the error sum of squares obtained when a model is fit to the data set
  $\hat{\sigma}^2$ is an estimate of pure error variance which is usually obtained from MSE from fitting the full model using all possible predictors specified by the researcher
  p = number of parameters estimated in the model
  n = the sample size

Candidate models that have a value of $C_p \approx p$, or less than p, may be considered to be good models, assuming the estimate of pure error variance in Equation 10.5 is unbiased.

Some model builders have used the minimum value of Mallows' C_p statistic as a rule for selecting the "best" model to describe the data. However, Mallows (1973) cautioned that "using the minimum C_p rule to select a subset of terms for least-squares fitting cannot be recommended universally". He pointed out that sometimes the model with the minimum C_p will have a worse prediction error than the other candidate models. He noted that the minimum C_p rule will give "bad results" for ambiguous cases where a large number of candidate models have C_p values similar to the minimum C_p value.

Instead of applying the minimum C_p rule, Mallows recommended examining a graphical display, known as the C_p plot, in which C_p values of all the candidate models under consideration are plotted against p (the number of predictors in the model including the intercept). He noted that a C_p plot can provide insight as to whether one or several candidate models are suggested by the data and emphasized the desirability of often considering two or more models, including models with the same number of predictors, as possible models that fit the data well.

10.4.6 Mean Square Error

Mean square error (MSE) may be used as a fit criterion in model selection by computing the MSE for all the candidate models and selecting the model with the smallest value of MSE. Using MSE as a selection criterion is equivalent to using adjusted R^2. The candidate model with the minimum value of MSE will always have the maximum value of adjusted R^2 among the candidate models for a given data set. Montgomery et al. (2012, p. 334) provided mathematical details showing why the fit criteria minimum MSE and maximum adjusted R^2 are equivalent.

10.4.7 The PRESS Statistic

The PRESS statistic is the predicted residual sum of squares. It is computed by totalling the values of the studentized deleted residuals (Definition 3.5) for all n cases in the data set. Recall the studentized deleted residual for case i is the difference between the observed and predicted Y value for case i when the predicted Y value for case i is based on the regression equation that is estimated from the data when case i is deleted from the sample. The PRESS statistic is therefore sometimes referred to as leave-one-out cross-validation. When the PRESS statistic is used as a fit criterion for comparing models, the model with the smallest PRESS statistic is considered to be the "best".

10.4.8 Schwarz's Bayesian Information Criterion

Schwarz (1978) developed a criterion for model selection based on principles of an important branch of statistics known as Bayesian statistics (see Burnham and Anderson (2002) for a technical explanation). In some literature, the Bayesian Information Criterion (Schwarz, 1978) is abbreviated as BIC. In SAS documentation, the Schwarz Bayesian criterion is identified by the keyword, SBC and is given by

$$\text{SBC} = n \log \left(\frac{\text{SSE}}{n} \right) + p \log(n) \tag{10.6}$$

where
n = the sample size
SSE = sum of squares error obtained when a model is fitted to a given data set
p = number of model parameters estimated, including the intercept

When SBC is used as the selection criterion, the model selected is the one with the minimum value of SBC as compared to other candidate models. The penalty for an increased number of predictors in SBC is $p \log(n)$, which is a stronger penalty than either AIC or AICC imposes. Burnham and Anderson (2004) noted that SBC tends to select an under-fitted model (too few predictors) in economics and in the biological and behavioural sciences where many factors may be involved in the prediction of an outcome variable. They suggested AIC or AICC may be more appropriate than SBC in these situations.

10.4.9 Significance Level (p-Value) Criterion

The significance level (p-value) of a partial F-test statistic is a criterion that can be used to assess an effect's contribution to the fit when that effect is added or dropped from the model. A partial F-test statistic is equivalent to the two-sided partial t-test statistic when an effect is represented in a multiple regression model by a single predictor as illustrated in previous chapters.

This significance level criterion, although widely used in the past and still currently used in some research disciplines, is generally not recommended for sequential predictor selection methods (Harrell, 2001; Burnham and Anderson, 2002). This is because this F-statistic computed to compare models throughout the selection process does not follow an F-distribution or a distribution even remotely similar to an F-distribution (Draper et al., 1971; Pope and Webster, 1972; Miller, 2002) and thus the p-value corresponding to this computed partial F-statistic is incorrect. Hence a wrong decision may be made as to whether a predictor should be added or deleted from the model, because this decision is based on whether an incorrect p-value is less than or equal to the significance level criterion specified for the procedure. The p-value computed at each step does not take into account the multiple testing that occurs over successive steps in the selection process. When more than one hypothesis test is applied there is a cumulative probability of Type I error, and thus the p-value over the many hypothesis tests in a model selection procedure is greater than the p-value associated with the hypothesis test computed for any one step. Therefore, when one uses a significance level (p-value) as a criterion, where many hypothesis tests are applied to select the "best" model, the probability of false positive results is inflated, often resulting in an incorrect model ultimately selected by the procedure as being the "best".

10.5   Post-Selection Inference Issues

In subset selection, inference issues result when the same data set is used for selection and estimation. Post-selection inference issues typically include:

- The usual conventional formula for a standard error of a partial regression coefficient (Equation 7.4) assumes a model has been prespecified and does not apply when a model has been obtained via a selection procedure. This conventional formula does not account for the search process involved in subset selection. The bootstrap method (Section 14.9) may be useful for estimating standard errors in such settings (Hastie et al., 2009). Berk et al. (2013) proposed an approach based on simultaneous inference of all possible subset models to account for the search process in subset selection. They reported their approach, although generally conservative for particular selection procedures, is less conservative than a "data snooping" method proposed by Scheffé (1959) that will be discussed in Section 12.10.

- The actual coverage probability of a *naive* confidence interval for a regression coefficient parameter from the final model after subset selection is less than the nominal: A naive confidence interval in this context refers to the usual confidence interval for model parameters where the model has been pre-specified and is not the result of a data-dependent search from a selection algorithm. It is important to realize that if one constructs a naive 95% confidence interval for the true regression coefficient corresponding to a predictor variable in the final model, the actual probability of that interval containing the true value of the regression coefficient will be less than 0.95. For example, Kabaila (2009) demonstrated for his real-life data that after variable selection, the minimum coverage probability of the nominal 95% confidence interval for a regression coefficient parameter in a final model selected was 0.79 using the Akaike criterion for selection and 0.70 using Bayes information criterion.

- The prediction performance of the "best" model selected tends to be optimistic (Leeb and Pötscher, 2005).

10.6   Predicting Percentage Body Fat: Naval Academy Example

We will use the following naval academy example to illustrate predictor subset capabilities of the REG and GLMSELECT procedures in Sections 10.7 and 10.8.

A researcher decided to use a subset selection procedure to determine if percentage body fat of an individual could be predicted with reasonable accuracy using fewer predictors than a full set of predictors that may be associated with this outcome variable. The data were collected from 86 volunteers at the U.S. Naval Academy (Hocking, 2013). The variables in the data set that may be used as predictors of percentage of body fat (FAT) are:

1. HT = height in centimetres
2. LOGWT = log base 10 of weight in kilograms
3. HIP = hip circumference in centimetres
4. FORE = forearm circumference in centimetres
5. NECK = neck circumference in centimetres
6. WRST = wrist circumference in centimetres
7. TRI = triceps skinfold thickness in millimetres
8. SCAP = scapula skinfold thickness in millimetres
9. SUP = suprailiac skinfold thickness in millimetres

The full regression model for these data is:

$$\begin{aligned} \text{FAT}_i = \ &\beta_0 + \beta_1 \text{HT}_i + \beta_2 \text{LOGWT}_i + \beta_3 \text{HIP}_i + \beta_4 \text{FORE}_i \\ &+ \beta_5 \text{NECK}_i + \beta_6 \text{WRST}_i + \beta_7 \text{TRI}_i + \beta_8 \text{SCAP}_i + \beta_9 \text{SUP}_i + \epsilon_i \end{aligned} \qquad (10.7)$$

where i = 1,2,...,86.

In Program 10.A, given in this chapter's appendix, we fit a full multiple linear regression additive model to these data using the SAS commands:

```
proc reg data=academy plots=none;
  *FULL MODEL;
  model FAT = HT LOGWT HIP FORE NECK WRST TRI SCAP SUP/vif collin;
```

These previous SAS statements generated the following results:

Source	DF	Squares	Square	F Value	Pr > F
Model	9	2189.25035	243.25004	22.36	<.0001
Error	76	826.87954	10.87999		
Corrected Total	85	3016.12988			

Root MSE	3.29848	R-Square	0.7258
Dependent Mean	28.09884	Adj R-Sq	0.6934
Coeff Var	11.73886		

Parameter Estimates

Variable	DF	Parameter Estimate	Standard Error	t Value	Pr > |t|	Variance Inflation
Intercept	1	-4.50263	15.65439	-0.29	0.7744	0
HT	1	-0.21220	0.09154	-2.32	0.0231	2.56189
LOGWT	1	37.66793	18.66890	2.02	0.0472	10.03156
HIP	1	0.07357	0.09830	0.75	0.4565	4.38299

FORE	1	-0.18599	0.30526	-0.61	0.5442	2.27263
NECK	1	-0.21044	0.20104	-1.05	0.2985	1.43230
WRST	1	-0.35453	0.18115	-1.96	0.0540	1.15720
TRI	1	0.33781	0.10783	3.13	0.0025	3.75919
SCAP	1	0.16587	0.10871	1.53	0.1312	2.53475
SUP	1	0.02893	0.05875	0.49	0.6239	2.32371

Collinearity Diagnostics

Number	Eigenvalue	Condition Index	--Proportion of Variation-- Intercept	HT
1	9.67448	1.00000	0.00000545	0.00000604
2	0.21668	6.68198	0.00011028	0.00013249
3	0.04590	14.51745	0.00001794	0.00003348
4	0.04073	15.41113	2.65072E-7	0.00000364
5	0.01375	26.52370	0.00164	0.00176
6	0.00312	55.69618	0.02690	0.01899
7	0.00272	59.59089	0.00387	0.00246
8	0.00185	72.34428	0.03895	0.00396

Number	Eigenvalue	Condition Index	--Proportion of Variation-- Intercept	HT
9	0.00067183	120.00063	0.28539	0.51350
10	0.00009020	327.50382	0.64312	0.45914

Collinearity Diagnostics

Number	----Proportion of Variation---- LOGWT	HIP	FORE	NECK
1	0.00000122	0.00001530	0.00002683	0.00003498
2	0.00001654	0.00008926	0.00034124	0.00054145
3	2.189316E-8	0.00005138	0.00004306	0.00013633
4	3.779496E-7	0.00018330	2.945205E-7	0.00024337
5	0.00022443	0.00125	0.00345	0.01103
6	0.00101	0.00054966	0.53928	0.04006
7	0.00094957	0.02026	0.11873	0.93662
8	0.00023541	0.60915	0.06561	0.00200
9	0.00005488	0.04952	0.00015275	0.00821
10	0.99751	0.31894	0.27238	0.00113

Collinearity Diagnostics

Number	----Proportion of Variation---- WRST	TRI	SCAP	SUP
1	0.00017393	0.00029193	0.00046398	0.00069996
2	0.00623	0.01971	0.03710	0.13379

3	0.01092	0.10499	0.29282	0.81322
4	0.00111	0.39652	0.49431	0.00430
5	0.95698	0.03436	0.02553	0.01227
6	0.00258	0.01047	0.03283	0.00578
7	0.00698	0.06710	0.00052071	0.02320
8	0.00564	0.18249	0.01222	0.00033780
9	0.00000498	0.04715	0.00594	0.00051852
10	0.00938	0.13692	0.09817	0.00588

Collinearity Among Predictors in the Full Model

The preceding results reveal that the value of the largest condition index for a collinear relationship in the sample which has two or more variance-decomposition proportions greater than 0.5 is 327, which indicates moderate collinearity. However, 327 is much less than the threshold value of 1000, suggested by Belsley (1991) and Myers (1990), as indicating severe collinearity that **may** be causing major problems in least squares estimation of regression coefficients (Section 6.2). Moreover, examination of the variation inflation factor (VIF) values for the predictors, given in the Parameter Estimates section of the preceding results, suggests that collinearity does not appear to be a severe problem in this sample as the VIF values for the predictors are all approximately ≤ 10 (subsection 6.2.3).

10.7   Model Selection and the REG Procedure

The REG procedure offers forward, backward, stepwise (Efromyson's algorithm), and all subsets model selection methods. This procedure has fewer options than a more recently developed model selection procedure, the GLMSELECT procedure which we will consider in Section 10.8. However, a feature available in the REG procedure but not in GLMSELECT is all subsets selection as described in subsection 10.3.4. As an example of specifying all subsets selection in the REG procedure, consider the following MODEL statement in Program 10.1 (Section 10.A) for the naval academy data of Section 10.6:

```
* All SUBSETS SELECTION USING Cp as SELECTION CRITERION;
model FAT = HT LOGWT HIP FORE NECK WRST TRI SCAP SUP
/selection=cp start=1 stop=6 best=10 adjrsq mse;
```

The preceding SAS statement specifies that all possible subsets of one-variable, two-variable up to six-variable models should be generated from the predictors in the full model given in the MODEL statement and from all these models generated, 10 models with the smallest Mallows' C_p statistic should be reported. The following results generated by this MODEL statement lists 10 models of all possible subsets of one-variable, two-variable, up to six-variable models in ascending order of each model's Mallows' C_p statistic.

```
The REG Procedure
Model: MODEL2
Dependent Variable: FAT


C(p) Selection Method

Number of Observations Read        86
```

Number of Observations Used   86

Number in Model	C(p)	R-Square	Adjusted R-Square	MSE	Variables in Model
5	4.5580	0.7166	0.6989	10.68389	HT LOGWT WRST TRI SCAP
6	5.3941	0.7208	0.6996	10.65882	HT LOGWT NECK WRST TRI SCAP
6	5.7238	0.7196	0.6983	10.70424	HT LOGWT FORE WRST TRI SCAP
4	5.8932	0.7046	0.6900	10.99997	HT LOGWT WRST TRI
6	5.9107	0.7190	0.6976	10.72997	HT LOGWT HIP WRST TRI SCAP
6	6.3543	0.7174	0.6959	10.79107	HT LOGWT WRST TRI SCAP SUP
5	6.4426	0.7098	0.6917	10.94019	HT LOGWT WRST TRI SUP
4	6.4773	0.7025	0.6878	11.07843	HT LOGWT TRI SCAP
5	6.8205	0.7085	0.6902	10.99158	HT LOGWT FORE WRST TRI
5	7.0454	0.7076	0.6894	11.02217	HT HIP WRST TRI SCAP

The rows of the preceding results are listed in ascending order of C_p. Recall, from subsection 10.4.5, that models which may be considered to be good fitting models are those which have Mallows' C_p statistic approximately equal or less than p, where p = number of estimated parameters, including the intercept, assuming MSE is an unbiased estimate of pure error variance. Thus, we see from the preceding output, the good fitting models based on Mallows' C_p statistic are the models reported in rows 1, 2, 3, and 5, viz., the models with predictor variables:

• HT LOGWT WRST TRI SCAP and a C_p of 4.56, which is less than p (the number of predictor variables plus the intercept), i.e., 4.56 is less than $p = 1 + 5$ predictor variables $= 6$

• HT LOGWT NECK WRST TRI SCAP and a C_p of 5.3941 (less than $p = 1 + 6$ predictor variables $= 7$)

• HT LOGWT FORE WRST TRI SCAP and a C_p of 5.7238 (less than $p = 1 + 6$ predictor variables $= 7$)

• HT LOGWT HIP WRST TRI SCAP and a C_p of 5.9107 (less than $p = 1 + 6$ predictor variables $= 7$)

These models will be considered again in Section 10.8.1. The remaining models all have a C_p greater than p.

10.8 Model Selection and the GLMSELECT Procedure

The GLMSELECT procedure offers forward, backward, and stepwise (Efromyson's algorithm) model selection methods and can support selection from a huge number of effects (up to tens of thousands). The selection process can be readily customized in GLMSELECT by specifying options and suboptions on the MODEL statement. The model selection method is specified by the SELECTION = option on GLMSELECT's MODEL statement using the keywords, FORWARD, BACKWARD, or STEPWISE. A selection method can be customized by specifying fit criteria for the SELECT = , STOP = , and CHOOSE = suboptions of the SELECTION = option. SAS keywords for specifying the fit criteria are given in Section 10.4.

Details of Options and Suboptions for Customizing Forward, Backward, and Stepwise Selection

The SELECT Suboption
You can use the SELECT suboption of the SELECTION option to specify the fit criterion to be used to determine the order in which predictors are selected or eliminated at each step of the selection process. Criteria that can be specified using SELECT = include ADJRSQ, AIC, AICC, CP, PRESS, RSQUARE, SBC, SL (defined in Section 10.4). If the SELECT = suboption is not specified, then by default, the SBC is used as the criterion which determines the selection or elimination of predictors.

The STOP Suboption
You can use the STOP suboption of the SELECTION option to specify a criterion for terminating the selection process. If you specify a criterion for the SELECT = suboption but do not specify the STOP = suboption, then the criterion specified for SELECT = is also used for the STOP criterion. If you neither specify the SELECT = nor the STOP = suboptions, then by default SBC is used as the criterion for terminating the selection process. If you specify a number for STOP = , for example STOP = 5, this directs GLMSELECT to stop the selection process at the first step for which the selected model has that number of effects.

The CHOOSE Suboption
You can use the CHOOSE suboption of the SELECTION = option to specify a criterion for choosing a model as will be illustrated in Section 10.8.1.

The INCLUDE Option
You can use the INCLUDE=s option to force the first s effects specified in the MODEL statement to always be included in all models with selections performed on only the remaining predictors in the MODEL statement. For example, the SAS commands that would force HT and LOGWT to be included in all models being fitted when using the STEPWISE procedure are:

```
proc glmselect data=academy;
MODEL FAT = HT LOGWT HIP FORE NECK WRST TRI SCAP SUP/
selection=stepwise(aicc) include=2;
run;
```

The INCLUDE option in GLMSELECT is available only when forward, backward, or stepwise selection is requested.

10.8.1   Example of Subset Selection in GLMSELECT

The following statements from Program 10.2 (Section 10.B) illustrate how predictor subset selection can be customized in GLMSELECT:

```
proc glmselect data=academy plots=(asePlot Criteria);
  title Backward Elimination;
MODEL FAT = HT LOGWT HIP FORE NECK WRST TRI SCAP SUP/
selection=backward(select=aicc choose=press stop=none) details=steps(fit)
stats=(ase adjrsq cp);
  run;
```

In the preceding code:

- The PROC GLMSELECT statement option, **plots = (asePlot Criteria)** requests that an ASE plot and a Criteria Panel plot be generated.

  An ASE plot displays the progression of the average square error at each step of the selection process.

  A Criteria Panel plot displays at each step of the selection process the following criteria: ADJRSQ, AIC, AICC, and SBC, as well as any other criteria that are named in the CHOOSE =, SELECT =, STOP =, or STATS = option in the MODEL statement.

- The MODEL statement, **MODEL FAT = HT LOGWT HIP FORE NECK WRST TRI SCAP SUP** specifies that the predictor variables of the full model from which predictor subselection should be made are HT LOGWT HIP FORE NECK WRST TRI SCAP SUP.

  (selection = backward) specifies that backward elimination should be used for the selection method.

  (select = aicc) specifies that the corrected Akaike's information criterion should be used as the fit criterion for determining the order in which a predictor variable is eliminated at each step of the selection process.

  (choose = press) specifies the PRESS statistic should be used as the criterion for choosing the final model.

  (stop = none) specifies that backward elimination should not stop until all predictor variables in the full model have been eliminated to provide insight into the selection process.

  details = steps(fit) stats = (ase adjrsq cp) specifies that average square error, adjusted R^2, and Mallows' C_p in addition to AICC and PRESS should be reported in the output at each step.

10.8.2   Model Information: Program 10.2

The GLMSELECT Procedure

Data Set	WORK.ACADEMY
Dependent Variable	FAT
Selection Method	Backward
Select Criterion	AICC
Stop Criterion	None
Choose Criterion	PRESS
Effect Hierarchy Enforced	None

```
Number of Observations Read        86
Number of Observations Used        86
```

```
             Dimensions

    Number of Effects     10
    Number of Parameters  10
```

10.8.3   Model Building Summary Results: Program 10.2

Backward Selection Summary

Step	Effect Removed	Number Effects In	Adjusted R-Square	AICC	CP
0		10	0.6934	306.2124	10.0000
1	SUP	9	0.6964	303.8520	8.2424
2	FORE	8	0.6987	301.7530	6.6565
3	HIP	7	0.6996*	300.0781	5.3941
4	NECK	6	0.6989	298.9277*	4.5580*
5	SCAP	5	0.6900	300.1308	5.8932
6	WRST	4	0.6746	303.0497	9.0302
7	HT	3	0.6450	309.3203	16.1014
8	LOGWT	2	0.6358	310.3493	17.7796
9	TRI	1	-.0000	396.0795	193.2180

Backward Selection Summary continued

Step	Effect Removed	PRESS	ASE
0		1083.8994	9.6149
1	SUP	1062.6169	9.6455
2	FORE	1024.6739	9.6979
3	HIP	980.9957	9.7912
4	NECK	978.3435*	9.9385
5	SCAP	985.0362	10.3604
6	WRST	1041.2456	11.0103
7	HT	1129.0496	12.1579
8	LOGWT	1149.2675	12.6233
9	TRI	3087.5151	35.0713

* Optimal Value of Criterion

10.8.4   Comments Regarding Model Building Summary Results

The model building summary for backward elimination given in the previous subsection shows that at:

* Step 0

  Ten effects are in the full model, viz., the intercept and the nine predictor variables specified in the MODEL statement.

* Step 1

  As AICC was the fit criterion specified for the SELECT suboption, SUP was the first predictor to be removed from the full model. This was because the model with SUP excluded has the smallest value of AICC compared to any of the other models with a single predictor variable removed. Recall AICC is a measure of information lost when an approximating model is used to approximate a full model. Thus, the least information may be lost when the model with SUP excluded is used to approximate the full model as compared to the other eight-variable models which have a single predictor excluded. The model at the end of Step 1 consisted of the remaining eight variables and the intercept.

* Step 2

  At Step 2, the AICC values of seven-variable models are compared where each model was formed by excluding SUP and one of the other remaining predictor variables. The model with SUP and FORE excluded had the smallest AICC value indicating that the least information may be lost when the model with SUP and FORE excluded is used to approximate the full model. Therefore FORE was removed from the model and at the end of Step 2 the model consisted of the intercept and the seven predictor variables, HIP NECK SCAP WRST HT LOGWT TRI.

* Step 3

  At Step 3, the AICC values of six-variable models were compared where each model was formed by excluding SUP, FORE, and one of the other remaining predictor variables. The model with SUP, FORE, and HIP excluded had the smallest AICC value as compared to the other six-variable models. Therefore HIP was removed and the model selected at Step 3 consisted of the intercept and the six predictor variables, NECK SCAP WRST HT LOGWT TRI.

* Step 4

  At Step 4, the AICC values of five-variable models are compared where each model was formed by excluding SUP, FORE, HIP, and one of the other remaining predictor variables. The model with SUP, FORE, HIP, and NECK excluded had the smallest AICC value as compared to the other five-variable models. Therefore NECK was removed and the model selected at Step 4 consisted of the intercept and the five predictor variables, SCAP WRST HT LOGWT TRI.

* Steps 5–8

  Each successive step removes a single predictor variable from the model selected at the previous step based on the minimum AICC criterion.

* Step 9

  At Step 9, the last predictor variable, TRI, is removed from the model selected at Step 8, leaving only the intercept in the model.

The final model chosen is the one that has a global minimum value of PRESS. Therefore, the final selected model has a PRESS value of 978.34 (subsection 10.8.3) and is the five-variable model with SUP FORE HIP NECK excluded from the full model, i.e., the final selected model is the five-variable model comprised by the Intercept HT LOGWT WRST TRI SCAP. However, as **stop = none** was specified, the backward elimination process continues until all variables have been removed from the full model.

10.8.5 Selected Model Results: Program 10.2

```
                    The GLMSELECT Procedure
                        Selected Model

    The selected model, based on PRESS, is the model at Step 4.


        Effects: Intercept HT LOGWT WRST TRI SCAP


                      Analysis of Variance

                            Sum of           Mean
Source                DF    Squares         Square    F Value

Model                  5   2161.41879     432.28376     40.46
Error                 80    854.71110      10.68389
Corrected Total       85   3016.12988


            Root MSE              3.26862
            Dependent Mean       28.09884
            R-Square              0.7166
            Adj R-Sq              0.6989
            AIC                 297.49179
            AICC                298.92769
            BIC                 212.61063
            C(p)                  4.55805
            PRESS               978.34354
            SBC                 224.21788
            ASE                   9.93850



                    Backward Elimination
                  The GLMSELECT Procedure
                        Selected Model

                   Parameter Estimates

                                      Standard
    Parameter    DF      Estimate        Error    t Value
    Intercept     1     -4.703474    13.065192      -0.36
```

HT	1	-0.209616	0.085043	-2.46
LOGWT	1	34.978047	12.055941	2.90
WRST	1	-0.356712	0.178552	-2.00
TRI	1	0.407195	0.092995	4.38
SCAP	1	0.177534	0.096333	1.84

10.8.6   Comments Regarding Standard Errors of the Partial Regression Coefficients for the Selected Model

The standard errors of the partial regression coefficients (parameter estimates) for the selected model given in the preceding subsection have been computed using the conventional formula (Equation 7.4) which assumes the model has been prespecified and not obtained via a search process. You can verify that the conventional formula was used to compute the standard errors of the regression coefficients for the selected model by noting these standard errors are identical with their counterparts obtained from the REG procedure where the model is prespecified using the model statement HT LOGWT HIP WRST TRI SCAP. (Recall in the REG procedure the intercept is included by default in the model.) Estimation of standard errors of partial regression coefficients for a final model chosen by a selection procedure and other aspects of post-selection inference are an ongoing area of research (e.g., Berk et al., 2013; Fithian et al., 2014; Lee et al., 2016; Tibshirani et al., 2016; Bachoc et al., 2019).

10.8.7   Average Square Error Plot: Program 10.2

Figure 10.1, the average square errors plot generated by Program 10.2, allows us to evaluate the prediction performance of the current model at each step of the backward elimination procedure where AICC was used at each step to select a model and the PRESS statistic was used to choose the final model.

10.8.8   Criteria Panel Plot: Program 10.2

A Criteria Panel plot can answer a data analyst's question, "Would the final model selected be different if I had specified other fit criteria as suboptions for this analysis?"

Figure 10.2, the Criteria Panel plot generated by Program 10.2 for the Academy data, indicates that regardless of whether C_p, AIC, AICC, or the PRESS had been specified as the fit criterion for the CHOOSE suboption, the same final model would be identified as "best", i.e., the model comprised by the intercept and the five predictor variables, SCAP WRST HT LOGWT and TRI (as SUP, FORE, HIP, and NECK have been removed from the full model).

If SBC had been used as the fit criterion, SCAP would have been removed from the full model in addition to SUP, FORE, HIP, NECK, so that the "best" final model chosen by SBC would have been comprised by the intercept and the four predictor variables, WRST HT LOGWT and TRI. When SBC is the fit criterion used, we would expect the final selected model to have fewer predictor variables as compared to the final model selected by AIC or AICC because, as noted in subsection 10.4.8, SBC imposes a stronger penalty for adding more predictors to a model as compared to AIC or AICC.

If adjusted R^2 had been used as a criterion for the CHOOSE suboption, the "best" final model chosen would have been comprised by the intercept and the six predictor variables, NECK SCAP WRST HT LOGWT and TRI.

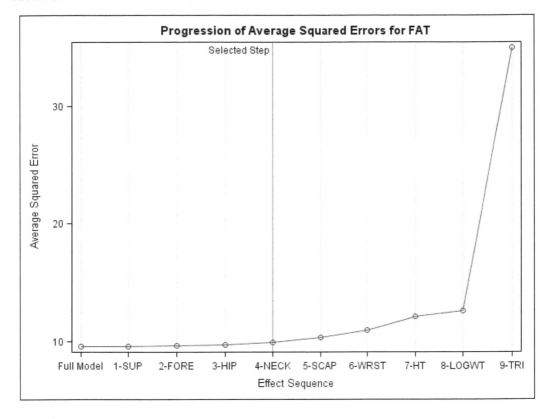

FIGURE 10.1
Plot of average square errors for Academy data at each step of backward elimination with the corrected Akaike's information criterion (AICC) for selecting models at each step and the PRESS statistic for choosing the final selected model.

Table 10.1 summarizes results from the ASE plot (Figure 10.1), the Criteria Panel (Figure 10.2), and the Model Building Summary output (subsection 10.8.3), obtained from the backward elimination procedure, which can inform the decision as to what models might be useful for predicting the outcome in the Academy example.

TABLE 10.1
Final models selected via backward elimination for academy data based on various fit criteria for the CHOOSE suboption in GLMSELECT where k = the number of predictor variables.

Choose Criterion	k	Predictor Variables in Model	ASE
SBC	4	WRST HT LOGWT TRI	10.3604
PRESS, C_p, AIC, AICC	5	SCAP WRST HT LOGWT TRI	9.9385
R^2	6	NECK SCAP WRST HT LOGWT TRI	9.7912

As summarized in Table 10.1, the five-variable model that would have been chosen by C_p, AIC, AICC, or PRESS has an ASE of 9.9385 as compared to the ASE of 9.7912 for the six-variable model that would have been selected by specifying adjusted R^2 as the criterion for choosing the final model. It is interesting to note that the five-variable model chosen by the backward elimination procedure is identical to that chosen by the all subsets selection procedure (Section 10.7) when both procedures utilize Mallows' C_p fit criterion.

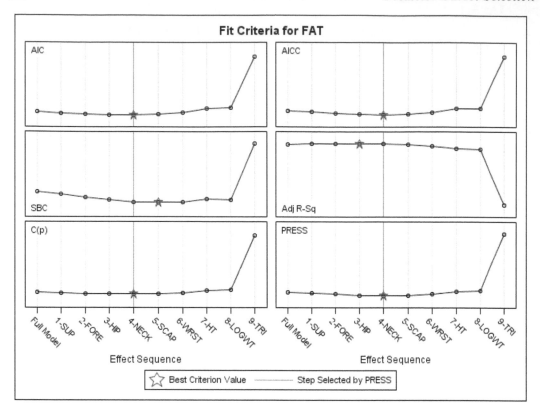

FIGURE 10.2
Panel plot of fit criteria at each step of backward elimination using Academy data.

Unless expert opinion would suggest otherwise, the five-variable model might be considered preferable to the six-variable model, as the former is more parsimonious and appears to have comparable prediction performance based on the average squared error observed in these data. The five-variable model might also be considered preferable to the four-variable model because the ASE for the five-variable model is slightly less than that of the four-variable model. Moreover, the four-variable model was the final model selected by SBC, a fit criterion which Burnham and Anderson (2004) pointed out tends to select an under-fitted model (too few predictors) in the biological sciences where many factors may be involved in the prediction of an outcome variable. They suggested AIC or AICC may be more appropriate as a fit criterion in these situations. From Table 10.1 it may be seen that AICC was one of the fit criteria that chose the five-variable model.

Based on the parameter estimates for the Selected Model (subsection 10.8.5), the estimated five-variable model is

$$\text{FAT} = -4.703 \quad\quad -0.210 \text{ HT} \quad\quad +34.978 \text{ LOGWT} \quad -0.357 \text{ WRST}$$
$$+0.407 \text{ TRI} \quad +0.178 \text{ SCAP}$$

However, if for some reason, it is inconvenient to measure SCAP, researchers may have a strong substantive justification to override these statistical considerations and choose the four-variable model, which apart from excluding SCAP, has the same predictors as the five-variable model. This example underscores the art and science of model selection where good decisions emerge from the interplay of substantive expertise and statistical results.

10.9   Other Features Available in GLMSELECT

Predictor subset selection methods available in GLMSELECT other than those discussed in this chapter are penalized regression procedures, also referred to as shrinkage methods. Although these methods are beyond the scope of this book, it is important to be aware of their availability as they address a problem that may occur in traditional sequential selection procedures, viz., the problem of unstable estimation of a selected model's regression coefficients. This problem tends to be exacerbated when there is collinearity among the predictors and/or when there is a large number of predictors relative to the sample size.

The following penalized regression methods are available in the GLMSELECT procedure:

- least angle regression (Efron et al., 2004)

- lasso selection (**L**east **A**bsolute **S**hrinkage and **S**election **O**perator) (Tibshirani, 1996; Efron et al., 2004)

- adaptive lasso selection (Zou, 2006)

- elastic net selection (Zou and Hastie, 2005)

A brief review of these methods with examples illustrating their implementation in GLM-SELECT are given in Rodriguez (2016), Gunes (2005), and Cohen (2006; 2009). Although some problems associated with traditional sequential selection can be mitigated with penalized regression methods, these methods are nevertheless automated statistical selection methods which can never be a substitute for substantive expertise.

A generally recommended approach to assess a selected model is to evaluate its predictive performance on a new data set other than the *training* data set, i.e., the data set used by the model selection procedure to select a predictor subset from the full model. For situations where this recommended approach is not feasible, GLMSELECT offers a variety of alternative approaches which involve partitioning a data set into disjoint subsets for the purpose of validation and testing. Cohen (2006) has provided illustrative examples. For a discussion of how common sense can be an important motivating factor in data partitioning see Myers (1990).

10.10   Chapter Summary

This chapter:

- describes model selection procedures for choosing a subset from a full set of predictors for a multiple linear regression model

- discusses why relying solely on automated selection procedures is contentious if the primary research objective is to understand the underlying relationship between an outcome variable and its predictors

- suggests that a model selection procedure may work well if the primary research objective is prediction

- cites reasons that researchers may want to select a subset of predictors from a full set

- discusses under-fitting, over-fitting, and the bias-variance trade-off in model building

- describes forward, backward, stepwise (Efromyson's algorithm), and all subsets selection methods and discusses the impact of collinearity on these methods

- discusses various fit criteria involved in predictor selection

- indicates data-driven inference issues that occur when the same data set is used for selection and estimation

- illustrates model selection methods in the REG and GLMSELECT procedures

- shows how ODS graphics in GLMSELECT can provide researchers with an overview of the selection process

- indicates the availability of penalized regression selection methods and a variety of model assessment approaches in GLMSELECT

- emphasizes that an automated statistical selection method can never be a substitute for substantive expertise

Appendix

10.A   Program 10.1

Program 10.1 reads in the data for the Naval Academy example described in Section 10.6 and stores it in the **temporary** SAS data set arbitrarily named ACADEMY. Program 10.A illustrates customization of predictor subset selection in the REG procedure.

```
*Program 10.1;
data Academy;
input CASE HT  LOGWT HIP  FORE  NECK   WRST  TRI SCAP  SUP FAT;
datalines;
  1  173.4 1.762   91.8  21.1  29.7  14.6  12.2  12.1  22.9  26.9
  2  173.6 1.848   99.3  27.5  31.5  15.2  16.5  16.4  26.4  28.0
  3  164.1 1.825  100.8  24.5  32.4  15.5  20.3  23.9  34.6  36.4
  4  160.1 1.775   96.2  22.5  31.1  14.1  12.9  11.4  25.3  20.9
  5  160.3 1.843  102.2  25.1  31.7  14.9  28.2  17.8  33.7  32.4
  6  166.9 1.782   91.4  24.6  31.5  15.2  18.1  13.8  17.9  21.7
  7  165.1 1.808   98.3  23.7  32.1  14.8  24.0  16.6  35.0  34.3
  8  167.3 1.860  103.7  25.2  37.2  15.2  27.9  20.7  40.0  34.3
  9  147.9 1.698   89.3  20.1  28.6  13.0  17.1   7.4  17.7  27.7
 10  163.3 1.663   87.1  21.3  27.0  14.3   9.4   8.2   9.1  21.4
 11  160.2 1.776   96.2  22.4  33.9  13.9  27.2  21.2  26.4  30.3
 12  159.0 1.777   83.9  21.6  29.9  14.2  15.2  17.7  24.5  29.9
 13  150.7 1.730   90.5  22.9  30.4  14.6  19.1  18.4  31.2  34.1
 14  160.7 1.804   97.8  24.8  32.4  25.8  17.7  12.4  15.0  25.1
 15  158.4 1.763   95.3  22.7  29.2  13.7  22.3  15.7  21.3  27.4
 16  158.7 1.788   96.8  24.5  28.9  14.7  30.2  13.0  23.5  31.5
 17  160.9 1.770   93.9  22.4  29.0  14.0  20.3  17.3  26.6  35.4
 18  176.3 1.912  111.5  24.7  31.5  15.5  22.5  21.5  40.0  33.5
 19  162.6 1.773   95.3  22.5  30.9  15.0  21.0  16.6  33.5  26.2
 20  160.2 1.760   91.5  23.1  30.5  15.0  21.0  13.7  20.2  31.0
 21  154.7 1.677   85.6  21.6  28.3  14.4  16.5   7.3  13.6  28.7
 22  152.4 1.781  102.6  22.9  29.6  14.5  22.8  19.0  30.8  38.4
 23  166.2 1.795   92.6  27.4  32.3  23.9   8.6   9.7  12.5  18.8
 24  160.1 1.766   92.3  22.5  29.2  14.4  14.4  11.8  18.7  27.7
 25  162.4 1.805   99.5  22.8  32.6  14.5  17.0  13.8  16.8  31.5
 26  170.4 1.840   97.8  24.5  34.8  16.0  18.1  13.8  32.5  26.6
 27  155.9 1.780   94.5  23.9  31.4  15.1  26.2   9.8  26.5  29.3
 28  162.8 1.828  121.0  23.6  30.4  15.0  22.4  12.8  17.8  28.6
 29  156.6 1.754   92.1  24.1  30.6  14.1  24.2  23.3   8.8  26.8
 30  160.9 1.825   98.3  25.1  32.7  14.7  12.8  21.7  16.4  29.9
 31  158.8 1.769   94.5  23.6  32.6  14.1  18.1  11.8  17.3  25.2
 32  154.5 1.699   83.5  23.2  31.3  14.7   8.0   9.2  10.9  14.9
```

223

33	157.5	1.725	93.1	22.4	30.6	14.3	13.6	10.0	14.4	22.1
34	171.9	1.816	95.6	22.9	32.5	14.8	16.0	10.2	15.6	26.8
35	161.2	1.797	102.9	23.0	32.6	14.1	23.3	15.3	27.1	31.7
36	156.3	1.892	114.4	24.3	32.3	14.6	34.1	26.0	32.3	42.9
37	165.2	1.835	101.5	25.2	31.1	15.0	14.8	20.2	22.8	31.5
38	160.7	1.723	85.1	21.3	31.8	13.4	12.2	13.0	8.2	23.5
39	166.8	1.786	94.4	22.1	31.9	12.9	18.0	8.8	9.7	26.4
40	154.3	1.695	87.7	20.9	28.5	13.2	14.9	15.5	16.4	32.0
41	158.2	1.741	93.7	21.9	29.6	14.0	17.0	9.8	10.0	29.3
42	166.9	1.710	89.7	21.0	31.5	13.8	13.2	13.3	22.1	24.8
43	166.1	1.812	98.4	24.8	31.8	14.4	17.2	10.2	6.6	28.7
44	164.4	1.823	97.2	24.6	31.8	13.9	15.6	23.2	14.2	30.6
45	172.8	1.833	98.1	23.8	31.8	15.0	18.2	13.4	8.2	28.1
46	167.8	1.823	100.6	25.5	31.9	15.5	15.2	15.6	13.0	32.5
47	161.2	1.686	86.6	21.2	30.6	14.7	10.0	7.0	12.9	17.9
48	176.3	1.759	89.0	21.7	30.7	15.2	13.4	9.8	17.8	21.2
49	166.7	1.784	91.9	23.0	31.9	15.1	15.2	12.8	16.4	28.2
50	159.0	1.700	91.4	21.1	29.4	14.1	16.0	8.6	15.6	26.8
51	168.3	1.732	89.6	22.6	33.7	16.2	14.1	9.2	6.0	21.7
52	159.1	1.788	94.0	23.7	22.1	14.7	24.2	13.0	21.7	32.0
53	154.1	1.668	85.2	21.7	29.4	13.9	13.5	9.4	16.3	22.2
54	158.3	1.721	93.5	22.4	30.4	14.0	18.7	14.6	31.2	31.3
55	165.2	1.858	102.7	24.6	34.9	15.3	14.8	14.0	8.8	28.3
56	160.8	1.735	77.5	22.2	30.2	14.8	11.6	9.1	5.2	28.5
57	168.9	1.780	89.2	22.2	31.7	14.5	13.9	12.6	24.2	23.6
58	162.2	1.741	89.0	23.5	30.4	14.4	7.3	9.6	3.8	23.3
59	163.1	1.804	90.9	24.0	24.1	15.5	19.4	13.2	9.5	26.9
60	162.8	1.746	89.3	23.3	32.0	14.9	14.2	9.0	17.7	18.6
61	156.7	1.741	87.9	22.6	29.1	14.5	14.5	12.8	16.6	23.8
62	163.2	1.749	88.2	22.2	30.6	15.1	10.7	9.1	20.2	19.2
63	162.4	1.672	84.9	14.1	27.4	21.6	9.1	9.5	11.0	16.6
64	156.0	1.857	107.9	24.3	32.1	15.2	32.4	28.0	38.0	42.2
65	174.2	1.884	103.8	24.2	33.3	15.2	22.4	12.3	25.8	29.7
66	151.2	1.652	82.2	21.3	28.0	14.2	12.0	10.7	12.9	24.1
67	170.8	1.819	98.5	22.5	28.6	14.6	20.0	21.5	30.9	25.6
68	158.4	1.703	86.9	21.4	28.7	14.2	10.7	8.5	10.0	22.2
69	161.7	1.797	96.9	23.4	30.5	14.4	21.4	24.5	29.3	27.5
70	173.4	1.847	98.2	23.4	29.8	15.1	22.1	8.6	18.0	33.6
71	160.0	1.761	94.6	21.8	29.7	14.3	18.3	9.7	17.5	29.5
72	166.1	1.798	97.5	23.7	30.4	15.6	24.8	14.7	23.7	30.3
73	149.3	1.653	82.9	21.1	28.6	13.2	9.4	11.3	23.3	24.5
74	161.4	1.875	104.6	23.8	31.4	14.4	29.9	27.5	36.5	29.2
75	152.5	1.800	102.5	23.5	29.9	13.4	31.5	20.2	40.0	38.1
76	161.0	1.748	92.4	23.3	29.1	14.6	14.7	7.4	7.5	21.5
77	159.7	1.776	91.2	24.3	31.7	14.6	14.9	13.5	18.8	22.4
78	158.8	1.752	96.8	20.2	29.2	13.5	21.4	17.7	16.5	31.1
79	160.7	1.829	103.7	25.8	30.2	15.2	20.0	11.6	15.5	32.3
80	171.1	1.828	99.1	24.4	32.1	14.9	23.2	18.3	22.7	31.7
81	174.6	1.946	113.2	24.9	33.5	15.7	26.5	17.4	22.6	36.5
82	162.0	1.778	93.5	22.7	28.6	14.2	20.5	11.1	19.8	26.1
83	155.5	1.892	109.9	24.5	30.2	14.8	39.7	23.7	40.0	50.1

```
84   158.0   1.741    90.6    22.7    30.8    14.5    16.5    21.7    31.1    29.8
85   165.1   1.776    96.6    23.9    31.3    24.7    10.8     8.4    17.9    21.5
86   165.9   1.731    92.1    21.5    27.4    14.4     9.0     8.0     7.4    21.2
;
proc reg data=academy plots=none;
    *FULL MODEL;
    model FAT = HT LOGWT HIP FORE NECK WRST TRI SCAP SUP/vif collin;
    * All SUBSETS SELECTION USING Cp as SELECTION CRITERION;
    model FAT = HT LOGWT HIP FORE NECK WRST TRI SCAP SUP/selection=cp
    start=1 stop=6 best=10 adjrsq mse;
quit;
```

10.B   Program 10.2

Using the temporary SAS data set ACADEMY created in Program 10.1, Program 10.2 illustrates customization of predictor subset selection in the GLMSELECT procedure as described in subsection 10.8.1.

Please note:
If you try to execute Program 10.2 after having logged out from the SAS session in which you executed Program 10.1, you will get error messages. To avoid this, you could create the SAS data set ACADEMY again by executing Program 10.1 in your current SAS session or alternatively by inserting the DATA STEP and the SORT procedure statements from Program 10.1 at the beginning of Program 10.2.

```
* Program 10.2;
proc glmselect data=academy plots=(asePlot Criteria);
    title Backward Elimination;
    MODEL FAT = HT LOGWT HIP FORE NECK WRST TRI SCAP SUP/
    selection=backward(select=aicc choose=press stop=none)
    details=steps(fit) stats=(ase adjrsq cp);
run;
```

11

Evaluating Equality of Group Means with a One-Way Analysis of Variance

11.1 Characteristic Features of a One-Way Analysis of Variance Model

Some characteristic features of a one-way analysis of variance (ANOVA) model are:

- There is only one response variable in the model.

- The response variable is continuous.

- The model is a linear model.

- There is only one explanatory variable represented by the model and it is treated as a qualitative (unordered categorical) variable in the analysis, i.e., the categories (values) of the explanatory variable represented in the model are not related to any sort of ranking. This explanatory variable represented in an ANOVA model is called **a factor** or **a classification variable**. For this reason a one-way ANOVA is also referred to as a **single-factor ANOVA**.

The features of the one-way ANOVA model cited in the first three bullets are also characteristic of the simple linear regression and multiple linear regression models described in previous chapters. However, the characteristic feature of the fourth bullet indicates that a one-way ANOVA model is a special case of a linear model where the explanatory variable is represented in the model as a qualitative variable.

11.2 Fixed Effects vs. Random Effects in One-Way ANOVA

A factor (i.e., classification variable or qualitative explanatory variable) represented in a one-way ANOVA model is referred to as a **fixed effect** when categories of the factor in the model have been deliberately chosen by the researcher. For example, a researcher wanted to compare the mean birth weight of babies delivered in 2020 at three particular hospitals (Health Sciences Center, Gimli, and St. Boniface) in the province of Manitoba, Canada. To investigate birth weights at these hospitals, the researcher took a random sample of 100 records from each of these three hospitals and applied a one-way ANOVA with birth weight as the response variable and the factor "HOSPITAL" with three categories (levels) represented in the model. In this situation, the factor "HOSPITAL", is a fixed effect as the researcher specifically chose those particular hospitals for the study. The ANOVA applied was a **one-way fixed effects ANOVA** where the primary question investigated was whether the average birth weight was the same in all three hospitals in 2020.

DOI: 10.1201/9780429107368-11

Conclusions from a one-way fixed effects ANOVA pertain only to those specific categories of the factor that were deliberately chosen by the researcher for the study (only to Health Sciences, Gimli, and St. Boniface hospitals in our example).

A factor is a **random effect** when the categories of the factor in the model have not been deliberately chosen by the researcher for the study but rather the categories have been randomly selected from the population of categories for the factor of interest. For example, a researcher was interested in the variability among mean birth weights of babies delivered in 2020 in **all** Manitoba hospitals. The researcher's budget allowed enough resources to study only three hospitals. Therefore, the researcher used a formal mechanism to randomly choose three hospitals from the population of all Manitoba hospitals and then took a random sample of 100 records from each of the three randomly chosen hospitals. The researcher applied a **random effects one-way ANOVA** to generalize the findings from these three randomly selected hospitals to all Manitoba hospitals in 2020. In this situation, the factor, "HOSPITAL", is a random effect in the one-way ANOVA model and the primary question investigated was whether or not mean birth weights vary among Manitoba hospitals in 2020.

In contrast to a fixed effects model, conclusions from a random effects one-way ANOVA can be generalized to the entire population of categories of the factor. Thus, in the example where the researcher randomly selected three hospitals from the specified population of all the hospitals in Manitoba, the conclusions based on the analysis of these three hospitals can be extended to all Manitoba hospitals in 2020.

In summary, a fixed effects model and a random effects one-way ANOVA can be distinguished on the basis of:

- how the categories of the factor (classification variable) have been chosen for the study by the researcher

- whether the conclusions of the ANOVA can be extended beyond the particular categories of the factor investigated in the study

- the researcher's primary question of interest

  - In a one-way fixed effects ANOVA, the primary question of interest is whether or not the true means of the response variable in the specifically chosen categories of the factor are equal to each other.

  - In a one-way random effects ANOVA, the primary question of interest is whether the true response variable means vary among **all** categories of the factor represented in the model.

- the robustness of a fixed effects vs. a random effects ANOVA to non-normal distributions

  - A fixed effects ANOVA has been shown to be more robust to departures from normality than a random effects ANOVA. Non-normality may have serious consequences on the validity of inferences in the random effects ANOVA (Scheffé, 1959; Kendall and Stuart, 1976). For a detailed example of a one-way random effects ANOVA, see Kutner et al. (2005).

Although we will discuss only the more frequently used one-way fixed effects ANOVA in this book, we nevertheless think it is important to be aware of fundamental differences between these two types of one-way ANOVAS. If you should encounter a random effects one-way ANOVA in a description of a research study, then you will be aware that the validity of the inferences from this ANOVA may be highly dependent on normal distributions.

11.3   Why Use a One-Way Fixed Effects Analysis of Variance?

You would apply a one-way fixed effects ANOVA when you have only sample data and you want to evaluate the equality of population means of a response (Y) variable in two or more groups which are classified on the basis of a single qualitative variable. In other words, you want to test the global hypothesis whether a single classification variable (factor) which defines your groups is of any use in explaining the response variable.

11.4   Task Study Example

A completely randomized experiment was carried out to investigate whether three different kinds of tasks could affect pulse rate. Thirty-nine volunteers were assigned at random to one of three groups so that there were 13 individuals in each group. On a selected day after training, the pulse rate (number of pulses in 20 seconds) of each individual was measured after performing the assigned task for one hour. Unfortunately some individuals withdrew from the training process so that there were unequal sample sizes in the groups. The data, a subset from Milliken and Johnson (2009), are given in Table 11.1.

To investigate whether these tasks affected pulse rates in their study, the researchers applied a one-way ANOVA in which pulse rate in 20 seconds was the response variable and the type of task assigned was the factor. Type of task was a fixed factor as the researchers deliberately chose these three particular tasks and did not randomly choose these three tasks from a specified population of tasks. The researchers were interested only in how these three particular tasks affected pulse rate. They were not interested in generalizing their results from these three tasks to a specified population of tasks.

TABLE 11.1

Pulse rate data for task experiment: n_i is the group sample size

	Task		
	A	**B**	**C**
	34	28	28
	34	28	26
	43	26	29
	44	35	25
	40	31	35
	47	30	34
	34	34	37
	31	34	28
	45	26	21
	28	20	28
		41	26
		21	
n_i	10	12	11
mean	38.00	29.50	28.82

We arbitrarily labelled the three tasks in this study A, B, C. The null hypothesis the researchers initially tested in this ANOVA was H_0: $\mu_A = \mu_B = \mu_C$ vs. H_a (not H_0) where μ_A, μ_B, μ_C are respectively the population mean pulse rates in the groups assigned tasks A, B, or C.

11.5   The Means Model for a One-Way Fixed Effects ANOVA

There are several ways to write a model for a one-way fixed effects ANOVA. One way is the **means model** (also sometimes called **the cell means model**), which is written as

$$Y_{ij} = \mu_i + \epsilon_{ij} \tag{11.1}$$

for $i = 1, 2, \ldots, g$; $j = 1, 2, \ldots, n_i$ where g is the number of groups in the analysis and n_i is the sample size in the jth group
Y_{ij} is the response for individual j in group i
μ_i is the population mean of the Y values in group i
ϵ_{ij} is the model error component for individual j in group i and is given by

$$\epsilon_{ij} = Y_{ij} - \mu_i \tag{11.2}$$

i.e., in the means model representation of one-way ANOVA (Equation 11.1) ϵ_{ij} is the difference between individual j's observed value for Y and the population mean of Y in group i (the group to which individual j belongs).

Stating null hypotheses regarding the group population means is straightforward when a one-way fixed effects ANOVA is represented by a means model. For example, the null hypothesis that all the group population means of the response variable are equal is given as:

H_0: $\mu_1 = \mu_2 = \ldots = \mu_g$ where g indicates the number of groups in the analysis.
H_a: Not H_0, i.e., not all group population means are equal.

Another way of writing a one-way ANOVA model other than as a means model is as a multiple regression model where the single qualitative predictor variable (factor) is represented in the model using reference cell coding (Sections 9.2, 9.3). However, although there are other alternative formulations for expressing a one-way ANOVA model,[1] our focus is on the means model because as Muller and Fetterman (2012) have pointed out there is a natural correspondence of the parameters of the means model and most discussions of the ANOVA procedure.

11.6   Basic Concepts Underlying a One-Way Fixed Effects ANOVA

In this section we provide an introduction to some of the basic concepts underlying a one-way fixed effects ANOVA which are helpful in understanding the ideal inference conditions

---

[1]Muller and Fetterman (2012) provided a discussion of other commonly used formulations of the ANOVA model in matrix algebra terms.

for this procedure (Section 11.7). Numerical details of a one-way ANOVA analysis will be given in Section 11.18.

In a one-way ANOVA, the Mean Square Among Means is the variance of the sample group means, weighted by the sample size of each group.

$$\text{Mean Square Among Group Means} = \frac{\text{Sum of Squares Among Group Means}}{g - 1} \tag{11.3}$$

where

Sum of Squares Among Means is given by $\sum_{i=1}^{g} n_i (\bar{Y}_{i.} - \bar{Y}_{..})^2$

$\bar{Y}_{i.}$ is the sample mean of the response variable for group i, i=1,...,g groups

$\bar{Y}_{..}$ is the overall response variable mean ignoring group and is given by

$\sum_{i=1}^{g} \sum_{j=1}^{n_i} Y_{ij} / (n_1 + \ldots + n_g)$ where Y_{ij} = the value of the response for individual j in group i.

The Mean Square Among Means is compared to the variance of the observations on the response variable **within** those groups. If the null hypothesis is true, i.e., if it is true that the population means of the response variable in the groups are all equal, then the variance among the sample group means is expected to be an estimate of the same population value as is the variance of the observations on the response variable within the groups.

The one-way ANOVA procedure assumes that the within group population variance of the response variable (σ^2) is the same value for all the groups and that any differences observed in the sample variances of the response variable (the s^2) of the groups in the study are simply due to sampling variation, i.e., the variation you would expect when taking samples from the same population. This assumption is referred to as the equality of variances ideal inference condition in Section 11.7. Assuming this ideal inference condition holds, the one-way ANOVA procedure averages the sample variances (the s^2) of all the groups in the analysis to obtain a more precise estimate of σ^2, the within group population variance, which is referred to as Mean Square Within Groups (or Mean Square Error) in a one-way ANOVA. For example, if there are three groups with equal sample sizes in the one-way ANOVA, the sample variance of each group is averaged to give a pooled estimate of the common within group population variance as follows

$$\text{Mean Square Within Groups} = \frac{s_1^2 + s_2^2 + s_3^2}{3} \tag{11.4}$$

If the sample sizes are not equal, then a pooled estimate of the common within group population variance is obtained by taking a weighted average of the sample variances where the weights are the degrees of freedom, i.e., functions of the sample sizes. For example, if there are three groups with unequal sample sizes, n_1, n_2, n_3, in the ANOVA, then

$$\text{Mean Square Within Groups} = \frac{(n_1 - 1)s_1^2 + (n_2 - 1)s_2^2 + (n_3 - 1)s_3^2}{(n_1 - 1) + (n_2 - 1) + (n_3 - 1)} \tag{11.5}$$

The ANOVA procedure uses weights that are a function of the sample size for this weighted average, so that any estimate of s^2 which is less reliable because it is based on a smaller sample size will be down weighted as compared to estimates of s^2 from groups with larger sample sizes which will be given larger weights.

The one-way ANOVA procedure calculates a test statistic to compare the variance among sample group means to the variance within groups, by forming an F ratio as follows:

$$F = \frac{\text{Mean Square Among Group Means}}{\text{Mean Square Within Groups}} \tag{11.6}$$

The procedure compares the value of this F test statistic of Equation 11.6 to a critical value from the central F distribution to decide whether the variance observed among sample group means is estimating something more than the variance within groups. The central F distribution is a ratio of two sample variances from a normal distribution which has the same mean and the same standard deviation. If there is a low probability that the F test statistic belongs to the central F distribution (i.e., a low p-value associated with the test statistic), then this suggests that the null hypothesis of mean equality is unlikely assuming ideal inference conditions are reasonable. The referral of the value of the F test statistic to the central F distribution underlies why the ANOVA procedure makes the assumption that the observations within each group comes from a normal distribution. This assumption is referred to as the ideal inference condition of normality in Section 11.7.

11.7   Ideal   Inference   Conditions   for   One-Way   Fixed   Effects ANOVA

1. INDEPENDENCE OF ERRORS

   The ϵ_{ij} (Equation 11.1), i.e., the model error components, are all independent and therefore uncorrelated. In a one-way fixed effects ANOVA, this condition implies that the observations on the response variable Y_{ij} are all independent and uncorrelated.

2. EQUALITY OF VARIANCES

   The ϵ_{ij} have the same variance in each population considered in the analysis, which implies that the population variances of the observations on the response variable are equal in the groups considered in the one-way fixed effects ANOVA.

3. NO MISCLASSIFICATIONS

   Each individual (sampling unit) in the study has been correctly classified as to the level (category) of the factor to which the individual belongs.

4. NORMALITY

   The ϵ_{ij} are normally distributed in the population of each group considered in the one-way fixed effects ANOVA which implies that the response values in each population considered are normally distributed.

11.8   Potential Consequences of Violating Ideal Inference Conditions for One-Way Fixed Effects ANOVA Test of Equality of Group Means

Much has been written about the consequences of violating ideal inference conditions for the one-way fixed effects ANOVA test of mean equality (e.g., Pearson, 1931; Cochran, 1947; Welch, 1951; Norton, 1952; Lindquist, 1953; Box, 1954; Scheffé, 1959; Boneau, 1960; van der Vaart, 1961; Glass et al., 1972; Brown and Forsythe, 1974b; Clinch and Keselman, 1982; Harwell et al., 1992; Miller, 1998).

The following are some of the broad general guidelines that have emerged from the literature:

- Violation of independence of the model errors can have serious consequences and can invalidate conclusions from a one-way ANOVA (Greenhouse and Geisser, 1959; Lana and Lubin, 1963).

- Equal sample sizes can mitigate the effect of unequal variances on Type I error rate of the Model F-test for the one-way fixed effects ANOVA. This beneficial effect of equal sample sizes has been shown mathematically for two groups by van der Vaart (1961) and since then many simulation studies have confirmed this effect of equal sample sizes. For example, Clinch and Keselman (1982) found that with equal group sample sizes (n=12 or n=36), Type I error rates were not seriously affected even for a skewed distribution, except for cases of extreme heterogeneity of variances. However, Box (1954) showed that with equal sample sizes in three or more groups and relatively small variance heterogeneity among the groups, Type I error rates were distorted: For example the actual Type I error rate observed was 0.07 for a nominal significance level of 0.05 in his simulation for five groups with variances in the ratio of 1: 1: 1: 1:3.

- When sample sizes are unequal, it has been found (Horsnell, 1953; Box, 1954; Glass et al., 1972; Harwell et al., 1992) that when **larger population variances** occur in the groups with **smaller sample sizes**, the one-way fixed effects ANOVA for detecting mean differences may have an inflated probability of Type I error, even when the normality assumption is met.

- When sample sizes are unequal, it has also been found (Horsnell, 1953; Box, 1954; Glass et al., 1972; Harwell et al., 1992) that when **larger population variances** occur in the groups with **larger sample sizes**, the one-way fixed effects ANOVA may be less powerful in detecting true differences among group population means as compared to other tests that adjust for heterogeneous variances.

- Misclassification of the categories of a factor can have a huge impact on the validity of a statistical analysis (Gustafson, 2004). Misclassification is especially a problem in a one-way ANOVA where typically more than two groups (i.e., categories of a factor) are considered, as it is difficult to determine whether a bias towards the null or alternative hypothesis will be caused by the classification errors.

- Departures from the normality assumption have little effect on an actual Type I error of a Model F-test in a one-way fixed effects ANOVA, provided the samples sizes are not too small. The early work of Pearson (1931) and the multivariate central limit theorem (see (Miller, 1998) provide a mathematical basis for the robustness of Type I errors to non-normality. This robustness improves not only with increasing sample sizes in the groups but also with an increasing number of groups in the analysis. These mathematical findings have been corroborated by numerous simulation studies (reviewed by Glass et al. (1972) and Harwell et al. (1992). However, it is important to note if the nature of departure from normality in any of the groups is such that a population mean is not the centre of location, then applying a Model F-test of mean equality can give a misleading substantive conclusion (e.g., in a left-skewed distribution the mean is less than the median and is not the centre of location).

- Simulations have shown that mild departures from normality have little effect on the power of a Model F-test in a one-way fixed effects ANOVA. However, moderate departures from normality may reduce the power of this F-test even when there are equal population variances (David and Johnson, 1951; Srivastava, 1959; Tiku, 1971; Harwell et al., 1992; Miller, 1998).

11.9   Overview of General Approaches for Evaluating Ideal Inference Conditions for a One-Way Fixed Effects ANOVA

General approaches for evaluating ideal inference conditions regarding the ϵ_{ij} in a one-way fixed effects ANOVA include:

* considering prior expert knowledge of the populations in the research area

* examining the distribution of internally studentized residuals pooled from the entire data set using methods for model checking described in Chapter 3

    Recall an internally studentized residual is a raw residual divided by its standard error where the raw residual is the sample analogue of ϵ_{ij}. In the means model for the one-way fixed effects ANOVA, the raw residual, e_{ij}, for individual j in group i is given by

$$e_{ij} = \text{observed } Y_{ij} - \text{predicted } Y_{ij}$$

    where predicted Y_{ij} is the sample mean of the response variable in the group to which individual j belongs.

    For example, consider the first individual in Group A in the task study in Table 11.1. The raw residual for this individual is 34 minus 38 = −4, where 34 is the observed Y value for this individual and 38 is the sample mean of Y in Group A. It may be noted that raw residuals are output by the PRINT procedure of Program 11.2 (Section 11.B) which applies a one-way ANOVA to the task study data. The following are results for the first individual in Group A from the PRINT procedure of Program 11.2, where the observed Y variable is labelled pulse20, the predicted Y is labelled as p, and the raw residual as r.

```
Obs        pulse20          p                r

 1           34          38.0000         -4.0000
```

* examining the distribution of the response variable observed in each group, provided the sample sizes are adequate, say greater than 20 in each group. When sample sizes are as small as 10 or 20 per group, very misleading conclusions may be reached by examining the distribution of the response variable in each group's sample (Muller and Fetterman, 2012).

    Based on the general approaches discussed in this section, we will provide a detailed description of a nine-step approach for evaluating ideal inference conditions for a one-way fixed effects ANOVA in Section11.16 with an illustrative example in Section 11.17 along with SAS programs that facilitate this evaluation in Sections 11.A and 11.B.

11.10   Suggestions for Alternative Approaches When Ideal Inference Conditions are Not Reasonable

Chapter 14 provides a general discussion of alternative approaches that may be employed when there is concern about violations of ideal inferences of the linear models presented in this book. Several procedures that have been specifically developed for testing equality of means in a one-way fixed effects ANOVA when variances are unequal are discussed in Section 11.19.

11.11   Testing Equality of Group Variances

In this section we consider methods for testing the equality of group variances hypothesis,

$$\text{H}_0\colon \sigma_1^2 = \dots, \sigma_g^2 \tag{11.7}$$

versus the alternative that the variances in the g populations are not all equal. Relying on an equality of variance test to evaluate the impact of violating the ideal inference condition of equality of variances for a one-way fixed effects ANOVA test of population means can be problematical as

- a test may not have sufficient power to detect population variance inequality which may distort the conclusions of a one-way fixed effects ANOVA test of means

- a test may indicate that population variance inequality exists but this inequality is such that it does not have an important impact on the conclusions of the one-way fixed effects ANOVA test of means

We would like to emphasize, however, an equality of variances test can be quite useful in those situations where the variability of an outcome variable is in itself of substantive interest. For example, in evolutionary biology, variation of traits is important as it is related to the strength of a selective force affecting those traits while, in population ecology, variation of traits can be used to study niche width (Nelson and Moodie, 1985; Moodie, 1985).

In medicine, much emphasis is placed on investigating average treatment effects with tests for equality of means. However, the tests for equality of variance described in this section can also be quite informative when comparing efficacy of treatment protocols. For example, Treatment A may have a better average outcome than Treatment B but if Treatment A has a more variable effect on the outcome than Treatment B, this should be taken into consideration when evaluating the relative merits of the two treatments. Other research areas where investigating variability (i.e., assessing uniformity and risk) is of substantive interest are stock market volatility, education methodologies, intervention programs in public health or crime prevention, as well as quality control studies in manufacturing, agriculture, and aquaculture.

11.12   Some Earlier Tests for Equality of Variances

As will be discussed in the following subsections, Bartlett's test and Hartley's $F_{\max}$ test are tests of variance equality that are no longer recommended.

11.12.1   Bartlett's Test

Bartlett's test (Bartlett, 1937), although still available in various statistical software packages, is not recommended for testing equality of population variances as this test has been found to be highly sensitive to departures from normality (Box, 1953; Brown and Forsythe, 1974a; Miller, 1968; Sharma and Kibria, 2013). Box (1953) has shown mathematically that for any sample size, Bartlett's test is biased unless the kurtosis of the parent population is exactly zero as is the case for a normal distribution. Thus, when Bartlett's test has a small p-value, it is impossible to tell whether the population variances are unequal or whether the kurtosis of the populations are not equal to zero, i.e., real differences in population variances are confounded with non-normality. For populations with leptokurtic (peaked) distributions, Bartlett's test will tend to suggest that there are differences in population variances where none exist, while for populations with platykurtic (flat) distributions, real differences in the population variances will be not be detected by Bartlett's test. The sensitivity of Bartlett's test to the kurtosis of the populations becomes progressively worse as the number of groups in the test increases.

Sharma and Kibria (2013) observed in their simulations of various non-normal distributions that Bartlett's test, at the 0.05 level of significance, had empirical Type I error rates that always exceeded 0.10. In one of their experiments, Bartlett's test, at the 0.05 level of significance, had an empirical Type I error rate as high as 0.95!

11.12.2   Hartley's F_{max} Test

Hartley's F_{max} test (Hartley, 1950) is not recommended as a test of variances because it is highly sensitive to slight departures from normality (Box, 1953; Brown and Forsythe, 1974a; Van Valen, 1978). The test statistic, which is calculated by computing the ratio of the largest sample variance to the smallest sample variance, is compared to an appropriate F_{max} critical value under the assumption of equal sample sizes.

11.13   ANOVA-Based Tests for Equality of Variances

In this section we will consider ANOVA-based tests for testing equality of population variances. An advantage of the ANOVA-based tests for equality of variances is that the null hypothesis of equal variances is tested by evaluating **mean** deviations from a measure of central location. It is well known that tests on means tend to be robust to moderate departures from normality because of the central limit theorem.

11.13.1   Levene's Approach for Testing Equality of Variances

Levene (1960) proposed testing the null hypothesis of equal variances by applying a one-way ANOVA to the absolute values of residuals from the means model (Equation 11.1). The absolute values of these residuals are given by $Z_{ij} = |Y_{ij} - \bar{Y}_{i.}|$, $i = 1, 2, \ldots, g$; $j = 1, \ldots, n_i$ where $n_i =$ the number of subjects in group i and $g =$ the number of groups in the analysis. For example, in the task study, the Z_{ij} for subject 1 who has a pulse rate of 34 and who is in the first group (Group A) is given by

$$\begin{aligned} Z_{11} &= |34 - \text{sample mean of Group A}| \\ &= |34 - 38| \\ &= 4 \end{aligned}$$

When the group sample sizes are equal this test evaluates the correct null hypothesis (Equation 11.7) for g groups.

However, when the group sample sizes are unequal, misleading conclusions regarding equality of group variances arise from this test as it evaluates a null hypothesis which is a function of the population variances and the sample sizes. Keyes and Levy (1997).

$$H_0: (1 - 1/n_1)\sigma_1^2 = \ldots, (1 - 1/n_t)\sigma_g^2 \tag{11.8}$$

Levene (1960) also suggested using the squared values rather than the absolute values of the residuals from the group sample means as an alternative spread variable in his ANOVA-based approach to test equality of group variances. The squared values of these residuals are given by $Z_{ij}^2 = (y_{ij} - \bar{y}_i)^2, i = 1, 2, \ldots, g; \ j = 1, \ldots, n_i$ where n_i = the number of subjects in group i and g= the number of groups in the analysis. The versions of Levene's test using either Z_{ij} or Z_{ij}^2 as spread variables are available as options in the MEANS statement in the GLM procedure (e.g. Program 11.2, Section 11.C).

11.13.2 Brown's and Forsythe's Tests of Variances

Brown and Forsythe (1974a) found in their simulations that Levene's test, using Z_{ij} as the spread variable, had elevated Type I error rates when the parent distributions were skewed. They therefore suggested two tests which were modifications of Levene's ANOVA-based approach. In their proposed tests they used a more robust estimate of the centre of location than the simple group mean in the calculation of the spread variable.

In one of their tests they used the absolute value of deviations from a group's **median** instead of a mean as a spread variable in a one-way ANOVA. For example, in the task study, the value of Brown's and Forsythe's spread variable, denoted here as Z_{med}, for the first subject whose pulse rate was 34 in Group A which had a median pulse rate of 37 is

$$Z_{\text{med}} = |34 - 37|$$
$$= 3$$

This test is available in the MEANS statement in the GLM procedure (e.g. Program 11.2, Section 11.C).

The other variance test proposed by Brown and Forsythe (1974a) involved applying a one-way ANOVA to a spread variable which consisted of absolute deviations from a 10% symmetrically trimmed group mean, which they defined as the mean of the observations after deleting 10% of the largest and 10% of the smallest values in the group. They stated that in their quest for a more robust estimate of a group's sample mean, their choice of 10% was arbitrary. We denote this spread variable as Z_{10}.

Based on their simulation research, Brown and Forsythe (1974a) recommended applying one-way ANOVA to Z_{med} to test equality of variances in skewed distributions. For long-tailed distributions they recommended testing variance equality by applying a one-way ANOVA to a spread variable which is based on deviations from a symmetrically trimmed mean (for example Z_{10}).

11.13.3 O'Brien's Test

O'Brien (1979) proposed using $r_{ij}(w)$ as a spread variable in the one-way fixed effects ANOVA as a modification to Levene's approach where

$$r_{ij}(w) = [(w + n_i - 2)n_i(Y_{ij} - \bar{Y}_{i.})^2 - ws_i^2(n_i - 1)]/[(n_i - 1)(n_i - 2)] \tag{11.9}$$

where n_i, $\bar{Y}_{i.}$, s_i^2 are the sample size, sample mean, and sample variance of the ith group and w is a weight parameter which adjusts O'Brien's test for the effect of kurtosis in the underlying parent population.

When w=0, $r_{ij}(0)$ is a slight modification of Levene's Z^2 variable for unequal n. When w=1, $r_{ij}(1)$ is equal to the jackknife spread variable used in Miller's test of equal variances (Miller, 1968), which has been shown to always be conservative so we will not consider it here. Although O'Brien (1979) provided a formula for calculating w which is a function of sample sizes and the kurtosis of the parent distribution, he stated that precise calculation of w would be impractical for most researchers, who typically do not have information about the kurtosis of the parent distributions. He therefore suggested that using $w=0.5$ in $r_{ij}(w)$ would provide a test that was a compromise between Levene's test, which often is too liberal, and Miller's test, which is always conservative. O'Brien's test is available in the MEANS statement in the GLM procedure (Section 11.C). By default SAS uses $w = 0.5$ as the weighting parameter for this test. However, you can set the weighting parameter to another value if you prefer using the (w=) option (Section 11.C).

11.14   Simulation Studies on ANOVA-Based Equality of Variances Tests

In an investigation of robust measures of variability, Keselman et al. (2008) compared 633 equality of variances tests based on modifications of procedures proposed by Levene (1960), Brown and Forsythe (1974a) and O'Brien (1981). Based on this research, Kesselman et al. (2008) proposed, as a general recommendation, an approach involving application of an ANOVA *F*- test to Levene's type measure of spread based upon empirically determined 20% trimmed asymmetric means.

Earlier simulation studies (Brown and Forsythe, 1974a; O'Brien, 1978; Algina et al., 1989) indicated that Brown and Forsythe's equality of variances test using the spread variable, Z_{med} is useful when avoidance of a Type I error is the more critical research objective rather than avoidance of a Type II error. These studies, investigating a wide variety of factors known to affect equality of variances tests, found that this test always met Bradley's criterion of a robust test. Bradley (1978) classified a test as robust if the test's observed Type I error rate in a simulation study falls within the interval (0.025, 0.075) when the test is carried out at the 0.05 level of significance. These studies also showed that Brown and Forsythe's equality of variances test at the 0.05 level of significance typically has an observed probability of Type I error rate that falls well below the upper limit of Bradley's criterion of 0.075 for an observed Type I error rate. A caveat regarding Brown and Forsythe's test using Z_{med} is that this test was shown to have low power to detect moderate differences in variances when any of the sample sizes are less than or equal to 10. Examples of this low power are:

- Brown and Forsythe (1974a) found that at the 5% level of significance their test using Z_{med} had a power of 0.23 of correctly rejecting the null hypothesis of equal variances in two groups where the ratio of the parent population variances was 4:1, the sample sizes, n1=n2=10 and the parent distributions were skewed (χ^2, 4 df).

- Their test had a power of 0.14 when the ratio of the parent population variances was 2:1, for the same conditions (i.e., skewed parent distribution and sample sizes, n1=n2=10).

- Even when one of the sample sizes was 20, i.e., n1=10 and n2=20 and variances ratios 4:1 with group 1 having the smallest variance, Brown and Forsythe found that the power of their test using Z_{med} was 0.40.

As with many statistical simulation studies, the research described in this section underscores the difficulty of recommending a single hypothesis test that is uniformly best for all situations.

11.15   Additional Comments about ANOVA-Based Equality of Variances Tests

We do not think it is necessary to apply an equality of variances test to evaluate the ideal inference condition of equal group variances, especially when the group sample sizes are equal or almost equal. However, provided the sample sizes are not too small (say greater than 10), the results of an ANOVA-based equality of variances test, reported to have reasonable Type I error rates and power, can provide useful substantive information to researchers about variability in the parent populations.

An advantage of ANOVA-based tests of equality of variances is that they can be readily extended to more complicated ANOVA models such as an analysis of covariance (Milliken and Johnson, 2002) and a multivariate ANOVA where the equality of variances of more than one response variable in several groups are considered (Nelson and Moodie, 1988). Simulations have shown these extensions of ANOVA-based tests of equality of variances performed well in terms of empirical Type I error rates and power (Nelson and Moodie, 1988; Milliken and Johnson, 2002).

11.16   Overview of a Step-by-Step Approach for Checking Ideal Inference Conditions

We suggest the following nine-step approach for checking ideal inference conditions for a one-way fixed effects ANOVA:

Step (1) Screen for any data input errors and correct any errors found.

Step (2) Screen the data for recording and measurement errors. A scatter plot, a box plot, and a listing of extreme values of the response variable for each group can reveal unusual cases which should be investigated for errors.

Step (3) Confirm that the response variable is continuous or if quantitative discrete, that it has more than a few values (as previously discussed for Step (3) Section 3.13): otherwise consider a categorical data analysis method.

Step (4) Review the details of the sampling design for the study to see if there may be any obvious aspects of the design that would suggest that a more complicated model is required.

Step (5) Apply a one-way ANOVA to the data to obtain model checking diagnostics.

Step (6) Evaluate the ideal inference condition of equal variances of the ϵ_{ij} (the model errors) in the parent populations of the groups by informally comparing the groups' scatter plots of internally studentized residuals obtained in Step (5). If sample sizes are adequate, say greater than 20 in each group, you can also informally compare the sizes of the sample variances of the response variable in each group. Depending on the group sample sizes or your philosophy towards hypothesis tests, or both, you may choose to formally test this ideal inference condition by applying an ANOVA-based test for equality of variances (subsection 11.13).

Note regarding omitting Step (7): You can proceed directly to Step (8), omitting Step (7) which further investigates independence of errors, if your study was a well designed randomized controlled experiment without any withdrawals from the "treatment" groups (as discussed in "Note regarding omitting Step (7)" in Section 3.13).

Step (7) Use graphical displays of internally studentized residuals obtained from Step (5) to informally evaluate whether the ideal inference condition of independence of errors has been violated due to the model errors being systematically related to the group categories in the model or to any other variable not included in the model (as previously discussed for Step (7) in Section 3.13). You can informally evaluate independence of errors in these situations by examining scatter plots of internally studentized residuals against any time, space, or other key variable related to the sampling and measurement details of the study. If the internally studentized residuals appear to be exhibiting a systematic pattern in any of these plots (subsection 3.9.1), you should stop checking the appropriateness of the one-way fixed effects ANOVA model at this step and consider other approaches discussed in Chapter 14.

Step (8) Check if there is evidence of serious departure from normality in each group's parent population. Prior expert knowledge of the research area may provide insight when evaluating this ideal inference condition. A normal Q-Q plot of internally studentized residuals from the entire data set can be helpful in detecting serious non-normality (Section 3.9.3). If the sample size in each group is reasonably large (say greater than 20) in a one-way fixed effects ANOVA, the distribution of the response variable in each group's sample can be examined informally to check for serious non-normality via:

- a box plot of the response values in each group
- a normal QQ plot of the response values in each group
- a comparison of the sample median and mean response value in each group to see how close they are in value (Recall the population median and mean have the same value in a normally distributed population.)
- noting the values of the sample skewness and kurtosis statistics in each group, with the realization that these statistics are sensitive to sampling variation, so the values of skewness and kurtosis observed in the samples may not necessarily precisely inform us about the shape of the distributions in the populations

Step (9) Identify and investigate model outliers using studentized deleted residuals obtained from Step (5), provided ideal inference conditions of the one-way fixed effects ANOVA were found to be reasonable in Steps (5) to (8) and hence studentized deleted residuals are appropriate for use in outlier detection.

11.17 Nine-Step Approach for Evaluating Ideal Inference Conditions: Task Study Example

Steps (1) and (2) Screen and correct any data input errors as well as recording and measurement errors. No such errors were found using the PRINT and UNIVARIATE procedures in Program 11.1 (Section 11.A).

Step (3) Confirm that the response variable is continuous or if quantitative discrete, that it has more than a few values. The response variable, pulse rate in 20s is continuous.

Step (4) Review the details of the sampling design for the study. There were no obvious aspects of the sampling design that would suggest that a model with a more complicated error structure was necessary. The subjects were drawn at random from a single population of volunteers. There were no repeated measures on any of the individuals.

Note: Steps (5)–(9) were facilitated by applying Program 11.2 (Section 11.B to the task study data.

Step (5) Apply a one-way ANOVA to the data to obtain model checking diagnostics. A one-way ANOVA test of means was applied to the data using the GLM procedure.

Step (6) Evaluate the ideal inference condition of equal variances of the ϵ_{ij} (the model errors) in the parent populations of the groups. Figure 11.1, the graph of the internally studentized residuals for each group, combined with the fact that sample sizes

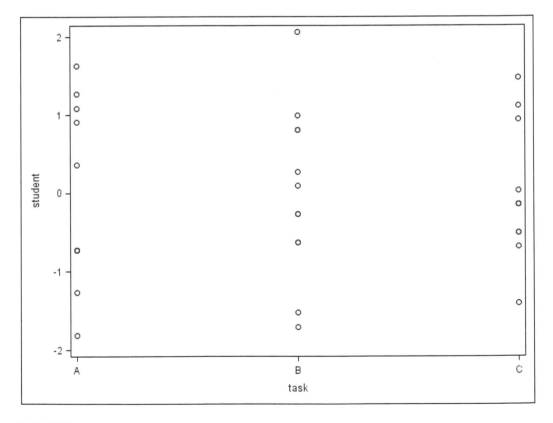

FIGURE 11.1
A scatter plot of internally studentized residuals from one-way fixed effects ANOVA of task example data.

are almost equal, suggests that violation of the equal variances ideal inference condition may not be a serious problem for this example. The following results from Brown and Forsythe's test of equal variances in Program 11.2 suggest there is not sufficient evidence to reject the null hypothesis of equal population variances (p-value 0.2397). However, given the small group sample sizes in this example, there may not be sufficient power to detect differences in the parent population variances.

```
                        The GLM Procedure

      Brown and Forsythe's Test for Homogeneity of pulse20 Variance
             ANOVA of Absolute Deviations from Group Medians

                          Sum of        Mean
       Source      DF     Squares       Square     F Value    Pr > F

       task         2     31.2485      15.6242       1.50     0.2397
       Error       30       312.8      10.4271
```

Step (7) Although the study was carried out as a randomized experiment, there is always concern about potential violation of independence of errors when individuals withdraw from their "treatment" groups (the tasks in this example). The hope is that the individuals are withdrawing for some reason unrelated to the "treatment" itself (e.g., a family member got sick, moving to a different city). In this example we do not have information regarding the reasons for withdrawal from the "treatment" groups or data on the baseline characteristics of the subjects (i.e., characteristics of the individuals measured prior to assignment to the "treatment" groups) that might affect pulse rates, so we cannot evaluate whether the withdrawals resulted in violation of independence of errors.

Step (8) Check if there is evidence of serious departure from the normality ideal inference condition. Neither the QQ plot of the internally studentized residuals from the one-way ANOVA model (Figure 11.2) nor the values of the sample shape statistics given in the inset box of this figure indicate that there is a serious departure from a normal distribution.

Step (9) Identify and investigate model outliers using studentized deleted residuals obtained from Step (5), provided ideal inference conditions of the one-way fixed effects ANOVA were found to be reasonable. As ideal inference conditions evaluated in previous steps appear to be reasonable, studentized deleted residuals were used to identify presence of model outliers in Program 11.2. Only one case, Record 21, was found to have a studentized deleted residual greater than 2. However its studentized deleted residual, 2.19, was just slightly greater than 2 and upon investigation nothing was found noteworthy about this case.

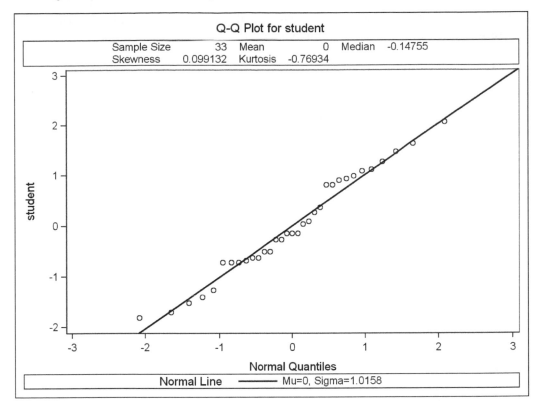

FIGURE 11.2
A normal QQ plot of internally studentized residuals from one-way fixed effects ANOVA of task example data.

11.18 One-Way ANOVA Model F-Test of Means

11.18.1 Task Study Example

Source	DF	Sum of Squares	Mean Square	F Value	Pr > F
Model	2	545.606061	272.803030	8.07	0.0016
Error	30	1014.636364	33.821212		
Corrected Total	32	1560.242424			

The preceding ANOVA table shows that the observed p-value for the ANOVA Model F-test of the null hypothesis of mean equality for the task example is less than 0.0016. This small p-value suggests that it is unlikely to observe an F value of 8.07 if the null hypothesis that the group population means are all equal is true. However, this small observed p-value for the Model F-test does not tell us whether the specific comparisons of the population means in which we are interested are different. This small p-value suggests only that there is **at least one** *linear contrast* of the group population means that may be different from zero.

11.18.2   What is a Linear Contrast?

Definition 11.1. *A linear contrast* (often simply called a contrast) in the context of a one-way fixed effects ANOVA is a linear combination of two or more group means with the restriction that the coefficients for the means sum to zero.

Let L denote a contrast in a fixed one-way fixed effects ANOVA

$$L = c_1\mu_1 + c_2\mu_2 + \ldots + c_g\mu_g \tag{11.10}$$

where $c_1, c_2, \ldots, c_g$, the coefficients of the group population means, sum to zero.

A null hypothesis about group population means in a one-way fixed effects ANOVA can be expressed via a linear contrast. For example, in the task study where there are three groups (A, B, and C), the null hypothesis that the population mean of Group A is equal to the population mean of Group B is evaluated by testing whether the following linear contrast[2] is equal to zero:

$$L = 1\mu_A - 1\mu_B + 0\mu_C$$

There are an infinite number of possibilities for linear contrasts of group means as the coefficients of a linear contrast may be positive, negative, an integer, or a fraction. A small p-value from the Model F-test suggests only that at least one out of an infinite number of possible linear contrasts of the group population means in the ANOVA may be different from zero. In Chapter 12 we will describe multiple comparison tests that can be used to evaluate specific hypotheses about group means of interest to researchers (e.g., whether the population mean of Group A is the same as the population mean of Group B). For information about how meaningful null hypotheses can be specifically tested using contrasts or any linear combination of the mean parameters, see Chapter 1, Milliken and Johnson (2009).

11.18.3   Comments on Model F-Test in One-Way Fixed Effects ANOVA

Although the Model F-test applied to your data may not have a low p-value, you may nevertheless find that a particular contrast of group means does have a low p-value when you apply a multiple comparison procedure. This is because multiple comparison procedures are often more powerful than the Model F-test for a one-way fixed effects ANOVA. For example, a multiple comparison procedure testing the equality of populations means in Groups A and B may have a p-value less than 0.05, even though the Model F-test from the one-way fixed effects ANOVA may have a p-value greater than 0.05.

You may ask whether you should bother applying an overall Model F-test in a one-way fixed effects ANOVA, if you are interested in testing specific comparisons of group means using multiple comparisons procedures described in Chapter 12. The answer is that it is typically not necessary to apply an overall Model F-test and that multiple comparison procedures can be used instead when you want to test specific comparisons of the group means. However, as the ideal inference conditions for some multiple comparison procedures are identical to those for the Model F-test, it is helpful to apply a one-way fixed effects ANOVA model to the data in SAS to obtain model residuals which can be informative for evaluation of these ideal inference conditions.

---

[2]Note c_1, c_2, c_3, the coefficients for the population means, A, B, and C (μ_A, μ_B, and μ_C) are respectively 1, -1, 0 thus satisfying the definition of a contrast that the coefficients sum to zero.

If you are interested in the general question, does the factor in the one-way ANOVA have any effect at all on the response variable, then a Model F-test can be considered as a method to shed light on this question. Moreover, in the ANOVA-based tests of variances (Section 11.13), a Model F-test with a low p-value can indicate that the population variances are unequal, which can be of substantive interest to researchers. Thus, the Model F-test can be useful as a research tool.

11.19   Testing Equality of Population Means When Population Variances Are Unequal

11.19.1   Welch's Test

Welch's test (Welch, 1951), sometimes referred to as Welch's ANOVA, is a widely used test of the equality of population means for the situation where population variances are unequal.

Welch's test is a modification of the F-test for the one-way ANOVA. Recall from Section 11.6, that, in the F-test, the sample variances are all added together and averaged (i.e., pooled) to estimate a common population variance, assuming the population variances are all equal. However, in Welch's test, the sample variances of the groups are not assumed to be from populations with equal variances, and so are not pooled. Instead, in Welch's test, a weighting factor for each group is used to adjust the F-test for unequal population variances.

Welch's test was developed under the assumption that the response variable is normally distributed in the parent population of each group. The weighting factor for the ith group, w_i, that Welch uses in his test, i.e.,

$$w_i = n_i/s_i^2 \tag{11.11}$$

is appropriate for a normally distributed population where the size of the mean and variance are independent of each other. Welch's test is not recommended for non-normal distributions where the size of the mean and the variance are dependent. In these situations, Welch's weighting factor not only adjusts for unequal variances but also ends up adjusting for unequal means because the size of the means is dependent on the variances. The take away message is to avoid Welch's test when it appears that the mean and variance are related in any of the parent population distributions.

Welch's test is available as an option on the MEANS statement in the GLM procedure. If we would like to apply Welch's ANOVA to the data stored in the temporary SAS dataset TASKDTA, we simply have to specify the keyword welch to the MEANS statement of the GLM procedure, as follows:

```
proc glm data=taskdta;
   class task;
   model pulse20=task;
   means task / welch;
quit;
```

Equation for Welch's Test Statistic

Welch's test statistic, which we denote as F_{Welch} tests the null hypothesis of mean equality in g groups, i.e., H_0: $\mu_1 = \mu_2 = \ldots, \mu_g$ vs. H_a (not H_0) and is given by

$$F_{\text{Welch}} = \frac{\sum_{i=1}^{g} w_i \frac{(\bar{Y}_{i.} - \bar{Y}^*)}{(g-1)}}{1 + 2(g-1)\Lambda/(g^2-1)} \tag{11.12}$$

where

> g = the total number of groups in the analysis
> w_i, the weighting factor for Group i, is equal to n_i/s_i^2
> n_i and s_i^2 are respectively the sample size and sample variance for Group i
> $\bar{Y}_{i.}$ = the sample mean of the response variable for Group i
> $\bar{Y}^*$ is the weighted average of the sample means where the weights are the w_i

i.e., $\bar{Y}^* = \sum_{i=1}^{g} w_i \bar{Y}_{i.} / \sum_{i=1}^{g} w_i$

$\Lambda = \sum_{i=1}^{g} \frac{1 - w_i/w_.)^2}{n_i - 1}$

> where $w_. = \sum_{i=1}^{g} w_i$.

Welch's test statistic, F_{Welch}, has an approximate F distribution with numerator and denominator degrees of freedom, $v_1 = g - 1$, $v_2 = (g^2 - 1)/3\Lambda$, respectively. In Welch's test the null hypothesis H_0: $\mu_1 = \mu_2 = \ldots, \mu_g$ is considered unlikely if $F_{\text{Welch}} > F_{\alpha, v_1, v_2}$.

11.19.2   Fitting Unequal Variance ANOVA Models

Fitting an unequal variance ANOVA model is an approach that can be considered when heterogeneity of population variances may be a problem in testing mean equality. Such an approach is beyond the scope of this book. For those interested in the underlying mathematical details and an illustrative SAS program for fitting an unequal variances one-way ANOVA model, see pages 36–38, Milliken and Johnson (2009).

11.19.3   Other Suggestions

Section 14.6 provides a discussion of general approaches that can be considered when violation of the equal variances ideal inference condition may be a problem.

11.20   Chapter Summary

This chapter:

- describes the characteristic features of a one-way analysis of variance (ANOVA) and explains why you would want to use this procedure

- outlines the differences between a fixed effects and random effects one-way ANOVA

- discusses ideal inference conditions for a one-way fixed effects ANOVA, potential consequences when these conditions are not met, as well as suggestions for alternative approaches

- describes tests for equality of population variances and discusses when such tests can be useful for researchers

- suggests a nine-step approach for evaluating ideal inference conditions in a one-way fixed effects ANOVA and illustrates this approach with an example

- provides SAS programs which implement the suggested nine-step approach

- discusses the results of the one-way fixed effects ANOVA table for the illustrative example

- describes Welch's test of mean equality for situations in which the population variances are unequal and explains why Welch's test is not recommended when population variances are related to population means

- suggests other approaches for testing mean equality that could be also considered when population variances are unequal

Appendix

11.A Program 11.1: Data Screening

The statements in this program have been explained in previous data screening programs. The input data are a subset from Milliken and Johnson (2009) and are stored by Program 11.1 in the temporary SAS data set arbitrarily named TASKDTA.

```
data taskdta;
input record task $ pulse20;
datalines;
36 A 34
37 A 34
38 A 43
39 A 44
40 A 40
41 A 47
42 A 34
43 A 31
44 A 45
45 A 28
46 B 28
47 B 28
48 B 26
49 B 35
50 B 31
51 B 30
52 B 34
53 B 34
54 B 26
55 B 20
56 B 41
57 B 21
58 C 28
59 C 26
60 C 29
61 C 25
62 C 35
63 C 34
64 C 37
65 C 28
66 C 21
67 C 28
68 C 26
```

```
;
run;
*STEP 1;
proc print;
   title Pulse rate data;
run;
*STEP 2;
/*The following plot can be examined
to check for outliers*/
ods listing style=journal2 image_dpi=200;
proc sgplot data=taskdta;
   title Pulse rate data;
   scatter x=task y=pulse20;
run;
```

11.B Program 11.2

Program 11.2 provides output associated with Steps (5) to (9) of our suggested approach for checking ideal inference conditions in a one-way fixed effects ANOVA test of means (Section 11.16). This program uses the **temporary** SAS data set TASKDTA created by Program 11.1.

Please note:
You will get error messages if you try to execute Program 11.2 after having logged out from the SAS session in which you created the temporary SAS data set TASKDTA. To avoid this, you could create the SAS data set TASKDTA again by executing Program 11.1 in your current SAS session or alternatively by inserting the DATA step from Program 11.1 at the beginning of Program 11.2.

```
*Program 11.2;
/*STEP 5  Apply a One-way ANOVA to obtain model diagnostics*/
/*STEP 6  Evaluate Equality of Variances
with a Homogeneity of Variance Test*/
proc glm data=taskdta;
   /*default less than full rank parameterization*/
   class task;
   model pulse20=task;
   output out=diag r=r p=p rstudent=rstudent student=student;
   means task /hovtest=bf;
run;
proc print data=diag;
   var  record pulse20 p r student rstudent;
run;
/*Examine Distribution of Internally
Studentized Residuals in each Group
for Step 6 to informally evaluate Equality of Variances
for Step 7 to check Independence of Errors in an Observational Study */
```

```
proc sgscatter data=diag;
   plot student*task;
run;
*STEP 7;
/*Examine the distribution of internally
studentized residuals to see
if any serious departures from normality are evident */
proc univariate data = diag  normal;
   title Pulse rate data;
   ods select ExtremeValues Frequencies;
   ods exclude TestsForLocation;
   var  student ;
   qqplot student /normal (mu=est sigma=est);
   inset n='Sample Size' mean median skewness kurtosis  /pos=tm ;
run;
* STEP 9;
/* Identify model outliers */
proc print data=diag;
   where abs(rstudent) gt 2.;
run;
```

11.C   Explanation of Statements in Program 11.2

The statement **proc glm data=taskdta;** requests that the GLM procedure be applied to the temporary SAS data set called TASKDTA.

The **CLASS** statement, i.e., **class task;** specifies that the variable named "task" be treated as a qualitative (classification) predictor in the ANOVA model. When you use the CLASS statement, it must appear before the MODEL statement. It is very important to remember to include the CLASS statement when applying an ANOVA to test group means. If you do not include the CLASS statement, the GLM procedure will treat your predictor variable as a quantitative predictor variable in a regression.

The statement **model pulse20=task;** specifies the response variable and the predictor variable for the model in the analysis according to the following general format for the model statement,

MODEL response variable(s) = predictor variable(s);

The statement **output out=diag rstudent=rstudent student=student;** specifies that rstudent (the studentized deleted residual) and student (the internally studentized residual) for each case should be saved in a new temporary SAS data set called DIAG along with the values of variables for the records that were in the original data set TASKDTA that was input in the GLM procedure. We will need this data set DIAG in subsequent procedures in the program. The name a user specifies after the = sign for each of the quantities, i.e., rstudent=rstudent, student=student, are the names these quantities will be given in the new data set DIAG. To keep things simple we specified the same names as the keywords.

The statement **means task /hovtest=bf;** is an example of the MEANS statement in the GLM procedure which has the following general format, **MEANS effects / options ;**.

In a one-way ANOVA there is just one effect (i.e., just one factor or classification variable). In our example the effect or classification variable is called "task".

You can request a test of equal variances (homogeneity of variances) by specifying the keyword **hovtest** as an option on the MEANS statement, as illustrated in the statement **means task /hovtest=bf.**

The keywords for specifying the variance tests are:

- **levene (abs)** for Levene's test using the spread variable, z_{ij} (the absolute value of the deviations from a group mean) as described in subsection 11.13.1

- **levene** for Levene's test using the spread variable, z_{ij}^2 (the square of the deviations from a group mean) also described in subsection 11.13.1

- **bf** for Brown and Forsythe's test using the spread variable Z_{med}, described in subsection 11.13.2

- **obrien** for O'Brien's test using w=.5 as a default weighting parameter (subsection 11.13.3). To specify an alternative weighting parameter for O'Brien's test, you simply use the (w=) option. For example, if you wanted a weighting parameter of .6 you would specify hovtest=obrien (w=.6)

More than one variance test can be requested on the MEANS statement, for example,

means task /hovtest=bf hovtest=levene (type=abs) hovtest=obrien;

However, the two different Levene's test cannot be requested at the same time.

Comment

If you want a Model F-test of the null hypothesis H_0: $\mu_1 = \mu_2 = \ldots = \mu_g$ in the GLM procedure, it is very important **to not specify the NOINT option** on the MODEL statement. The NOINT option in GLM will result in a Model F-test of the null hypothesis that all the group means are equal to zero, viz, H_0: $\mu_1 = \mu_2 = \ldots = \mu_g = 0$, which is rarely a useful hypothesis to be testing. The reason this null hypothesis is tested by specifying the NOINT option is related to the default parametrization of linear models by the GLM procedure and is beyond the scope of this book.

12

Multiple Testing and Simultaneous Confidence Intervals

12.1   Multiple Testing and the Multiplicity Problem

Whenever we carry out more than one hypothesis test, the probability of making at least one Type I error (rejecting a null hypothesis when it is actually true) increases as the number of tests increases. This is referred to as a multiplicity (or a multiple comparison) problem. For example, if we performed three tests using a 0.05 significance level for each test, then the overall probability of rejecting the null hypothesis falsely in one or more of these three tests could be as high as 0.14. If we performed 20 tests of 0.05 using a 0.05 significance level for each test, then the overall probability of falsely rejecting the null hypothesis in one or more of these 20 tests could be as high as 0.64. This problem of increased probability of making a Type I error as the number of tests increases should not discourage researchers from investigating multiple hypotheses in a study. Gleaning as much valid information as possible from a study is usually an efficient use of research resources. However, when many hypothesis tests are applied, the concern is always whether a low p-value associated with any of the tests is actually due to a real effect or merely due to chance because of the large number of tests performed.

Many procedures have been developed which address this concern by reducing the overall Type I error rate over a set of multiple tests. Such procedures are known as **multiplicity-adjusted procedures**. However, a multiplicity-adjusted procedure, in reducing the Type I error rate over a set of tests, typically concomitantly decreases the power of each single test in the set, thereby increasing the probability of Type II error (i.e., increasing the probability of not detecting real effects). The challenge in developing these procedures is to achieve a reasonable balance between the probabilities of Type I and Type II errors.

We will discuss the multiplicity problem in the context of constructing multiple confidence intervals in Section 12.11. We discuss the multiplicity problem in the context of hypothesis testing first because researchers may have already been introduced to the multiplicity problem via hypothesis tests and we would like to build on this knowledge. By considering multiple hypothesis testing first we by no means want to imply that hypothesis tests are more important than confidence intervals with respect to the information conveyed. As we have emphasized throughout this book, confidence intervals always provide more information about the findings than hypothesis tests.

12.2   Measures of Error Rates

To understand multiplicity-adjusted procedures we need to distinguish the various measures of error rates defined in this section:

DOI: 10.1201/9780429107368-12

- **Definition 12.1.** A *comparisonwise (per comparison) error rate*, denoted as **CWER**, is the probability of making a false inferential claim when a single inference procedure is applied. Comparisonwise error rate is also abbreviated as **CER** in the literature.

In the context of hypothesis testing, the comparisonwise error rate is the probability of making a Type I error for a single test, (i.e., the probability in a single test of making the false claim that a null hypothesis is not true when it actually is true). Generally, a p-value for a test reported in the literature is the comparisonwise error rate for that single test, unless the authors specifically state that a multiplicity-adjusted procedure had been applied to adjust for multiple testing.

- **Definition 12.2.** A *familywise error rate*, denoted as **FWER**, is the probability of making one or more false inferential claims when a set of inference procedures are applied. Familywise error rate is also abbreviated as **FER** or as **FWR** in the literature.

In the context of multiple testing, the familywise error rate is the probability of making one or more false rejections of the null hypothesis (Type I errors) in the set of hypothesis tests under consideration.

The set of inference procedures under consideration is called a family because the set is comprised by the procedures which are typically all related to a particular overarching concept that the researcher is evaluating. For example, in the task study (Section 11.4), the family of hypothesis tests of interest to the researchers consisted of all pairwise comparisons between the mean pulse rate of the three task groups (i.e., there were three hypothesis tests in the family comparing the mean pulse rates of Task A vs. Task B, Task A vs. Task C, and Task B vs. Task C).

- **Definition 12.3.** A *false discovery rate*, denoted as **FDR**, is the **expected false discovery proportion**, where the **FDP** is the proportion of false rejections of null hypotheses out of the total number of rejected null hypotheses in a family of hypothesis tests, i.e.,

$$\text{FDR} = \text{expected}[\text{FDP}] \tag{12.1}$$

$$\text{FDP} = \frac{\text{number of false rejections of the null hypotheses in a set of tests}}{\text{total number of rejections of the null hypotheses in the set of tests}} \tag{12.2}$$

The term false discovery rate was coined because a false rejection of a null hypothesis would result in a "false discovery".

12.3   Overview of Multiple Testing Procedures

Multiple testing procedures can be classified as follows:

- **Multiplicity-unadjusted procedures** are those procedures which do not adjust for an increased probability of Type I error rate over multiple testing. A p-value from such procedures is called a **raw p-value or an unadjusted p-value**. An unadjusted or raw p-value is the observed probability of falsely rejecting the null hypothesis for a single test without adjusting for the number of other related hypothesis tests the researcher is making when investigating a particular concept.

When an investigation involving multiple testing is exploratory (hypothesis-generating) rather than confirmatory (hypothesis-validating), it has been argued that it is acceptable to use procedures that do not directly adjust for increased overall Type I error due to multiple testing (Saville, 1990; van Belle et al., 2004; Bender and Lange, 2001; Westfall et al., 2011)

However, if you decide to carry out many hypothesis tests without adjustment for multiplicity in an exploratory study, we recommend that you report how many hypothesis tests you applied and the p-value associated with each test. This allows your readers the possibility of adjusting each p-value for multiplicity (e.g., using Equation 12.5 to compute a Bonferroni adjusted p-value). We also recommend when reporting results which have not been adjusted for multiplicity that you explicitly state your study was exploratory in nature and that other studies are needed to confirm the conclusions in your study because the p-values reported are unadjusted for multiple testing.

- **Familywise error rate (FWER) controlling procedures** are those procedures in which the primary focus is on controlling the familywise error rate (Definition 12.2, Section 12.2). Such procedures are used when it is extremely important not to make even a single Type I error (false rejection of a null hypothesis) in a set of tests. For example, FWER controlling procedures are used for analysis of primary outcomes in confirmatory clinical trials where strong control of the familywise error is mandated by regulatory agencies (Dmitrienko et al., 2010). FWER controlling procedures are also used in hypothesis-generating investigations where researchers do not want to get any false leads (due to false rejections of null hypotheses in their set of multiple tests) as follow-up investigations from false leads may have ethical implications and/or may be costly in terms of effort and resources.

- **False discovery rate (FDR) controlling procedures** are those procedures that focus on controlling the FDR (Definition 12.3, Section 12.2). These procedures are potentially more powerful than familywise error rate controlling procedures. This is because FDR-controlling procedures use a less stringent criterion for overall Type I error rate control. FDR-controlling procedures allow a **specified proportion** of rejected null hypotheses to be falsely rejected **on average** in their control of overall Type I error rate in a set of tests as compared to familywise error rate controlling procedures which control the probability of **no** false rejections in a set of tests.

FDR-controlling procedures are often used to control Type I error rate over multiple tests in studies where the number of hypothesis tests in a family is quite large (sometimes even in the millions) and where familywise error rate controlling procedures would typically have low power due to their controlling a more stringent measure of Type I error rate. Investigations in microarray gene expression, neuro-imaging, astrophysics, and clinical safety are examples of studies where FDR-controlling procedures are often applied.

Benjamini and Hochberg (1995), who first introduced FDR-controlling procedures, also cited screening studies as a type of study in which controlling the FDR is relevant, and controlling the familywise error rate is not necessarily needed. As an example, they described a study involving screening many chemicals for drug development. In such a study, researchers would want to obtain as many possible discoveries (candidates for potential drug development) so they are willing to accept a proportion of false discoveries (falsely rejected null hypotheses). However, they would not want to use a procedure that has no adjustment at all for multiple testing. They would want to have some control on the number of falsely rejected null hypotheses to avoid the cost incurred in following up and investigating too many false leads as potential candidates for drug development in the confirmatory phase of their research. Given the study's objective, a FDR-controlling procedure would provide an appropriate compromise between multiplicity-unadjusted procedures (in which no multiplicity adjustment for Type I error is made) and familywise error rate controlling procedures (in which adjustment for Type I error rate is too stringent).

12.4 The Least Significant Difference: A Multiplicity-Unadjusted Procedure

12.4.1 Introduction

The least significant difference (LSD) procedure does not adjust for the increased probability of Type I error rate which results from multiple testing and therefore is not recommended for hypothesis-confirming studies. This procedure is often used to test all pairwise comparisons in a set of group means.

The LSD procedure for a two-tailed test of population means, μ_i, μ_j, involves computing a least significance difference using the following equation:

$$\text{LSD}_{ij} = t_{\alpha/2,v} \sqrt{\hat{\sigma}^2(1/n_i + 1/n_j)} \tag{12.3}$$

where

t_α is the critical value from the Student's t-distribution corresponding to the specified comparisonwise level of significance for a two-tailed test with v degrees of freedom, where v is the number of degrees of freedom associated with $\hat{\sigma}^2$ (e.g. $v = N - g$ for a one-way ANOVA where N is the total sample size, i.e., $N = n_1 + n_2 + \ldots + n_g$, and g is the total number of groups

$\hat{\sigma}^2$ is an estimate of the common within group variance of Y, e.g., the Mean Square Error from the one-way ANOVA, in the task study under the assumption that the ideal inferences conditions of independence of error and equal group population variances hold

n_i and n_j are the sample sizes of groups i and j, respectively

If the observed absolute difference in a pair of group sample means exceeds this least significance difference, i.e.,

$$|\bar{Y}_{i.} - \bar{Y}_{j.}| > \text{LSD}_{ij}$$

then the null hypothesis of the equality of the population means of Group i and j is rejected at the specified comparisonwise level of significance for this test. We provide an example of implementing this procedure in SAS in subsection 12.4.4.

12.4.2 Ideal Inference Conditions

The ideal inference conditions for the LSD procedure applied to pairwise comparisons of group means considered in a single factor fixed effects ANOVA are identical to the ideal inference conditions for the Model F-test of this ANOVA (Section 11.7). Having established in Section 11.17 that these ideal inference conditions are reasonable for the task study, we will use this study as an illustrative example for the LSD procedure.

12.4.3 Fisher's LSD

Fisher (1935) recognized the problem of increased probability of Type I error rate over multiple tests of group means associated with the LSD procedure. He suggested that this multiplicity problem could be mitigated, if the LSD procedure were used **only if** the Model F-test from the one-way ANOVA of the group means being considered in the multiple tests was significant at the same level of significance desired for each single hypothesis test, usually 0.05. This suggestion became known as **Fisher's LSD or Fisher's least significant difference procedure or Fisher's protected LSD**.

It can be shown (Meier, 2006; Westfall et al., 2011) that Fisher's LSD is guaranteed to maintain the familywise error rate at the specified desired level for all pairwise comparisons of group means, **if and only if there are just three groups in the family**. When there are four or more groups, Fisher's LSD is not guaranteed to maintain the familywise error rate for pairwise comparisons of the group means at the specified level and therefore is not recommended as a multiple testing procedure (Milliken and Johnson, 2009; Westfall et al., 2011).

12.4.4 SAS and the LSD Procedure

One way the LSD procedure can be obtained in the GLM procedure is by specifying the keyword T as an option on the LSMEANS statement. The following code for the task study provides an example.

Please note: The following code specifies the **temporary** SAS data set TASKDTA as the input data set. You will get error messages, if you try to execute this code after having logged out from the SAS session in which you created the temporary SAS data set TASKDTA. To avoid this, you could create the SAS data set TASKDTA again by executing Program 11.1 in your current SAS session or alternatively by inserting the DATA step from Program 11.1 at the beginning of this code.

```
proc glm data=taskdta;
   class task;
   model pulse20=task;
   lsmeans task /adjust=t;
quit;
```

The following results for the task example were generated by the LSMEANS statement in the preceding SAS code.

```
                 The GLM Procedure
              Least Squares Means

                      pulse20        LSMEAN
          task        LSMEAN         Number

          A        38.0000000          1
          B        29.5000000          2
          C        28.8181818          3


       Least Squares Means for effect task
       Pr > |t| for H0: LSMean(i)=LSMean(j)
```

Dependent Variable: pulse20

i/j	1	2	3
1		0.0019	0.0011
2	0.0019		0.7807
3	0.0011	0.7807	

The preceding results for this example are actually results from Fisher's LSD procedure (subsection 12.4.3) because the test statistic from the overall Model F-test from one-way ANOVA for the task study has an observed p-value of 0.0016 (as reported in Section 11.18). In this task study example the familywise error rate over all pairwise comparisons of the group means is no greater than 0.05 because, as discussed in subsection 12.4.3, there are just three groups in the family and the Model F-test from the corresponding one-way ANOVA has a p-value no greater than 0.05.

These preceding results from Fisher's LSD procedure are summarized as follows:

- 0.0019 is the p-value for the test of the null hypothesis that the absolute difference in population mean pulse rate between Group A and Group B is equal to zero. As ideal inference conditions of the procedure seem reasonable for the task study, this small p-value suggests that the null hypothesis of zero difference in population means of Groups A and B is unlikely. Therefore based on the Fisher's LSD procedure, we would conclude that the population mean pulse rates in Groups A and B are not equal.

- 0.0011 is the p-value for the test of the null hypothesis that the absolute difference in population mean pulse rate between Group A and Group C is equal to zero. This small p-value from Fisher's LSD procedure suggests that the population mean pulse rates in Groups A and C are not equal.

- 0.7807 is the p-value for the test of the null hypothesis that the absolute difference in population mean pulse rate between Group B and Group C is equal to zero. This large p-value suggests there is insufficient evidence from Fisher's LSD procedure to conclude that the population mean pulse rates in Groups B and C are not equal

12.5 Examples of Familywise Error Rate Controlling Procedures

As discussed in Section 12.3, familywise error rate (FWER) controlling procedures are typically used in situations where researchers want to restrict the maximum probability of making one or more Type I errors in a set of tests. In the sections that follow we will discuss some of the widely used familywise error rate controlling procedures:

- The Bonferroni method – a general procedure for adjusting raw p-values from various multiple testing situations (Section 12.6)

- The Tukey-Kramer method for all pairwise comparisons of group means (Section 12.7)

- Dunnett's method for all "treatments" vs. a "control" comparisons (Section 12.9)

- Scheffé's method for "data snooping" (Section 12.10)

- The Holm-Bonferroni Step-down method – a general sequential procedure for adjusting raw p-values from various multiple testing situations (Section 12.15)

We will also discuss the relatively recent simulation-based parametric resampling method (Westfall et al., 2011) which can be used to control familywise error rate for all pairwise comparisons of group means when sample sizes are unequal (Section 12.8).

With the exception of the Bonferroni and Holm-Bonferroni methods, all of the FWER controlling procedures which we will discuss (viz., Tukey-Kramer, Scheffé, Dunnett, and the parametric resampling method) have the same general ideal inference conditions:

- independence of errors

- normality of errors

- equality of sampled population error variances

- no misclassification of the sample data in the groups being compared

When any of these procedures (other than Bonferroni and Holm-Bonferroni) are used to test all pairwise comparisons of group means where the groups are classified on the basis of a single factor, the ideal inference conditions are identical to those for the Model F-test in one-way (single factor) fixed effects ANOVA (Section 11.7). Having already established that these ideal inference conditions are reasonable for the task study (Section 11.17), we will continue to use this study as an illustrative example.

The Bonferroni and Holm-Bonferroni methods assume only that the raw p-values which they adjust for multiplicity are valid.

12.6   The Bonferroni Method

12.6.1   Introduction

The Bonferroni method is a general procedure that can be used to control the familywise error rate in various multiple testing situations (e.g., partial t-tests of multiple regression predictors; pairwise comparisons of group means from a one-way fixed effects ANOVA).

The Bonferroni method controls the familywise error rate in the **strong sense**, which means it controls the familywise error rate for any combination of false and true null hypotheses being considered in the family. A misunderstanding about the Bonferroni method is that it provides only **weak control** of the familywise error rate (Goeman and Solari, 2014). Weak controlling familywise error rate methods guarantee control of the familywise error rate only if all the null hypotheses of the tests in the family are true.

The Bonferroni method can be conservative resulting in decreased probability of being able to detect true effects under any of the following circumstances:

- There are a large number of tests in a family.

- There is high positive correlation among test statistics in a family (Goeman and Solari, 2014).

- Many tests in the family have null hypotheses that are actually false (Goeman and Solari, 2014).

We will further discuss why the Bonferroni method can be conservative in subsection 12.6.3. However, to facilitate this discussion we will first describe a probability inequality underlying the Bonferroni method for multiple hypothesis tests.

It can be shown that

$$\text{FWER} \leq (m)(\alpha) \tag{12.4}$$

FWER refers to the familywise error rate for m tests carried out at the α level of significance

where m is the total number of tests in the family

α is the level of significance used for each test

On the basis of Equation 12.4 the Bonferroni method can be used to adjust for multiple hypothesis tests by multiplying the raw p-value for each test obtained from a multiplicity-unadjusted procedure by m, the number of tests in the family, i.e.,

$$\text{Bonferroni adjusted p-value} = \text{raw p-value} \times m \tag{12.5}$$

If the product obtained from multiplying the raw p-value by the number of tests in the family is greater than 1.00, the Bonferroni adjusted p-value is assigned a value of 1.00.

In the context of significance testing, the Bonferroni method can be used to control the FWER for m tests **to be no greater** than a pre-specified threshold, typically 0.05, by using $0.05/m$ as the level of significance for each test in the family. However as discussed in Section 1.9, making research decisions solely on the basis of whether a test result is labelled "statistically significant", (having a p-value no greater than a pre-specified threshold level) is contentious.

12.6.2   Example 12.1: An Application of the Bonferroni Method

This example illustrates using the Bonferroni method to control the familywise error rate to be no greater than 0.05 for a set (family) of partial t-tests from a multiple regression model. As previously discussed in Section 7.9, raw p-values from partial t-tests of multiple regression predictors are not adjusted for multiplicity and so the probability of Type I error increases as the number of partial t-tests of predictors increases. The multiple regression model in this example is based on a hypothetical data set where the response variable was oxygen consumption (a measure of aerobic fitness) and the four predictor variables in the model were resting pulse rate, running pulse rate, age, and running time. The raw (multiplicity-unadjusted) p-values from these partial t-tests are given in Table 12.1. Previous evaluation of ideal inference conditions for the partial t-tests in this multiple regression had indicated that these raw p-values are reasonably accurate. The Bonferroni adjusted p-values for the partial tests are respectively 1.00, 0.020, 0.080, and 0.004 (i.e., the number of tests in the family times the raw p-value: $4 \times 0.7810 = 3.12$, which is capped at 1.00; $4 \times 0.005 = 0.020$; $4 \times 0.02 = 0.08$; $4 \times 0.001 = 0.004$).

In the context of significance testing, suppose researchers in this study wanted to have a familywise error rate that was no greater than 0.05 for the four partial t-tests of the predictors. They applied the Bonferroni method which guarantees that the FWER is no greater than 0.05 for four tests by using $0.05/4 = 0.0125$ as the comparisonwise significance level for each test. From Table 12.1 it can be seen that the null hypotheses of the partial t-tests for running pulse rate and running time are rejected at the 0.05 familywise level of significance by the Bonferroni method as their raw p-values are less than 0.0125. However, the null hypotheses of the partial t-tests of age and resting pulse rate are not rejected by the Bonferroni method at the 0.05 familywise level of significance as their raw p-values are greater than 0.0125. However as previously discussed, rejecting null hypotheses solely on the basis of significance tests is no longer recommended.

TABLE 12.1
Raw and Bonferroni-adjusted p-values for Example 12.1

Model predictor	Raw p-value	Bonferroni adjusted p-value
Resting Pulse Rate	0.781	1.00
Running Pulse Rate	0.005	0.020
Age	0.02	0.080
Running Time	0.001	0.004

12.6.3 Why the Bonferroni Method Can Be Conservative

The Bonferroni method can be conservative because it adjusts for a familywise error rate for the worst case scenario which is unnecessary in most applications. The worst case scenario, which rarely occurs in multiple testing, is the scenario where false rejections of the null hypotheses by the tests in the family are mutually exclusive, i.e. perfectly negatively correlated (Westfall et al., 2011; Goeman and Solari, 2014). For example, consider the situation where there are two tests in the family: If false rejections of the tests' null hypotheses are mutually exclusive, this means that if the null hypothesis of Test 1 is falsely rejected, then this precludes the possibility that the null hypothesis of Test 2 is falsely rejected or vice versa.

The Bonferroni method always guarantees that the familywise error rate is **not greater than the specified** α, **say** $\alpha=0.05$. However, Goeman and Solari (2014) showed that when some of the null hypotheses of the tests in the family are not true, the Bonferroni method is conservative because it does not actually control the maximum familywise error rate at the specified nominal level α but rather at a **stricter level** equal to $(m_0/m)\alpha$, where m_0 is the number of tests in which the null hypothesis is actually true in the family and m is the total number of tests in the family.

Example: Consider the situation in which there are four tests in a family ($m = 4$) and the null hypothesis is actually true in just one of these tests ($m_0 = 1$). Although the specified nominal α for the Bonferroni method is 0.05, the Bonferroni method would be actually controlling for a maximum familywise error rate of $[1/4(0.05) = 0.0125]$ resulting in a more conservative test than if the null hypotheses for all the tests in the family were actually true.

Of course, we do not know the number of true null hypotheses in a family. However, current research involves investigating methods which attempt to estimate the number of true null hypotheses in a family in the effort to develop more powerful multiplicity-adjusted procedures.

12.6.4 The Bonferroni Method and All Possible Pairwise Comparisons of Group Means

The Bonferroni method is available in the GLM procedure for all pairwise comparisons between group means. For example, the following code specifies that Bonferroni's method should be applied to all pairwise comparisons of the group means in the task study using as input the **temporary**[1] SAS data set TASKDTA, created by Program 11.1.

---

[1]Recall the note regarding temporary SAS data sets in subsection 12.4.4.

```
proc glm data=taskdta;
   class task;
   model pulse20=task;
   lsmeans task/ adjust=bon;
quit;
```

In the previous code, a listing of Bonferroni adjusted p-values for each pairwise comparison is obtained by specifying the keywords ADJUST=BON as an option on the LSMEANS statement. The following are the results generated by this LSMEANS statement:

```
                    The GLM Procedure
                    Least Squares Means
           Adjustment for Multiple Comparisons: Bonferroni


                             pulse20        LSMEAN
            task             LSMEAN         Number

            A             38.0000000          1
            B             29.5000000          2
            C             28.8181818          3



            Least Squares Means for effect task
            Pr > |t| for H0: LSMean(i)=LSMean(j)

                Dependent Variable: pulse20

     i/j            1               2               3

      1                          0.0056          0.0033
      2          0.0056                          1.0000
      3          0.0033          1.0000
```

These preceding results suggest that:

- the Task A population mean pulse rate is different from Task B population mean pulse rate (Bonferroni adjusted p-value = 0.0056)

- Task A population mean pulse rate is different from that of Task C (Bonferroni adjusted p-value = 0.0033)

- no difference between Task B and Task C population mean pulse rates could be detected in this example

12.7   The Tukey-Kramer Method for All Pairwise Comparisons

12.7.1   Introduction

Tukey's original method maintains the familywise error rate at a desired level α for all pairwise comparisons of a set of means for groups with equal sample sizes. Tukey's method involves utilizing a value from the studentized range distribution to adjust the p-value of each t-statistic which tests H_0: $\mu_i = \mu_j$ vs. H_a $\mu_i \neq \mu_j$, $1 \leq i, j \leq g$, so that the probability of falsely rejecting the null hypothesis over all pairwise comparisons in the family is equal to the specified α. For those who are interested, Westfall et al. (2011) provide a theoretical discussion of how the studentized range distribution is based on the multivariate t-distribution when classical linear model assumptions are valid.

The Tukey-Kramer method is a modification of the original method proposed by Tukey (1953) and Kramer (1956) which accommodates unequal sample sizes in the groups. When the sample sizes are equal, the Tukey-Kramer method yields identical results as the original Tukey method. When the sample sizes are not equal, the Tukey-Kramer method can be conservative (Hayter, 1984; Westfall et al., 2011). When there are large differences in group sample sizes, the Tukey-Kramer method is not recommended (Miller, 1981). However, slightly unequal sample sizes may yield acceptable results.

12.7.2   SAS and the Tukey-Kramer Method

The Tukey-Kramer method is available in the GLM procedure for testing all pairwise comparisons of group means by specifying the keywords ADJUST=TUKEY as an option on the LSMEANS statement. This is illustrated by the following SAS code which uses as input the **temporary**[2] SAS data set TASKDTA, created by Program 11.1.

```
proc glm data=taskdta;
   class task;
   model pulse20=task;
   lsmeans task/ adjust=tukey;
quit;
```

The LSMEANS statement in the preceding code generates the following Tukey-Kramer's adjusted p-values which maintain a familywise error rate of 0.05 over all the pairwise comparisons in the task example.

```
                  The GLM Procedure
               Least Squares Means
     Adjustment for Multiple Comparisons: Tukey-Kramer
```

task	pulse20 LSMEAN	LSMEAN Number
A	38.0000000	1
B	29.5000000	2
C	28.8181818	3

```
        Least Squares Means for effect task
        Pr > |t| for H0: LSMean(i)=LSMean(j)

           Dependent Variable: pulse20
```

i/j	1	2	3
1		0.0051	0.0030
2	0.0051		0.9575
3	0.0030	0.9575	

---

[2]Recall the note regarding temporary SAS data sets in subsection 12.4.4.

The preceding results from the Tukey-Kramer procedure, where the familywise error rate is maintained at 0.05 may be summarized as follows:

- The Tukey-Kramer adjusted p-value for the test of the null hypothesis Group A population mean pulse rate (labelled as LSMEAN 1) = Group B population mean pulse rate (labelled as LSMEAN 2) is 0.0051, which suggests these population mean pulse rates are different.

- The Tukey-Kramer adjusted p-value for the test of the null hypothesis Group A population mean pulse rate (labelled as LSMEAN 1) = Group C population mean pulse rate (labelled as LSMEAN 3) is 0.003, which suggests these population mean pulse rates are different.

- The Tukey-Kramer adjusted p-value for the test of the null hypothesis Group B population mean pulse rate = Group C population mean pulse rate is 0.9575, which indicates on the basis of Tukey-Kramer's method applied to the task study data, there is insufficient evidence to indicate that these population mean pulse rates are different.

12.8   A Simulation-Based Method for All Pairwise Comparisons

12.8.1   Introduction

When group sample sizes are unequal, simulated estimates of adjusted p-values for all pairwise comparisons of means may be considered as an alternative to Tukey-Kramer's adjusted p-values. These simulated estimates are obtained via a parametric resampling algorithm described by Westfall et al. (2011).

12.8.2   SAS and Simulation-Based Adjusted p-Value Estimates

These simulation-based estimates of adjusted p-values for all pairwise comparisons of group means can be obtained in the GLM procedure by specifying the option AD-JUST=SIMULATE on the LSMEANS statement, as illustrated in the following code. This code uses as input the **temporary**[3] SAS data set TASKDTA, created by Program 11.1.

```
proc glm data=taskdta;
   class task;
   model pulse20=task;
   lsmeans task/adjust=simulate (seed=10653299);
quit;
```

The (SEED=*number*) option specifies an integer to be used to start the pseudo-random number generator for the simulation. The number specified can be any number greater than zero. If you do not specify the (SEED= *number*) option or if you specify a number for the seed that is less than or equal to zero, then by default the seed will be generated by reading the time of day from the computer clock which can result in slightly different values for the simulated-based p-values for the same data set.

The accuracy of the simulation-based adjusted p-value estimates can be increased by increasing the number of simulations, which is accomplished in the GLM procedure by the NSAMP= option, e.g.,

```
lsmeans task/adjust = simulate (nsamp = 20000000 seed = 10653299);
```

---

[3]Recall the note regarding temporary SAS data sets in subsection 12.4.4.

However, for most practical purposes the default accuracy provided by GLM should be adequate. Westfall et al. (2011) reported that, by default, a simulation-based adjusted p-value around 0.05 for all pairwise comparisons will be estimated between 0.045 and 0.055 with 99% confidence by the GLM procedure and that the number of simulations used to attain this default level of accuracy is 12,604.

12.8.3 Task Study Example of Simulation-Based Adjusted p-Values Estimates

The results generated by the SAS code given in the previous subsection are as follows:

```
                     The GLM Procedure
                   Least Squares Means
         Adjustment for Multiple Comparisons: Simulated


                         pulse20      LSMEAN
             task        LSMEAN       Number


             A         38.0000000       1
             B         29.5000000       2
             C         28.8181818       3



         Least Squares Means for effect task
         Pr > |t| for H0: LSMean(i)=LSMean(j)

            Dependent Variable: pulse20


     i/j           1            2            3

      1                      0.0044       0.0031
      2         0.0044                    0.9553
      3         0.0031       0.9553
```

The preceding results reveal that, in this example the simulation-based adjusted p-value estimates for all pairwise comparisons of the task study means using the default settings for accuracy are similar to the corresponding Tukey-Kramer adjusted p-values. This is not surprising as the group sample sizes are close in value (10, 12, 11, respectively for Groups A, B, and C). When the group sample sizes are quite different, the simulation-based adjusted p-value estimates are typically smaller than their counterparts from the Tukey-Kramer method, provided a sufficient number of simulations are used to estimate the simulation-based adjusted p-values (Westfall et al., 2011).

12.9 Dunnett's Method for "Treatment" vs. "Control" Comparisons

12.9.1 Introduction

Dunnett's method (Dunnett, 1955) maintains familywise error rate at a desired level specifically for making multiple "treatment" vs. "control" comparisons, assuming the ideal inference conditions (Section 12.5) for this method hold. For example, in the task study where there are three groups, Groups A, B, and C, if Group A is the control group, then Dunnett's method would control the familywise error rate at a desired level for two comparisons, Group B vs. Group A and Group C vs. Group A. Although this method is described as a treatment vs. control multiple comparison method, it can be used in any situation where each of several groups, not necessarily treatment groups, is compared to a common single group (e.g., in an investigation of dinosaur extinction two comparisons of interest were (i) an upper stratigraphic level vs. the lowest stratigraphic level and (ii) a middle vs. the lowest stratigraphic level with each stratigraphic level representing a period of approximately 730,000 years (Rogers and Hsu, 2001). Miller (1981) referred to Dunnett's method as "many to one t procedure".

Dunnett's method does not require equal group sample sizes as an ideal inference condition. In SAS, essentially exact critical values and adjusted p-values are calculated for Dunnett's method for unbalanced one-way ANOVA (Westfall et al., 2011) based on a numerical method described in Hochberg and Tamhane (1987). Dunnett originally developed his method for equal group sample sizes, and in his 1955 paper, provided a table of critical values for detecting statistical significance between a treatment and common control for equal sample sizes. He noted that these critical values would be approximate for unbalanced designs (i.e., unequal group sample sizes). Textbooks that use Dunnett's original table caution that results are approximate for unequal group sample sizes. However, with the advance of computer software since 1955, essentially exact distributions for Dunnett's critical value can be computed numerically for unbalanced designs.

12.9.2   SAS and Dunnett's Test

A listing of the DUNNETT adjusted p-values for each treatment vs. control two-sided test of mean equality is obtained by specifying the keywords ADJUST=DUNNETT as an option on the LSMEANS statement of GLM. The following code, which applies Dunnett's two-sided test to the **temporary**[4] SAS data set TASKDTA, provides an illustration:

```
proc glm data=taskdta;
   class task;
   model pulse20=task;
   lsmeans task/adjust=dunnett;
quit;
```

Specifying the Control Group: By default, GLM considers the group that is labelled with the lowest value for the classification variable as the control. So in the task study example where the classification variable is "task" and each group is labelled with the alphanumeric values of A, B, or C, the group labelled as A is identified by GLM as the control group because it has the lowest alphanumeric value for the task variable in the data.

Suppose you want to specify that the group coded as B in the data input is the control group. We can specify this in GLM by the following code:

```
proc glm data=taskdta;
   class task;
   model pulse20=task;
   lsmeans task/pdiff=control('B') adjust=dunnett ;
quit;
```

---

[4]Recall the note regarding temporary SAS data sets in subsection 12.4.4.

This code illustrates that an alternate way to specify the control group in GLM is to enclose the formatted value of the classification variable for your control group in quotes and parentheses as an option for pdiff=control.

Suppose we had used numeric values to code the task classification variable, say the values 1, 2, and 3 to respectively indicate whether an individual belonged to Task A, B, or C groups and we wanted to indicate Task B was the control group, we would simply enclose the number 2 in quotes and parentheses as follows:

```
lsmeans task/pdiff=control('2') adjust=dunnett ;
```

Adjusted p-Values from Dunnett's Two-sided Test where Task A was Specified as the Control Group for the Task Study:

```
                 The GLM Procedure
               Least Squares Means
        Adjustment for Multiple Comparisons: Dunnett


                                   H0:LSMean=
                         pulse20     Control
            task         LSMEAN      Pr > |t|

             A        38.0000000
             B        29.5000000     0.0035
             C        28.8181818     0.0021
```

From the preceding results, it can be seen that Dunnett's adjusted p-values which maintain the familywise error rate at 0.05 for "treatment vs. control" comparisons suggest that:

- population mean Group B is different from that of control Group A (Dunnett's adjusted p-value = 0.0035)

- population mean Group C is different from that of control Group A (Dunnett's adjusted p-value = 0.0021)

12.10   Scheffé's Method for "Data Snooping"

12.10.1   Introduction

Scheffé's method (Scheffé, 1953, 1969) controls the familywise error rate at the desired level but this method is often conservative as it is controlling the probability of making at least one Type I error over all the infinitely possible linear contrasts of the group means being considered. Thus, this method is over-adjusting for the multiplicity effect as it is adjusting for an infinite number of comparisons of no interest to the researcher. However, Scheffé's method may be considered whenever researchers want to make a large number of unplanned comparisons, i.e., comparisons that were not planned before collecting the data. Scheffé's method will hold the familywise error rate at the desired level, even when it is used to test contrasts of the means that are suggested by the data (Westfall et al., 2011). As Scheffé (1959) wrote, his method allows "data snooping".

The ideal inference conditions for Scheffé's method are given in Section 12.5. This method, which is based on an underlying F-distribution, is consistent with the Model F-test from the corresponding one-way ANOVA of the group means. If the corresponding Model F-test has a small p-value (e.g., less than 0.05), Scheffé's method will yield the same small p-value for at least one (not necessarily substantively meaningful) contrast of the infinitely possible linear contrasts of the group means. If the corresponding Model F-test does not have a small p-value (e.g., greater than 0.05) then Scheffé's method likewise will not yield a small p-value for any of the infinitely possible linear contrasts of the group means.

12.10.2   Application of Scheffé's Method

We apply Scheffé's method to test all pairwise comparisons of the task study group means solely to illustrate that this method is conservative compared to the familywise error rate controlling methods we have discussed up to now. Milliken and Johnson (2009) provide an example of applying Scheffé's method to evaluate linear contrasts other than pairwise comparisons using the MIXED procedure.

12.10.3   SAS and Scheffé's method

Adjusted p-values from Scheffé's method are generated for all pairwise comparisons of group means by specifying the keywords ADJUST= SCHEFFE as an option on the LSMEANS statement. This is illustrated by the following code, which uses as input the **temporary**[5] SAS data set TASKDTA, created by Program 11.1:

```
proc glm data=taskdta;
   class task;
   model pulse20=task;
   lsmeans task/ adjust=scheffe;
quit;
```

The results generated by the preceding SAS code are as follows::

```
                    The SAS System


                  The GLM Procedure
                Least Squares Means
      Adjustment for Multiple Comparisons: Scheffe


                          pulse20        LSMEAN
            task          LSMEAN         Number

             A          38.0000000          1
             B          29.5000000          2
             C          28.8181818          3



        Least Squares Means for effect task
        Pr > |t| for H0: LSMean(i)=LSMean(j)
```

---

[5]Recall the note regarding temporary SAS data sets in subsection 12.4.1.

```
                   Dependent Variable: pulse20

     i/j              1               2               3

      1                            0.0073          0.0044
      2            0.0073                          0.9614
      3            0.0044          0.9614
```

Controlling for a familywise error rate of 0.05, where the family consists of the infinite number of linear contrasts of population means for Tasks A, B, and C, the preceding adjusted p-values from Scheffé's method suggest that:

- population mean pulse rates for Task A and Task B are not equal (adjusted p-value = 0.0073)

- population mean pulse rates for Task A and Task C are not equal(adjusted p-value = 0.0044)

- the data do not provide evidence to conclude population mean pulse rates for Task B and Task C are different (adjusted p-value = 0.9614)

12.11   Ordinary Confidence Intervals and the Multiplicity Issue

12.11.1   Introduction

The multiplicity problem encountered when you estimate multiple ordinary (multiplicity-unadjusted) confidence intervals for various parameters and/or parameter functions is that the greater the number of these ordinary confidence intervals you estimate, the lower the level of confidence you can have that all of these intervals will contain the true values of the parameters and/or parameter functions being estimated.

For example, when you estimate three ordinary 95% confidence intervals, the level of confidence could be as low as 86% that all three intervals contain the true values of the parameters or parameter functions being estimated. If you were to estimate 10 ordinary 95% confidence intervals, the level of confidence could be as low as 60% that all 10 intervals contain the true values of the parameters or parameter functions being estimated.

12.11.2   Why Does the Overall Confidence Level Decrease as the Number of Ordinary Confidence Intervals Increases?

To facilitate the explanation to this question, we first define *familywise confidence level* and consider *familywise error rate* in the context of confidence intervals.

Definition 12.4. The **familywise confidence level**, which we denote as **FWCL**, is the probability that **all** intervals being estimated by a procedure will, under repeated random sampling of the population(s) studied, contain the true values of the parameters or parameter functions being estimated. In other words FWCL is the coverage probability of a procedure for a family of confidence intervals.

The familywise error rate, FWER, in the context of multiple confidence intervals is the probability (proportion of times) that **not all** intervals being estimated by a procedure will, under repeated random sampling of the population(s) studied, contain the true value of the parameter(s) or parameter functions being estimated.

The familywise confidence level and familywise error rate are clearly related. When the same procedure is used to estimate the same family of confidence intervals, the familywise confidence level of this procedure and its familywise error rate are probabilities of mutually exclusive events and hence their probabilities are complementary. Therefore

$$\text{FWCL} = 1 - \text{FWER} \tag{12.6}$$

It can be shown that when the number of confidence intervals in a family increases, the familywise error rate increases, which explains why its complement, the FWCL decreases. When confidence intervals in the family are all statistically independent of each other, the familywise error rate is

$$\text{FWER} = 1 - (1 - \text{CWER})^k \tag{12.7}$$

where

k is the number of statistically independent confidence intervals in the family

CWER, the comparisonwise error rate (Definition 12.1), in the context of confidence intervals, is the probability that a procedure, under repeated random sampling of the study population(s), will produce a confidence interval that does not contain the true value of the parameter or parameter function being estimated. For example, the comparisonwise error rate is equal to 0.05 for a single ordinary 95% confidence interval, assuming assumptions in estimating the confidence interval are met.

As an example, consider three statistically independent 95% ordinary confidence intervals (i.e., CWER=0.05) in a family. From Equation 12.7, we compute the familywise error rate for this example to be 14% and its complement the familywise confidence level to be 86%.

However, often confidence intervals in a family are not statistically independent due to their sharing overlapping information (e.g., the same estimate of σ^2 when computing each confidence interval; the same treatment mean used for some of the confidence intervals as occurs in the family of confidence intervals for the pairwise absolute difference between the group means in the task study, i.e., the three confidence intervals: $|\mu_A - \mu_B|$, $|\mu_A - \mu_C|$, $|\mu_B - \mu_C|$.

In these situations where confidence intervals in a family are not statistically independent, the familywise error rate will be less than that given by Equation 12.7 but will be greater than the comparisonwise error rate, CWER. Hence if you construct three ordinary 95% confidence intervals for the pairwise difference between means in the task study where there is some statistical dependence among the confidence intervals, the familywise error rate will be greater than 0.05 but less than 0.14 (the value obtained from Equation 12.7 which assumes independent confidence intervals) and the familywise confidence level will be less than 0.95 but not as low as 0.86. The actual familywise error rate and familywise confidence level for any particular family depend on the degree of statistical dependence among the confidence intervals in that family.

12.12 Controlling Familywise Error Rate for Confidence Intervals

12.12.1 Introduction

The familywise error rate, and hence its complement, the familywise level of confidence, can be controlled for a family (set) of two-sided confidence intervals by constructing each confidence interval in the family using the following general formula:

$$\text{estimate of a parameter or a contrast} \pm c_\alpha(\text{standard error of estimate}) \qquad (12.8)$$

where c_α is a critical value that is selected to control the familywise error rate to be no greater than α and hence the familywise confidence level not less than $1 - \alpha$.

For example, the general formula for constructing each confidence limit estimate in a family of pairwise mean differences among g groups so that the familywise error rate is controlled to be no greater than 0.05 is given by:

$$\bar{Y}_i - \bar{Y}_j \pm c_{0.05}\, \hat{\sigma}\sqrt{\frac{1}{n_i} + \frac{1}{n_j}} \qquad (12.9)$$

where

$\bar{Y}_i - \bar{Y}_j$ is the sample estimate of the difference between two population means, $\mu_i - \mu_j$, in the family

$c_{0.05}$ is a critical value that is selected to control the familywise error rate to be no greater than 0.05 and the familywise confidence level no less than 0.95

$\hat{\sigma}\sqrt{\frac{1}{n_i} + \frac{1}{n_j}}$ is the standard error of the estimate, $\bar{Y}_i - \bar{Y}_j$

Westfall et al. (2011, p. 86) summarized how c_α may be obtained in FWER controlling methods. When an exact analytic solution is used to obtain c_α, the method controls for a familywise error rate equal to α precisely, assuming the assumptions of the model hold. Otherwise, when an exact analytic solution for c_α is not feasible, a FWER controlling method controls for a familywise error rate that is less than or equal to the specified alpha. Examples of such methods that control the familywise error rate to be no greater than the specified α for all pairwise comparisons are the Bonferroni method (subsection 12.6.3), Tukey-Kramer's method for all pairwise comparisons in one-way ANOVA with unequal sample sizes (Section 12.7), and the simulation-based method described in Section 12.8. Such methods may be more conservative and result in a wider confidence interval than if the critical value for c_α is obtained via an exact analytic solution.

12.12.2   Task Study Example

We consider the task study to illustrate procedures that control familywise error rate for a set of confidence intervals. In this study our objective is estimate confidence intervals for the pairwise differences between the population means: i.e., between Group A and Group B; Group A and Group C; Group B and Group C.

Suppose it is desired that the inference procedure applied yields at least a 95% confidence level that **all** of these intervals will contain the true value of the difference being considered. We know that if we estimate the ordinary (multiplicity-unadjusted) 95% confidence intervals, the overall level of confidence will be less than 95% because of the multiplicity problem. For this reason we decided to estimate each confidence interval in the family using a method that ensures that the familiywise error rate is no greater than 0.05 and hence that the familywise confidence level is not less than 0.95.

The decision then arises which method should we use to control the familywise error rate. Tukey-Kramer and the simulation-based method are possible candidates. Also in this example where there are just three groups considered for pairwise comparisons, Fisher's Restricted LSD is a candidate method because it has been shown that Fisher's Restricted LSD will hold the familywise error rate at a specified level when there are three intervals in the family. (However, when there are more than three intervals in the family, Fisher's

Restricted LSD is not recommended because in these situations it does not maintain the familywise error rate at the specified nominal level.) We also include the simulation-based approach among the candidate methods because the group sample sizes are unequal. If the sample sizes had been equal, we would not consider the simulation-based method because Tukey's method would be preferred (Westfall et al., 2011).

12.12.3 SAS and Controlling Familywise Error Rate for Confidence Intervals

The following code illustrates how to request confidence interval estimates for the family of all pairwise comparisons of group mean differences using LSMEANS statements. This code specifies as input the **temporary**[6] SAS data set TASKDTA.

```
proc glm  data=taskdta;
   class task;
   model pulse20=task;
   lsmeans task/diff cl;
   lsmeans task/adjust=tukey cl;
   lsmeans task/adjust=simulate (seed=10653299) cl;
   lsmeans task/adjust=simulate (seed=10653299 nsamp=20000000) cl;
   lsmeans task/adjust=bon cl;
   lsmeans task/ adjust=scheffe cl;
quit;
```

In the preceding SAS code, the option **diff** and **cl** in the LSMEANS statement, i.e., lsmeans task/diff cl; specify LSD confidence intervals and adjusted p-values. In this task study example these LSD confidence intervals and adjusted p-values are in fact, as discussed in Section 12.4.3, Fisher's LSD simultaneous confidence intervals and adjusted p-values. By default α is set to 0.05 in the preceding code and familywise 95% confidence limits are requested. However, for the task example with unequal sample sizes and the particular methods specified in the preceding LSMEANS statements (Fisher's LSD, Tukey-Kramer, Simulation based, Bonferroni, and Scheffé's), the α option of 0.05 does not mean the familywise error rate is exactly equal to 0.05 even if all assumptions are met, but rather that the familywise error rate is no greater than 0.05.

Familywise confidence limits for the difference between Task A and Task B population means obtained from the LSMEANS statements in the preceding code are summarized in Table 12.2. Familywise p-values from the corresponding two-sided test of the null hypothesis of zero difference between Task A and Task B population means are also given. Similar trends were observed among these various methods when results for the other pairwise mean comparisons in the family (Task A vs. Task C and Task B vs. Task C) were compared.

In Table 12.2 we have listed the results from the various methods in ascending order based on confidence interval widths and corresponding p-values. From this table it can be seen that:

- Fisher's LSD confidence interval for the difference in population means between Groups A and B is the least wide and the familywise p-value for rejecting the null hypothesis, $|\mu_A - \mu_B| = 0$, is the smallest compared to the other methods. As previously emphasized, we have considered Fisher's LSD method only because there are three "treatment" groups in this family and the Model F-test from the corresponding one-way fixed effects ANOVA has a p-value less than 0.05, thereby guaranteeing the familywise error rate does not exceed 0.05.

---

[6]Recall the note regarding temporary SAS data sets in subsection 12.4.4.

TABLE 12.2
Familywise confidence limits for the difference between population means A vs. B in the task study and corresponding familywise p-values from a two-sided test of the null hypothesis of zero difference between Task A and Task B population means, obtained from methods which control FWER for all pairwise comparisons to be no greater than 0.05.

Method	Confidence Limits	Confidence Interval Width	p-Value
Fisher's LSD	3.4146 13.5854	10.1708	0.0019
Simulation-based (default number of samples)	2.3761 14.6239	12.2478	0.0044
Simulation-based (number of samples = 20,000,000)	2.3629 14.6371	12.2742	0.0051
Tukey-Kramer	2.3613 14.6387	12.2774	0.0051
Bonferroni	2.1858 14.8142	12.6284	0.0056
Scheffé	2.0875 14.9125	12.8250	0.0073

- The FWER controlled confidence interval and p-value from the Tukey-Kramer method are very similar to those from the simulation-based method for all pairwise comparisons. Although the simulation-based method typically estimates shorter confidence intervals and smaller adjusted p-values than those estimated by Tukey-Kramer's method when the sample sizes are unequal (Westfall et al., 2011), it is not surprising to observe similarity between the results from the two methods for this example where the sample sizes are very close ($n = 10$ in Group A, $n = 11$ in Group B).

- There are only slight differences in the results for the Tukey-Kramer and Bonferroni methods, which is also not surprising as there are only three comparisons in the family in this task example. The Bonferroni method becomes increasingly conservative compared to Tukey-Kramer's as the number of comparisons in a family increases.

- Scheffé's method gave the widest familywise confidence interval and the largest familywise p-value compared to the other methods considered in Table 12.2. Scheffé's method was included for comparison purposes only. This method is never recommended for preplanned pairwise comparisons as it is conservative compared to other methods (Section 12.10).

12.13 Confidence Bands for Simple Linear Regression

Often we may want to see at a glance how accurately an estimated simple linear regression model is predicting the outcome variable over a range of observed predictor values in our sample. This information may be obtained by a graphical display of a *confidence band* for the estimated simple linear regression model.

Definition 12.5. *A confidence band for an estimated simple linear regression model* is an infinite set of simultaneous confidence intervals for all the values of the outcome variable (Y) predicted by the model within a certain range of predictor values.

The x values usually considered for the confidence band are those in the range from the lowest to the largest x value in the sample, in order to avoid the perils of extrapolation. For example, suppose we were interested in constructing a confidence band for the simple linear regression described in Section 2.3 where the Y variable is systolic blood pressure (SBP) and the X variable is age (AGE). The minimum and maximum x values in the sample are 21 and 90 so we would want to construct a confidence band for the regression line for the range $21 \leq x \leq 90$. In other words, we would want to construct simultaneous confidence intervals for each predicted mean of Y (SBP) by the regression model for every value of X (AGE) where $21 \leq x \leq 90$.

The set of confidence intervals within any given range of X for a simple linear regression confidence band is infinite because the X variable in simple linear regression is continuous. This means there are infinitely many possible x values in any specified range of X for the confidence band. Thus, there are an infinite set of possible confidence intervals for predicting Y in the sampled population on the basis of an x value in any specified range using an estimated simple linear regression model. Therefore, the method, which produces a multiplicity-adjusted confidence band for a simple linear regression, has to adjust for a familywise error rate where the family is comprised by an infinite number of possible confidence intervals within the range of specified x values. In the next subsections we will discuss two methods for producing a multiplicity-adjusted confidence band for a simple linear regression – the Working-Hotelling method and a discrete simulation-based method.

12.13.1   The Working-Hotelling Method

Working-Hotelling's approach (Working and Hotelling, 1929) is similar to Scheffé's method in that it controls familywise error rate for an infinite set of inferences and like Scheffé's method is conservative. Working-Hotelling's method controls the familywise error rate for the infinite family of confidence intervals for predicted y values by the simple linear regression model in the range $-\infty < x < +\infty$. This approach is unnecessarily conservative since you want adjustment only for inferences in the range $a \leq x \leq b$ where a and b are x values observed in the sample, e.g., a and b could be respectively the minimum and maximum values in the sample.

At any x value, say x_0, the upper and lower y value which forms the boundary for the Working-Hotelling's confidence band for a simple linear regression line is given by

$$\hat{\mu}_{y|x_0} \pm c_{\text{WH},\alpha} \ \widehat{\text{SE}}(\hat{\mu}_{y|x_0}) \tag{12.10}$$

where

$\hat{\mu}_{y|x_0}$ is the mean of Y given x_0 in the sampled population that is predicted by the estimated simple linear regression model (Equation 2.9)

$c_{\text{WH},\alpha}$ is Working-Hotelling's critical value which is equal to $\sqrt{2F_{1-\alpha,2,n-2}}$

$\text{SE}(\hat{\mu}_{y|x_0})$, the standard error of $\hat{\mu}_{y|x_0}$, is given by Equation 4.12

12.13.2   Program 12.1: Working-Hotelling Confidence Band for a Simple Linear Regression Model

Program 12.1 generates a Working-Hotelling confidence band for the simple linear regression model of Example 2.1 (SBP vs. age).

Program 12.1

```
data SBPDATA;
title Program 12.1 Simple Linear Regression of Systolic Blood Pressure against Age;
input ID SBP AGE @@;
datalines;
```

```
1 124 82 2 114 34 3 115   40 4 121 51 5 120 34
6 128 25 7 94 54 8 120 54 9 148 64
10 118 67 11 120 42 12 149 90 13 106 56 14 114 69 15 79 66 16 128 73 17 123 66
18 112 25 19 148 51 20 131 22 21 102 70 22 122 65 23 135 57 24 112 41 25 99 48
26 112 27 27 100 34 28 134 83 29 123 69 30 140 36 31 110 62 32 117 42 33 121 66
34 127 89 35 114 78 36 154 71 37 131 50 38 135 26 39 103 54 40 116 72 41 136 85
42 88 34 43 120 46 44 118 37 45 151 57 46 122 40 47 134 60 48 136 58 49 123 47
50 115 85 51 134 41 52 98 39 53 119 68 54 129 65 55 151 71 56 122 53 57 114 55
58 156 51 59 124 51 60 127 22 61 91 64 62 131 55 63 125 30 64 97 35 65 122 32
66 128 47 67 148 90 68 133 59 69 126 33 70 112 45 71 121 61 72 111 73 73 127 48
74 136 39 75 110 82 76 135 82 77 140 83 78 101 59 79 130 73 80 110 59 81 116 33
82 129 72 83 105 21 84 110 55 85 113 54 86 123 85 87 130 36 88 121 64 89 118 42
90 97 33 91 109 42 92 116 33 93 136 74 94 101 26 95 137 62 96 119 51 97 . 29
;
proc orthoreg data=SBPDATA;
model SBP=AGE;
estimate
"21" Intercept 1 Age 21 ,
"30" Intercept 1 Age 30 ,
"40" Intercept 1 Age 40 ,
"50" Intercept 1 Age 50 ,
"60" Intercept 1 Age 60 ,
"70" Intercept 1 Age 70 ,
"80" Intercept 1 Age 80 ,
"90" Intercept 1 Age 90
/adjust=scheffe cl;
ods output Estimates=Estimates;
proc print data=Estimates noobs label;
var Label Estimate StdErr tValue probt Adjp AdjLower AdjUpper;
proc sgplot data =Estimates (rename=(Estimate=SBP label=AGE));
series x=AGE Y=SBP;
series x=AGE Y=AdjLower;
series x=AGE Y=AdjUpper;
title Working-Hotelling Confidence Bounds for Mean Systolic Blood Pressure (SBP);
run;
```

Figure 12.1, produced by Program 12.1, reaffirms visually what we discussed in sub-section 4.12.2, viz., that the mean of Y for a given X in the sampled population can be estimated more accurately by a simple linear regression model at those values of X that are closer to the sample mean of X. In Figure 12.1 we see that the confidence bounds are narrower at those values of AGE close to 54, the sample mean of AGE.

12.13.3   A Discrete Simulation-Based Method

Westfall et al. (2011) described an approach which they called the "discrete simulation-based" method that is less conservative than Working-Hotelling's method for estimating a multiplicity-adjusted confidence band for simple linear regression. This approach is particularly less conservative than Working-Hotelling's method when we want a confidence band for only a subset of the entire range of x values observed in the sample. For example, this approach would be less conservative for the scenario where we wanted a confidence band for the regression line between ages 65 and 90 in the SBP vs. age study (Example 2.1) in which the observed range of ages in the sample was between 21 and 90.

This method is called "discrete simulation-based" because it involves estimating simulation-consistent critical values to produce simultaneous confidence intervals for the

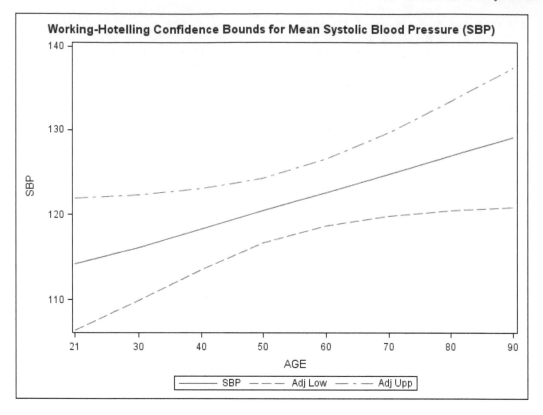

FIGURE 12.1
Working-Hotelling's simultaneous 95% confidence bounds for mean systolic blood pressure (SBP) in Example 2.1.

simple linear regression line at equally spaced x_i values, for $i = 1, 2, \ldots k$ in the specified range from a lower bound a to an upper bound b, i.e., $a < x_1 < x_2 \ldots < x_k < b$. Westfall et al. (2011) provided theoretical details of this method. They pointed out that a critical value estimated by this method is slightly smaller than the correct critical value. However, the difference between the critical value from the simulation-based method and the correct critical value becomes smaller as k, the number of discrete points increases.

Figure 12.2 gives discrete simulation-based approximate 95% confidence bounds for the simple linear regression model of Example 2.1 (SBP vs. AGE). This figure can be obtained by modifying Program 12.1 as follows:

1. Replace

   ```
   / adjust=scheffe cl;
   ```

   with

   ```
   / adjust=simulate (acc=.0002 seed=121211 report) cl;
   ```

2. Replace the title in the SGPLOT procedure of Program 12.1 with

   ```
   title Discrete Simulation-Based Approximate 95% Confidence Bounds
   for Systolic Blood Pressure (SBP) Predicted by a Simple Linear
   Regression of SBP vs. AGE;
   ```

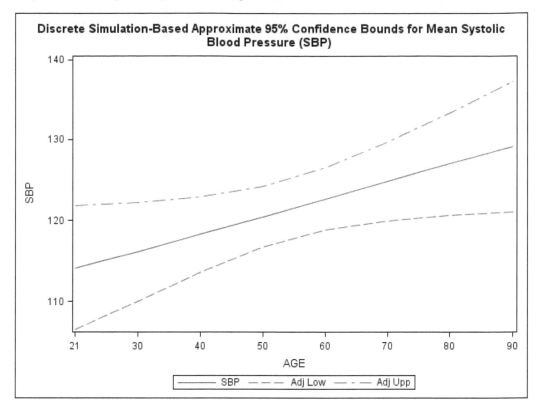

FIGURE 12.2
Simulation-based simultaneous 95% confidence bounds (discrete approximation) for mean systolic blood pressure (SBP) in Example 2.1.

The critical value of 2.448074 from the discrete simulation-based method for Example 2.1 (obtained by specifying the REPORT option for ADJUST=SIMULATE) is slightly smaller than the critical value of 2.487274 from Working-Hotelling's method (also obtained from this REPORT option). As Westfall et al. (2011) pointed out the discrete simulation-based method results in a slightly smaller (i.e., liberal) critical value for the simultaneous confidence bounds. Therefore, in this example, the correct critical value for the 95% simultaneous confidence intervals in the range of age bounded by 21 and 90 years is between 2.448074 (simulation-based method) and 2.487274 (Working-Hotelling's critical value). Thus, in this example where we want to produce simultaneous confidence bounds for mean SBP in the range of ages bounded by 21 and 90, we could choose either method depending as to whether we want to err on the slightly liberal or on the slightly conservative side.

Comparison of Figure 12.1 produced by Working-Hotelling method with Figure 12.2 shows there is very little difference between the confidence bounds for mean SBP produced by these two methods for this example. Moreover, comparison of the multiplicity-adjusted confidence intervals constructed by the two methods at the equally spaced x_i values, i.e., $21 < 30 < 40 \ldots < 90$ show very little difference. For example at AGE=21, the 95% confidence interval for the mean SBP estimated by the discrete simulation-based method is (106.50, 121.82) as compared to Working-Hotelling's 95% confidence interval, which is (106.38, 121.94) for AGE=21. The ordinary (multiplicity-unadjusted) 95% confidence interval for mean SBP at AGE=21 is (107.95, 120.37) (found in Program 12.1 output in the columns labelled "Upper" and "Lower" for age (label) 21). We do not recommend using many multiplicity-unadjusted confidence intervals to judge the accuracy of a linear regression model for a range of x values, as doing so will give the false impression that the model's predictive accuracy is better than it actually is.

12.14   Single Step vs. Sequential Multiplicity-Adjusted Procedures

The methods described in Sections 12.6 to 12.10 are examples of **single step** multiplicity-adjusted procedures. A single step multiplicity-adjusted procedure is so named because all the null hypotheses in the family are tested in a single step using the same value for the comparisonwise level of significance. In a single step multiplicity-adjusted procedure the rejection of a null hypothesis of any test in the family has no influence on the rejection of a null hypothesis of any other test in the family. Thus the order in which the hypothesis tests in the family are carried out is unimportant and therefore "one can think of the multiple inferences as being performed simultaneously in a single step" (Dmitrienko et al., 2010).

In contrast to single step are **sequential multiplicity-adjusted methods** (also referred to as **sequentially rejective** or as **stepwise multiplicity-adjusted methods** in the literature) in which the hypothesis tests in a family are evaluated in a sequence of steps such that the rejection of a null hypothesis of a given test depends on the results of previous steps in the sequence.

Sequential multiplicity-adjusted procedures are as powerful or more powerful than the single-step procedures on which they are based. For example, the Holm-Bonferroni sequential multiplicity-adjusted procedure is as powerful or more powerful than the single-step Bonferroni procedure. The reason sequential multiplicity-adjusted procedures are more powerful is that they use a less stringent adjustment for multiplicity at each step while controlling the familywise error rate at the same level as the single step. Sequential multiplicity-adjusted procedures are particularly likely to be more powerful than their single-step counterparts when many of the null hypotheses being considered in the family are false.

Single-step procedures have the advantage of allowing estimation of simultaneous confidence intervals (Section 12.12) which correspond to the hypotheses tested by these procedures whereas such simultaneous confidence intervals cannot be estimated using sequential multiplicity-adjusted procedures (Westfall et al., 2011). As previously discussed, estimated confidence intervals corresponding to the hypotheses tested can provide researchers with extremely useful information.

12.15   The Holm-Bonferroni Sequential Procedure

12.15.1   Introduction

In this section we consider the Holm-Bonferroni method (sometimes called the Stepdown Bonferroni method or simply Holm's method) as an example of one of the many sequential multiplicity-adjusted procedures available in SAS. Holm (1979) proposed his sequential method to increase the power of the Bonferroni method while maintaining the familywise error rate at the desired level.

A simple way to apply the Holm-Bonferroni method is to compute an adjusted Holm-Bonferroni p-value at each successive step in the sequence (details given in subsection

12.15.2) and reject the null hypothesis corresponding to the adjusted p-value if it is less than or equal to the desired familywise error rate, α_{FW}. Holm-Bonferroni multiplicity-adjusted p-values are readily obtained in the MULTTEST procedure as illustrated in subsection 12.15.3.

12.15.2 How to Compute Holm-Bonferroni Multiplicity-Adjusted p-Values

For those who might be interested, we describe the sequence of steps which are implemented to compute Holm-Bonferroni adjusted p-values.

Before computing Holm-Bonferroni adjusted p-values, the raw p-values should be sorted and ranked in ascending order. The smallest raw p-value is assigned a rank of 1 and the largest raw p-value is assigned rank m where there are m tests in the family.

Let $p(1) \leq p(2) \leq \ldots \leq p(m)$ denote the ordered raw p-values and $adjp(1)$ to $adjp(m)$ denote Holm-Bonferroni's adjusted p-values corresponding to $p(1)$ to $p(m)$.

Holm-Bonferroni adjusted p-values are computed as follows:

$$adjp(i) = \begin{cases} mp(i) & \text{for } i = 1 \\ \max[\ adjp(i-1)\ ,\ (m-i+1)p(i)\ ] & \text{for } i = 2, \ldots, m \end{cases}$$

If any adjusted p-value exceeds 1.00, it is assigned a value of 1.00.

We will illustrate with Example 12.1 (subsection 12.6.2) where researchers wanted to have the familywise error rate no greater than 0.05 for four partial t-tests of model predictors in a multiple regression analysis of aerobic fitness. The raw (multiplicity-unadjusted) p-values from the four partial t-tests are given in Table 12.1 (subsection 12.6.2). These raw p-values in ascending order are:

$p(1)$=0.001 (Running time)
$p(2)$=0.005 (Running Pulse Rate)
$p(3)$=0.02 (Age)
$p(4)$=0.781 (Resting Pulse Rate)

Step 1. Compute adjp(1).
$$\begin{aligned} adjp(1) &= mp(1) \\ &= 4(0.001) \\ &= 0.004 \end{aligned}$$

Step 2. Compute adjp(2).
$$\begin{aligned} adjp(2) &= max[adjp(1), (4-2+1)p(2)] \\ &= max[0.004, 3(0.005)] \\ &= max[0.004, 0.015] \\ &= 0.015 \end{aligned}$$

Step 3. Compute $adjp(3)$.
$$\begin{aligned} adjp(3) &= max[adjp(2), (4-3+1)p(3)] \\ &= max[0.015, 2(0.020)] \\ &= 0.040 \end{aligned}$$

Step 4. Compute $adjp(4)$.
$$\begin{aligned} adjp(4) &= max[adjp(3), (4-4+1)p(4)] \\ &= max[0.040, 0.781] \\ &= 0.781 \end{aligned}$$

Comment:

The Holm-Bonferroni method is sometimes referred to as a stepdown procedure because the steps in the sequence for this method involve going down from the most significant raw p-value (the smallest p-value) to the least significant raw p-value (the largest raw p-value).

12.15.3   Program 12.2: Holm-Bonferroni Method and the MULTTEST Procedure

Program 12.2 generates Holm-Bonferroni adjusted p-values for Example 12.1 using the MULTTEST procedure.

Program 12.2

```
data unadjp;
input Partial_t $ RawP @@;
datalines;
 RestPuls 0.781 RunPuls 0.005 Age 0.02 RunTime 0.001
;
proc multtest inpvalues(RawP)=unadjp bon holm out=new ;
run;
proc sort data=new out=sortnew;
by RawP;
run;
proc print data=sortnew label;
run;
```

Explanation of SAS statements in Program 12.2

Program 12.2 illustrates how raw p-values can be used as input to the MULTTEST procedure using the INPVALUES= option in the PROC MULTTEST statement.

As given in SAS documentation the option

$$\text{INPVALUES(pvalue-name)} = \text{SAS-data-set}$$

in the PROC MULTTEST statement names an input SAS data set that includes a variable containing raw p-values. In Program 12.1 "unadjp" is the name of the input SAS data set in the MULTTEST procedure. (The INPVALUES= and DATA= options cannot both be specified in the PROC MULTTEST statement.) The "pvalue-name" specifies the variable name you are using for the p-values in your data set. In Program 12.2 we arbitrarily used the name "RawP" as the variable name for the raw p values. If no "pvalue-name" is specified, the MULTTEST procedure uses the default name "raw_p" for raw p-values.

The keywords BON and HOLM in the PROC MULTTEST statement specify that we would like the Bonferroni method and Holm-Bonferroni method to adjust the input raw p-values. We requested the Bonferroni method so that this single step method could be compared to its sequential step counterpart in the output from Program 12.2.

The option OUT=SAS-data set allows you to save the adjusted p-values computed by the MULTTEST procedure along with the corresponding variable labels and raw p-values in a SAS data set with the name you specify after the = sign. In Program 12.2 we specify that we want this output to be saved in a SAS data set called "new".

We input the SAS data set we called "new" into the SORT procedure and sorted the records in "new" by the variable "RawP" in ascending order. (Ascending order is the default in the SORT procedure.) We specify the name of this sorted data set to be "sortnew" using the "OUT= " option in the PROC SORT statement.

A listing of the contents of the input SAS data set "sortnew" is requested in the PRINT procedure.

12.15.4 Familywise Adjusted p-Values for Partial t-Tests from Program 12.2

Obs	Partial_ t	RawP	Bonferroni p-value	Stepdown Bonferroni p-value
1	RunTime	0.001	0.004	0.004
2	RunPuls	0.005	0.020	0.015
3	Age	0.020	0.080	0.040
4	RestPuls	0.781	1.000	0.781

The preceding familywise adjusted p-values reveal that the Holm-Bonferroni method (labelled Stepdown Bonferroni in the MULTTEST procedure's output) is less conservative than the single step Bonferroni method. These results underline one of the reasons why rejecting a null hypothesis solely on the basis of a prespecified threshold p-value is not recommended. If researchers were using a threshold family-wise adjusted p-value of 0.05 and applied the Holm-Bonferroni method, the null hypothesis of the partial t-test for Age in this multiple regression example would be rejected. However, if the Bonferroni method had been applied in this situation, the null hypothesis of the partial t-test for Age would not be rejected.

12.16 Adjusting for Multiplicity Using Resampling Methods

When the groups do not have a normal distribution but have equal variances, resampling methods using either a bootstrap or permutation approach can be used to adjust for multiple testing involving these groups. Resampling methods can also be used to take dependence among test statistics into account. We will return to this topic when we discuss resampling methods in Chapter 14.

12.17 Benjamini-Hochberg's False Discovery Rate Method

12.17.1 Introduction

In 1995 Benjamini and Hochberg introduced a new approach for dealing with the multiple testing problem. This approach focuses on controlling the false discovery rate (FDR) in a set (family) of tests. They proved that their method is more powerful in detecting false null hypotheses than comparable FWER controlling procedures when there is at least one null hypothesis that is not true in the set of tests under consideration and the test statistics in the set are independent (Benjamini and Hochberg, 1995) or exhibit a special condition of positive dependence (Benjamini and Yekutieli, 2001).

Simulations have indicated that the Benjamini and Hochberg method is robust and controls the FDR for most applications that involve asymptotically normal two-sided tests

(Sarkar, 2004; Reiner-Benaim, 2007; Kim and van de Wiel, 2008; Romano et al., 2008; Yekutieli, 2008). However, when test statistics are highly correlated with one another, caution is advised in interpreting results from the Benjamin-Hochberg procedure (Heller, 2010; Goeman and Solari, 2014).

12.17.2 Benjamini-Hochberg's Adjusted p-Values

The Benjamini-Hochberg method involves computing FDR-adjusted p-values using a sequential Bonferroni-type approach. You can easily obtain Benjamini-Hochberg's FDR-adjusted p-values from the MULTTEST procedure as illustrated by Program 12.3 (subsection 12.17.3).

For those who may be interested, we describe the sequence of steps which are implemented to compute Benjamini-Hochberg's FDR-adjusted p-values.

Before computing Benjamini-Hochberg's FDR-adjusted p-values, the raw p-values should be sorted and ranked in ascending order. The smallest raw p-value is assigned a rank of 1 and the largest raw p-value is assigned rank m where there are m tests in the family.

Let $p(1) \le p(2) \le \ldots \le p(m)$ denote the ordered raw p-values and let $\mathrm{BH}p(1)$ to $\mathrm{BH}p(m)$ denote Benjamini-Hochberg's FDR-adjusted p-values corresponding to $p(1)$ to $p(m)$.

Benjamini-Hochberg's FDR-adjusted p-values are computed sequentially as follows:

$$\mathrm{BH}p(i) = \begin{cases} p(i) & \text{for } i = m \\ \min[\ \mathrm{BH}p(i+1)\ ,\ \frac{m}{i}p(i)\ ] & \text{for } i = m-1, m-2, \ldots, 1 \end{cases} \qquad (12.11)$$

From Equation 12.11 we see a Benjamini-Hochberg's FDR-adjusted p-value is computed by multiplying a raw p-value by the factor, m/i, where m is the number of tests in the set and i is the rank of the raw p-value where the smallest p-value has rank 1 and the largest p-value has rank m. To avoid the possibility of reversing the original order of the raw p-values, after Step 1 the minimum of $\mathrm{BH}p(i+1)$ and $\frac{m}{i}p(i)$ is taken at each step.

Example 12.2

We consider an illustrative example used by Benjamini and Hochberg (1995) in which a family of 15 hypothesis tests compared two treatment regimens for 15 outcomes (endpoints) in a randomized trial involving patients with acute myocardial infarction (Neuhaus et al., 1992). The raw p-values from these 15 tests are: 0.0001, 0.5719, 0.0298, 0.0004, 0.6528, 0.0019, 0.0095, 0.0201, 0.0278, 0.0344, 0.0459, 0.3240, 0.4262, 0.7590, 1.00.

These 15 raw p-values sorted in ascending order are 0.0001, 0.0004, 0.0019, 0.0095, 0.0201, 0.0278, 0.0298, 0.0344, 0.0459, 0.3240, 0.4262, 0.5719, 0.6528, 0.7590, 1.0000.

Step 1 Compute $\mathrm{BH}p(15) = p(15) = 1.00$.
Step 2 Compute $\mathrm{BH}p(14)$.
$\mathrm{BH}p(14) = \min[\mathrm{BH}p(15), \frac{15}{14}\ 0.7590] = \min[1.00, 0.8132] = 0.8132$.
$\vdots$

Step 15 Compute $\mathrm{BH}p(1)$.
$\mathrm{BH}p(1) = \min[\mathrm{BH}p(2), \frac{15}{1}\ 0.0001] = \min[0.003, 0.0015] = 0.0015$.

In Step 15 we obtained the result $\mathrm{BH}p(2)=0.003$ from the next subsection (see Program 12.3 Results, row 2 and the column labelled "False Discovery Rate p-value").

12.17.3   SAS and Benjamini-Hochberg's FDR-Controlling Method

Program 12.3 applies Benjamini-Hochberg's method to Example 12.2. Benjamini-Hochberg FDR-adjusted p-values are requested by specifying the keyword FDR as an option in the PROC MULTTEST statement. We have also requested FWER-adjusted p-values from the single-step and sequential Holm-Bonferroni methods for comparison by specifying the options BON and HOLM in the PROC MULTTEST statement. The input of the raw p-values to the MULTTEST procedure has been previously described for Program 12.2 in subsection 12.15.3.

Program 12.3: Benjamini-Hochberg's (1995) FDR-Controlling Procedure

```
data rawpvalues;
input TestID $ p@@;
datalines;
A 0.0001 B 0.5719 C  0.0298 D 0.0004
E 0.6528 F 0.0019 G 0.0095 H 0.0201
I 0.0278 J 0.0344 K 0.0459 L 0.3240
M 0.4262  N 0.7590 O 1.00
;
proc multtest inpvalues(p)=rawpvalues bon holm fdr out=new noprint;
run;
proc sort data=new out=sortnew;
by p;
proc print data=sortnew label;
run;
```

Program 12.3 Results

Obs	Test ID	p	Bonferroni p-value	Stepdown Bonferroni p-value	False Discovery Rate p-value
1	A	0.0001	0.0015	0.0015	0.00150
2	D	0.0004	0.0060	0.0056	0.00300
3	F	0.0019	0.0285	0.0247	0.00950
4	G	0.0095	0.1425	0.1140	0.03563
5	H	0.0201	0.3015	0.2211	0.06030
6	I	0.0278	0.4170	0.2780	0.06386
7	C	0.0298	0.4470	0.2780	0.06386
8	J	0.0344	0.5160	0.2780	0.06450
9	K	0.0459	0.6885	0.3213	0.07650
10	L	0.3240	1.0000	1.0000	0.48600
11	M	0.4262	1.0000	1.0000	0.58118
12	B	0.5719	1.0000	1.0000	0.71488
13	E	0.6528	1.0000	1.0000	0.75323
14	N	0.7590	1.0000	1.0000	0.81321
15	O	1.0000	1.0000	1.0000	1.00000

It can be seen from the preceding results that, for all tests (except the first and last) Benjamini-Hochberg's FDR-adjusted p-values are smaller than the corresponding

FWER-adjusted p-values from the single-step Bonferroni method and the sequential Holm-Bonferroni method (labelled as Stepdown Bonferroni by the MULTTEST procedure). In the next subsection we discuss how to interpret adjusted p-values from a procedure which controls the false discovery rate as compared to one which controls the familywise error rate, with special reference to these results.

12.17.4   Interpreting Results from a FDR-Controlling Procedure

When interpreting results, it is important to remember that a FDR-controlling procedure applied at a level, q, say 0.05, controls the false discovery proportion (the FDP) at 0.05 **only on average over hypothetical replications** of the set of tests under consideration in your study and does not guarantee that the proportion of false (rejections) discoveries in this set of tests in your study is 0.05. When there is non-negligible dependence among the test statistics in the set considered in your study, the actual false rejection proportion in this set may be much higher or lower than its expected value based on the FDR procedure (Ploner et al., 2006; Efron, 2007).

An FDR-controlling procedure can nevertheless be highly useful in exploratory research which involves testing many hypotheses. Such a procedure is useful as it can identify from among a huge set of tests, a subset of hypothesis tests that look promising for future hypothesis-confirming studies. By setting the nominal level of a FDR-controlling procedure to, for example 0.05, the procedure identifies a subset of hypothesis tests that on average has a false discovery proportion nominally equal to 0.05. The hypothesis tests in this subset are identified by having FDR-adjusted p-values that are not greater than 0.05. For example, the adjusted p-values from the Benjamini-Hochberg procedure, given in the results from Program 12.3, identify the hypotheses tested in Tests A, D, F, and G (which respectively had Benjamini-Hochberg FDR-adjusted p-values of 0.0015, 0.0030, 0.0095, 0.0356) as comprising the subset of hypotheses that look promising for future research.

It is extremely important, however, to realize that an adjusted p-value from an FDR-controlling procedure does not refer to the probability of a false discovery (Type I error) for an individual hypothesis test in the set under consideration. For example the FDR-adjusted p-value of 0.0030 for Test D from Benjamini-Hochberg's method (Program 12.3 Results) cannot be interpreted as the probability of making a mistake in rejecting the null hypothesis of Test D is 0.0030. Adjusted p-values from Benjamini-Hochberg's method and other FDR-controlling procedures should be used only to identify hypothesis tests that belong to the subset for which the false discovery rate is no greater than the specified level, e.g., 0.05. In contrast, an adjusted p-value for an individual test from a familywise error rate controlling procedure does provide information about the probability of Type I error for that particular individual test, taking into account the number of tests in the family. For example, the Bonferroni adjusted p-value of 0.0060 for Test D (Program 12.3 Results) can have the interpretation that the probability of making a mistake in rejecting the null hypothesis of Test D is 0.0060, taking into account there are 15 hypothesis tests considered in the family.

12.18   Recent Advances in FDR-Controlling Procedures

Much work has been done in FDR methodology since Benjamini and Hochberg's seminal paper in 1995. Goeman and Solari (2014) have provided an excellent review. Advances in FDR methodology are continuing to rapidly evolve. SAS has excelled in keeping apace with many of these advances and recently developed FDR-controlling methods are available for example in the MULTTEST, MIXED, and GLIMMIX procedures as described in SAS documentation for these procedures. The focus of recent statistical research in this area has been assessment of inaccuracies in FDR methods when the multiple test statistics are correlated, as well as the development of methods that specifically incorporate the dependence structure of the test statistics in the analysis. Although details from this theoretical research are beyond the scope of this book, we would like to emphasize several important points emerging from this research:

- Violation of the independence assumption for FDR-controlling procedures frequently occurs in many applications in high-throughput technologies.

- A FDR-controlling procedure may lead to inaccurate conclusions, unless dependence of the test statistics in the application has been taken into account.

- Even though many FDR-controlling procedures are available and easily executed in SAS, the help of a professional statistician should be enlisted so that the latest appropriate methods which consider dependence of the test statistics may be applied. The combined expertise of the statistician and the investigator(s) involved in the study are essential for sound and powerful data analysis in this rapidly evolving area of FDR-controlling procedures.

12.19 Chapter Summary

This chapter:

- describes the multiplicity problem in the context of multiple testing, viz., the greater the number of hypothesis tests you carry out, the greater the probability of your making at least one Type I error (falsely rejecting a null hypothesis) in your set of tests

- defines three measures of Type I error rate: comparisonwise, familywise, and false discovery error rate

- provides an overview of multiplicity-unadjusted procedures, familywise error rate controlling procedures, and FDR-controlling procedures

- describes the least significant difference (LSD) procedure as an example of a multiplicity-unadjusted procedure

- discusses and illustrates with SAS examples the following familywise error rate controlling procedures: Bonferroni, Tukey-Kramer for all pairwise comparisons, simulation-based method for all pairwise comparisons, Dunnett's method for "treatment" vs. control comparisons, and Scheffé's method for "data snooping"

- describes the multiplicity problem when multiple ordinary (multiplicity-unadjusted) confidence intervals are constructed

- discusses multiplicity-adjusted confidence intervals, illustrating with examples and SAS programs for simultaneous confidence intervals for all pairwise comparisons of group means from a one-way fixed effects ANOVA and simultaneous confidence bands for simple linear regression

- delineates the differences between single step and sequential step multiplicity-adjusted procedures describing Holm-Bonferroni's procedure as an example of a sequential method

- describes Benjamini-Hochberg method as an example of a FDR-controlling procedure

- discusses interpretation of results from FDR-controlling procedures

13

Analysis of Covariance: Adjusting Group Means for Nuisance Variables Using Regression

13.1 Introduction

Analysis of covariance, also known as ANCOVA, is a statistical procedure that:

- adjusts group means of a response variable for one or more nuisance variables (covariates) via regression

- compares the adjusted group means using confidence intervals and hypothesis tests based on the regression. These adjusted means of the response variable are the predicted values obtained from the regression.

You may recall from discussions in previous chapters that a variable is labelled a nuisance variable when it may influence the values of the response variable in a study and it is not directly of interest to the researcher's hypothesis. What a researcher would consider a nuisance variable depends on the objective of the analysis as illustrated in Sections 13.4 and 13.11.

In this chapter our focus will be on a one-way (single factor) analysis of covariance. The characteristic features of a one-way analysis of covariance model are described in the next section.

13.2 Characteristic Features of a One-Way Analysis of Covariance Linear Model

A one-way analysis of covariance model is a multiple linear regression model with the following characteristic features:

- There is only one response variable.

- The response variable is continuous.

- There are two or more explanatory (predictor) variables represented in the model.

- One of the explanatory variables is called **the study variable** which is always represented as a **qualitative** variable in the model via suitable coding (such as reference cell coding). The study variable is also referred to as the **treatment or group or focus variable**.

- One or more **covariates** variables are the other explanatory variables in the model which are known as **concomitant, or control or extraneous variables**. Covariates may be either **quantitative or qualitative variables**.

DOI: 10.1201/9780429107368-13

- A linear relationship is assumed to exist between the response variable and any continuous covariates in the model.

13.3   Why Apply an Analysis of Covariance?

In an observational or in a quasi-experimental study, you would want to apply an analysis of covariance as an attempt to control for non-random confounding (mixing up) of the effects of the study variable and one or more nuisance variables on the response variable.

In a randomized experiment (clinical trial), you would want to apply an analysis of covariance to reduce any bias due to a random baseline covariate imbalance that may occur between treatment groups (Koch et al., 1998; D'Agostino Jr and D'Agostino Sr, 2007).

In any type of study, whether observational, quasi-experimental, or a randomized experiment (clinical trial), you typically would want to apply an analysis of covariance to your data:

- as an attempt to **increase the efficiency of your analysis.** When you include a covariate in a model that is related to the response variable, you will reduce the mean square error (i.e., unexplained random variation of the response variable). Reducing mean square error in a linear model can increase the possibility of detecting a difference between the group population means of the response variable, decrease the width of the confidence interval for this difference, as well as increase the model's predictive accuracy of the response variable.

- to **estimate the mean value of a response variable for individuals with a specified covariate value in a focus (treatment) group**

13.4   Example 13.1: Exercise Programs and Heart Rate

The objective of this experiment was to compare the effect of two exercise programs on heart rate in males. A sample of male volunteers between 28 and 35 years of age was used for the subjects in this study. Eight males were randomly assigned to each exercise program. After completion of eight weeks of training in the assigned program, the response variable, heart rate after a six-minute run, was recorded for each male.

In this study, resting heart rate was considered a nuisance variable as it is known to affect heart rate after exercise and was not of direct interest to the researchers. They therefore decided to include baseline resting heart of each subject as a control variable in an analysis of covariance where baseline resting heart rate was the resting heart rate recorded for each subject at the time he volunteered for the experiment before he was randomly assigned to an exercise program. The investigators' objectives in including baseline resting heart rate as a covariate in the analysis were to adjust for any random imbalance in this variable between the exercise groups and to reduce the unexplained random variation of their response variable, thereby increasing their power to detect a difference between the exercise group population means of the response variable, heart rate after a six-minute run. The data for Example 13.1, given in Table 13.1, are a modified version of a data set from Table 3.15 given by Milliken and Johnson (2002). Section 13.8 provides details of the step-by-step approach

TABLE 13.1

Data for Example 13.1 – HR denotes heart rate after a six-minute run and BRHR denotes baseline resting heart rate of subject before randomization to Exercise Program A or B.

Subject	Exercise Program A HR	BRHR
1	131	58
2	138	59
3	142	62
4	147	68
5	160	71
6	166	76
7	165	83
8	171	87

Subject	Exercise Program B HR	BRHR
09	153	56
10	150	58
11	158	61
12	152	64
13	160	72
14	154	75
15	155	82
16	164	86

we implemented to apply an ANCOVA in which the response variable was heart rate after a six-minute test run and the two explanatory variables were exercise program (the group variable) and baseline resting heart rate (the covariate).

13.5 General Equation for a One-Way Analysis of Covariance with a Single Continuous Covariate

In this chapter we will focus on the simplest scenario for an ANCOVA where we want to adjust group means of a response variable, Y, for a single continuous covariate. It can be shown that a multiple regression model for this simplest scenario is based on the simple linear regression of Y vs. the covariate, X, for each of the g groups considered. The general model for this one-way ANCOVA can be written for $i = 1, \ldots, g$ as follows:

$$\mu_{Y_i|X} = \beta_{0i} + \beta_i X \tag{13.1}$$

where

$\mu_{Y_i|X}$ denotes the mean value of the response variable, Y, for all the individuals in population group i who have the same value of the explanatory variable, X

β_{0i} is the Y intercept of the straight line that describes the linear relationship between $\mu_{Y_i|X}$ and the explanatory variable X in the population of group i

β_i is the slope of the straight line that describes the linear relationship between $\mu_{Y_i|X}$ and the explanatory variable X in the population of group i. Often β_i is called the population regression coefficient for X in group i.

It may be helpful in subsequent sections in this chapter to keep in mind that a one-way ANCOVA involving a single continuous covariate is based on the general model of Equation 13.1, which encompasses a simple linear regression of Y vs. the covariate for each group in the analysis.

13.6   Two Critical Decisions in an Analysis of Covariance

If you want to adjust group response (outcome) variable means for a continuous covariate, you have to make two critical decisions before embarking on an analysis of covariance.

Decision Number 1: You have to decide whether it is reasonable to assume there is a linear relationship between the response variable (Y) and the covariate in each group.

Expert knowledge in the subject matter can be key in informing this decision. Also, as will be illustrated in Section 13.8, you can inform your decision regarding the plausibility of a linear relationship by applying a simple linear regression of Y vs. the covariate in each group so as to:

- informally evaluate the ideal inference condition of linearity using methods described in Chapter 3

- formally evaluate linearity, provided the other ideal inference conditions of this simple linear regression are reasonable for a group, via: estimation of a confidence interval for the covariate's regression coefficient; a Model F-test of the simple linear regression model or equivalently a test of the covariate's regression coefficient being equal to zero. Recall, however, from Step (8) Section 1.7 and Sections 1.8 to 1.10, that basing a decision solely on the basis of hypothesis tests and confidence interval estimates is not recommended.

The simple linear regression of the response variable vs. the covariate for each group and/or substantive prior research may suggest that:

- There is no relationship between the response variable and the covariate within any of the groups and hence a comparison of the group means, unadjusted for the covariate, based on a one-way ANOVA may be considered.

- There is a nonlinear relationship between the response variable and the covariate and hence a strategy to achieve linearity (Section 14.5) can be considered or a more complex nonlinear model (subsection 14.5.6) could be fitted.

- There is a relationship between the response variable and the covariate in each group that is optimally represented by a polynomial regression model, which includes the covariate, X as well as higher powers of X, e.g., X^2, X^3, in the model. Polynomial regression models are beyond the scope of this book but a brief description is given in subsection 14.5.5.

- There is a simple linear relationship between the response variable and the covariate in each group and hence an ANCOVA model is appropriate. Even if there is a linear relationship between the response variable and the covariate in at least one of the groups but no relationship between the response variable and covariate in the other groups being considered, an ANCOVA model may be considered.

Decision Number 2: If you decide that a linear model is appropriate for your analysis, you then have to decide which form of the model to apply, either **an unequal slopes model** or **an equal slopes model**.

An **unequal slopes model** is appropriate when the magnitude as well as the direction (sign) or just the magnitude of the difference between the study group means of the response variable depends on the value of the covariate.

Figure 13.1 is an example in which the slopes of the fitted regression line to the sample data for each group are unequal (i.e., the fitted regression lines for the two groups are clearly nonparallel) suggesting that an unequal slopes model is appropriate for the populations from which the data are drawn. If you compare the distance between the fitted lines in this figure, you will note that as you move along the X axis, you can see that the magnitude of the difference between the predicted means of Y for Groups (GRP) 1 and 2 increases with increasing values of the covariate, X, when X occurs in the range (1,20). However, the fitted regression lines displayed in Figure 13.1 (as well as those to be discussed in Figures 13.2 and 13.3) are based on sample data. Therefore we will need to apply statistical inferential methods (confidence intervals and/or hypothesis tests) to help us decide whether the slopes

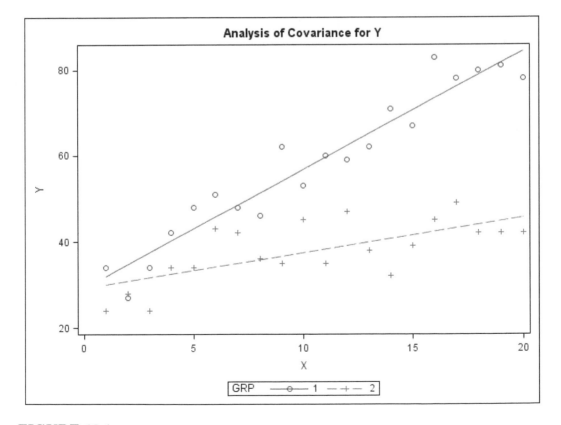

FIGURE 13.1
Fitted regression lines with unequal slopes where the magnitude of the difference in predicted mean Y between groups depends on the value of the covariate X.

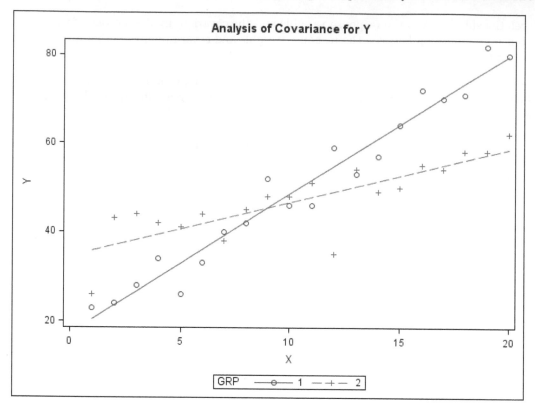

FIGURE 13.2

Fitted regression lines with unequal slopes where the direction and magnitude of the difference in predicted mean Y between groups depend on the value of the covariate X.

are actually unequal in the populations from which the sample data for Group 1 and 2 were drawn. We will describe these methods in later sections,

Figure 13.2 is another example in which the slopes of the fitted regression line to the sample data for each group are unequal suggesting that an unequal slopes model would be appropriate for the analysis. Examination of the fitted regression lines in this figure show that both the magnitude and direction of the difference in predicted mean Y between Groups 1 and 2 depend on the value of the covariate X, when X occurs in the range (1,20). At lower values of the covariate X, the predicted mean Y for Group 2 is greater than that predicted for Group 1 whereas at higher values of X the predicted mean Y for Group 1 is greater than that for Group 2.

An unequal slopes model can be written as a multiple linear regression interaction model which includes, in addition to the group indicator and covariate variables, **an interaction term representing an interaction effect between the group and covariate variables.** For example, one way you can request that an unequal slopes model be fitted to your data is by specifying the following CLASS and MODEL statement in the GLM procedure:

```
proc glm;
class grp;
model y=grp x grp*x ;
quit;
```

where

the CLASS statement specifies that the variable, grp, should be treated as a qualitative variable in the model

y denotes the response variable

grp denotes the predictor variable indicating the group

x denotes the continuous covariate

grp*x denotes the interaction term between the group and the covariate predictor variables which allows for unequal slopes in the fitted model

An **equal slopes model** assumes that the difference between the population means of the response variable in the two groups is the same regardless of the values of the covariate. This assumption is often referred to as **parallelism**. Figure 13.3 illustrates an example where the fitted regression lines for the sample data in the two groups have almost identical slopes in the observed range of the covariate, suggesting that an equal slopes model would be appropriate for the populations from which the samples were drawn.

An equal slopes model is an **additive** multiple linear regression model in which **no interaction is represented** between the group and covariate variable. An equal slopes model is also called a **common slopes model**. One way you can request an equal slopes model is by specifying the following CLASS and MODEL statements in the GLM procedure:

proc glm; class grp; model y=grp x; quit;

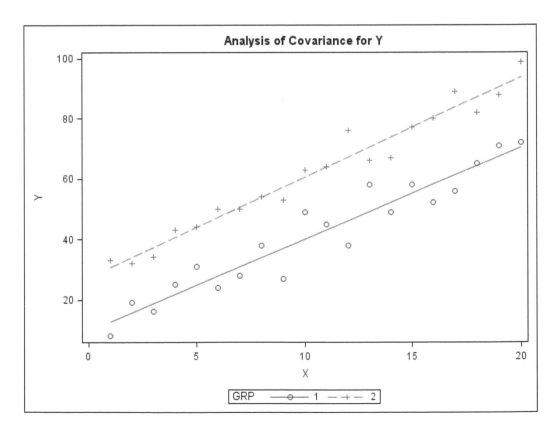

FIGURE 13.3

Fitted regression lines suggesting an equal slopes model would be appropriate for these data.

where

the CLASS statement specifies that the variable, grp, should be treated as a qualitative variable in the model

y denotes the response variable

grp denotes the predictor variable indicating the group

x denotes the continuous covariate

WARNING: When an interaction effect actually occurs between the group variable and a covariate in your data, but you incorrectly use an equal slopes model for your analysis, your conclusions may be entirely erroneous! Unfortunately, this important warning is often overlooked in many areas of research where results of an equal slopes model are routinely reported without evaluation of whether an equal or unequal slopes model is the appropriate model for the data.

We illustrate how you can be misled by application of an inappropriate model in Section 13.9 when we compare results from an unequal slopes and an equal slopes model applied to the data of Example 13.1 (the exercise experiment). In most research situations we suggest that you initially evaluate the fit of an unequal slopes model to your data. Details follow in Sections 13.7 and 13.8.

13.7   Implementing a One-Way ANCOVA A Step-by-Step Approach

In this section we suggest a step-by-step approach for implementing an ANCOVA for the simplest scenario, where you want to adjust the means of two groups for a single continuous covariate. Illustrative examples as well as SAS programs which implement these steps are given in Sections 13.8, 13.12, and in this chapter's appendix.

Steps A and B: Screen for data input and recording errors.

Step C: Review the details of your study's sampling design study to see if there are any obvious aspects of the design which would suggest that the proposed ordinary least squares model, which assumes an uncorrelated error structure, is inappropriate and that a more complicated model is required for your analysis.

If there is an obvious aspect of your sampling design, such as repeated measurements on the same individual, that would require a different model with a more complicated error structure, then stop at this step. Otherwise proceed to Step D.

Step D: Fit a simple linear regression model of Y (the response variable in the ANCOVA) vs. the covariate to the data in each group and evaluate whether ideal inference conditions of the simple linear regression are reasonable for each group using methods described in Section 3.13.

If any ideal inference conditions are not reasonable for either group, alternative approaches described in Chapter 14 may be considered.

Step E: Check whether any cases merit investigation due to their disproportionate influence in the simple linear regression applied to each group in Step D. Measures of influence described in Chapter 3 can be used for Step E.

Step F: Further evaluate whether ideal inference conditions of independence of errors and equality of variances are reasonable.

Independence of Errors Across Groups

Independence of errors across groups can be investigated by combining in a single graph **all** the internally studentized residuals from each group's simple linear regression of Y vs. the covariate against any key variable (e.g., a time or space variable) which may affect the response variable but has not been held constant or has not been part of a randomization protocol in the study.

By evaluating independence of errors across groups you may detect a dependence of errors that would not have been evident when you evaluated independence of errors solely within groups. This lack of detection of dependence of errors within groups could be due to small group sample sizes. Or there might be scenarios where a pattern of dependence between errors in different groups would be missed when independence of errors was evaluated only within groups. For example, suppose if in the mussel weight study (Example 9.1, Section 9.4), in the absence of instructions from the investigators, their research assistant weighed all the mussels from location A in the morning and all the mussels from location B in the afternoon of the same day. (The rationale of the research assistant in doing so was to avoid inadvertently mixing of specimens from the two locations.) Unfortunately by the afternoon the scales had drifted out of calibration such that they were underestimating the mussel weights. As a result, the errors, i.e., unexplained deviations from the model, were larger for group B than for group A (as group B were all measured in the afternoon). If a data analyst checked a plot of internally studentized residuals vs. time of weighing solely within each group, this dependence of errors between groups would be missed. Within each group the plot of internally studentized residuals vs. time of weighing would have exhibited a random pattern.

Equality of Variances Between Groups

Having evaluated the equality of error variances at each value of the covariate within each group in the simple linear regressions in Step D and found this to be reasonable, you can proceed to informally evaluate equality of error variances between groups by visually comparing the spread of internally studentized residuals from the simple linear regression of Y vs. the covariate for each group in a graphical display (as we illustrate in Section 13.8 with Figure 13.4).

You could also evaluate equality of error variances between groups by forming a ratio of the largest mean square error to the smallest mean square error from the simple linear regression of Y vs. the covariate for each group. It can be shown that a mean square error from a linear regression is an unbiased estimate of the population variance when the least squares ideal inference conditions hold. If the ratio of the largest to the smallest mean square error is no more than approximately four, and if other ideal inference conditions do not appear to be violated, then usually there is not concern regarding a serious impact on the analysis, unless research subject expertise suggests otherwise.

Proceed to Step G, if you have decided that the ideal inference conditions of independence of errors, equality of variances, and normality are reasonable. Otherwise, consider alternative approaches described in Chapter 14.

Step G: Evaluate whether an unequal slopes rather than an equal slopes model is appropriate for your analysis.

You can implement Step G by:

- drawing on subject matter expertise regarding the plausibility of an unequal slopes model

- estimating the slopes and standard errors of the population regression lines of Y vs. the covariate for each group

- informally comparing the fitted simple linear regression lines of Y vs. the covariate along with the original data points for each group in a single plot (i.e., a conditional effects plot) and checking if the lines fitted to the sample data in each group appear to be non-parallel, i.e., have unequal slopes

- fitting a multiple regression interaction (unequal slopes) model to the data from both groups and formally evaluating if the slopes for the simple linear regression line of Y vs. the covariate are different in each group's population by:

  - estimating a confidence interval[1] for the difference in these population slopes
  - testing the null hypothesis of no difference between these population slopes where a small p-value can be suggestive of inequality of slopes in the sampled populations

When inequality of the groups' population slopes in the simple linear regression line of Y vs. the covariate is indicated in Step G, we recommend that you proceed to Step H and complete your analysis using the unequal slopes model. Otherwise, go to Step I and apply an equal slopes model (additive multiple regression model) to the data. However, if prior substantive information and/or expert opinion suggest that the true population values of the slopes are actually unequal and your study lacked power to detect this inequality you may decide to go to Step H.

Step H: Using an unequal slopes model, evaluate group differences in covariate-adjusted means of the response variable (Y) at specific covariate values. As we will illustrate (Section 13.8), a covariate-adjusted mean for a group at a particular covariate value is estimated in an unequal slopes model by the **predicted value** for that particular covariate value from the group's simple linear regression of Y vs. the covariate. Various options for implementing Step H include:

- estimating multiplicity-adjusted confidence bounds (e.g. discrete simulation-based or Working-Hotelling confidence bounds) for the **difference between the predicted values** from each group's simple linear regression of Y vs. the covariate at each value of the covariate within the entire observed range of covariate values common to both groups

- applying multiplicity-adjusted (e.g., discrete simulation-based or Scheffé's) tests of the null hypothesis of **zero difference between the predicted values** from each group's simple linear regression of Y vs. the covariate at regular intervals within the entire observed range of covariate values common to both groups

Please note: You should implement the next step and apply an equal slopes model only if you were unable to detect a difference in the groups' population slopes in Step G, or if substantive information does not suggest a difference in the groups' population slopes.

Step I: Using an equal slopes model, evaluate group differences in covariate-adjusted means of the response variable.

---

[1]Typically 95% but the context of the research area may indicate a different level e.g., 90%, 99%.

13.8   An Example of Implementing an Analysis of Covariance

The suggested step-by-step approach outlined in Section 13.7 is illustrated in this section by applying an analysis of covariance to the exercise experiment (Example 13.1) where there are two groups (two categories of the qualitative explanatory variable, exercise program) and a single continuous covariate, baseline resting heart rate.

Steps A and B: Screen for data input and recording errors. No data input nor
   recording errors were found for the exercise experiment data given in Table 13.1. See
   Section 13.A for details.

**Step C: Review the details of your study's sampling design study to see if there
   are any obvious aspects of the design which would suggest that the proposed
   ordinary least squares model which assumes an uncorrelated error structure
   is inappropriate and that a more complicated model is required for your
   ANCOVA.**

The following known details suggest a model with a more complicated error structure
is not required to reflect the sampling design of the study:

- Each individual was represented only once in the study.

- The subjects in the study were assigned completely at random to exercise program A
  or B, regardless of their baseline resting heart rate or any other variable that might
  have an effect on the response variable.

- The subjects were from a single source, viz., males between the ages of 28 and 35
  who volunteered for this experiment.

- None of the subjects were related by birth or shared the same household.

- Measurement error of the predictor, baseline resting heart rate, was investigated at
  the planning stage of the study and found to be negligible.

**Step D: Fit a simple linear regression model of Y (the response variable in the
   ANCOVA) vs. the covariate to the data in each exercise group and evaluate
   whether ideal inference conditions of the simple linear regression are reasonable for each group using methods described in Section 3.13.**

A simple linear regression of heart rate after a six-minute test run (Y) vs. baseline resting
heart rate (the covariate) was applied in each exercise group using Program 13.2 (Section
13.D). Ideal inference conditions of this simple linear regression seem reasonable for each
group, based on results from applying the nine-step approach (Section 3.13) to the data
in each exercise program. Details are given in Section 13.A.

**Step E: Check whether any cases merit investigation due to their dispropor-
   tionate influence in the simple linear regression applied to each group in the
   previous step using methods described in Chapter 3.**

None of the cases in exercise group A merited investigation for their influence on the simple linear regression of heart rate after a six-minute test run (Y) vs. baseline resting heart
rate (the covariate) applied to this group's data. Examination of the output from Program
13.2, revealed that no case in exercise group A had a DFBETA or DFFITS greater than
1.0, a threshold value recommended for small to medium samples by Kutner et al. (2005)
and no case had a Cook's D greater than 0.5 (subsection 3.18.2). However, examination
of the output from Program 13.2 revealed that, in exercise group B, the case with ID

equal to 16 had a DFFITS value equal to 1.3 and a Cook's D equal 0.7, suggesting that this case might be slightly influential in the simple linear regression of heart rate after a six-minute test run vs. baseline resting heart rate in this exercise group. Investigation of this case revealed nothing in particular to distinguish this case from the others in exercise group B.

Step F: Further evaluate whether ideal inference conditions of independence of errors, equality of variances, and normality are reasonable.

Independence of Errors Between Groups

We concluded that the ideal inference condition of independence of errors between groups was reasonable as this study was an experiment with a completely randomized design structure, i.e., the subjects who volunteered for the experiment were assigned completely at random to the exercise programs. There were no withdrawals from the study. Moreover, aspects of the design related to measurement of the Y variable in both groups were either held constant or assigned at random to all subjects in the experiment.

Equality of Group Population Variances

Figure 13.4 (obtained from the SGPLOT procedure in Program 13.2, Section 13.D) suggests that the ideal inference condition of equal group population variances is reasonable.

We also investigated equality of group error variances by computing the ratio of the Mean Square Errors from the simple linear regression of heart rate after a six-minute test

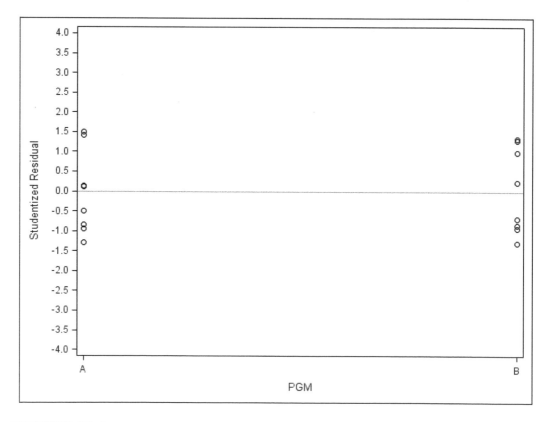

FIGURE 13.4

Comparison of internally studentized residuals from each group's simple linear regression of heart rate vs. baseline resting heart rate.

run vs. baseline resting heart rate for each group with the largest Mean Square Error as the numerator. The Mean Square Error from this simple linear regression for exercise group A is 23.85311, as given in the following Analysis of Variance table for "PGM = A" from Program 13.2:

PGM=A
Analysis of Variance

Source	DF	Sum of Squares	Mean Square	F Value	Pr > F
Model	1	1406.88136	1406.88136	58.98	0.0003
Error	6	143.11864	23.85311		

The Mean Square Error from this simple linear regression of heart rate after a six-minute test run vs. baseline resting heart rate in exercise group B is 14.69616, as reported in the following Analysis of Variance table for "PGM = B" from Program 13.2:

PGM=B
Analysis of Variance

Source	DF	Sum of Squares	Mean Square	F Value	Pr > F
Model	1	61.32303	61.32303	4.17	0.0871
Error	6	88.17697	14.69616		

The ratio of the Mean Square Errors from these simple linear regressions in exercise programs A and B is 23.85311/14.69616 = 1.62, The small size of this variance ratio combined with Figure 13.4 and the fact that the sample sizes were equal in each exercise group indicates it is reasonable to pool these two Mean Square Errors from the simple linear regression to form the Mean Square Error in the unequal slopes model. This pooled Mean Square Error is 19.275 as reported in the following Analysis of Variance table from Program 13.3A (Section 13.E).

Source	DF	Sum of Squares	Mean Square	F Value
Model	3	1510.454385	503.484795	26.12
Error	12	231.295615	19.274635	
Corrected Total	15	1741.750000		

Normality

No serious departure from normality was detected in examining normal Q-Q plots of internally studentized residuals from the separate simple linear regressions for each group (Section 13.A, Step 8). Similarly, no serious departure from normality was detected in Figure 13.5, a normal Q-Q plot of internally studentized residuals from the unequal slopes model generated by Program 13.3A. Comparison of regression diagnostics generated by Programs 13.2 and 13.3A show that internally studentized residuals from the unequal slopes model are comprised by the same raw residuals as their counterparts from the separate simple linear regressions for each group. However internally studentized residuals from the unequal slopes model are different from their counterparts in the separate linear regressions for each

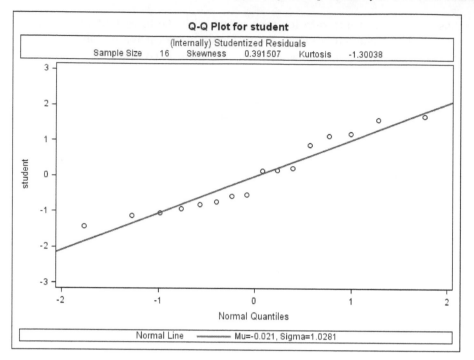

FIGURE 13.5
Normal Q-Q plot of internally studentized residuals from an unequal slopes model applied
to the exercise program data of Example 13.1.

group because in the unequal slopes model the pooled mean square error is utilized in
standardizing the raw residuals.

**Step G: Evaluate whether an unequal slopes rather than an equal slopes model
is appropriate for your analysis.**

The point estimates and associated estimated standard errors (given in parentheses)
of the slopes of the population simple linear regression lines of HR (Y) vs. BRHR (the
covariate) for each exercise group, 1.30 (0.17) for Group A and 0.26 (0.13) for Group B,
suggest that the true values of the population slopes for these regression lines are unequal.
These point estimates of the slopes and their standard errors are reported by the REG
procedure of Program 13.2 as the BRHR Parameter Estimate for PGM = A and PGM =
B as follows:

PGM=A

Variable	DF	Parameter Estimate	Standard Error	t Value	Pr > |t|
Intercept	1	60.49153	12.10421	5.00	0.0025
BRHR	1	1.30508	0.16993	7.68	0.0003

PGM=B

Variable	DF	Parameter Estimate	Standard Error	t Value	Pr > |t|
Intercept	1	137.48497	9.04365	15.20	<.0001
BRHR	1	0.26375	0.12912	2.04	0.0871

To continue our evaluation of the appropriateness of an unequal slopes model for this study, we examined Figure 13.6, a conditional effects plot of the fitted simple linear regression lines of HR vs. BRHR for each group generated by Program 13.3A. Figure 13.6 strongly suggests that an unequal slopes model is appropriate for the ANCOVA. The fitted simple linear regression lines of HR vs. BRHR for exercise programs A and B are clearly not parallel in this figure.

We then executed Program 13.3B (Section 13.F) to formally evaluate the appropriateness of an unequal slopes model by fitting this model to the data from both groups and estimating a 95% confidence interval for the true difference between the slopes and testing the null hypothesis that this difference is equal to zero. The following results were obtained from the GLM procedure in Program 13.3B:

```
                   Dependent Variable: HR


                                    Standard
Parameter                Estimate     Error    t Value   Pr > |t|

Slope A-Slope B        1.04132978  0.21260440     4.90     0.0004


            Parameter              95% Confidence Limits
            Slope A-Slope B        0.57810459   1.50455498
```

The preceding results suggest that there is a difference in slopes between the groups' population simple linear regression lines of HR vs. BRHR. The observed difference in slopes based on this study's sample data is 1.04 with a standard error of 0.21. The estimated

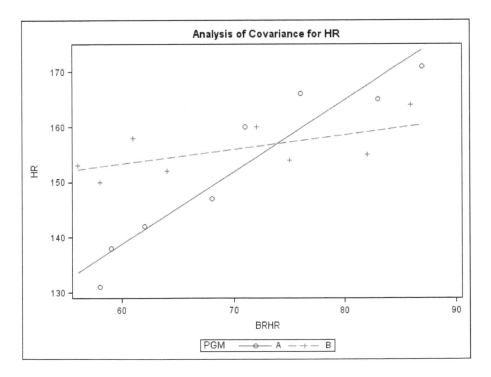

FIGURE 13.6

Fitted simple linear regression lines for heart rate after a six-minute test run (HR) vs. baseline resting heart rate (BRHR) for exercise program (PGM) A and B.

95% confidence limits for the difference in slopes in the sampled populations (0.58, 1.50) suggest that zero is not a plausible value for this difference in slopes that is compatible with the data. Moreover, the estimated upper limit for this interval indicates that the difference between population slopes could be as large as 1.50.

The p-value of 0.0004 reported from the t-test of the null hypothesis that the difference between slopes is equal to zero suggests that it is unlikely that the slopes for Groups A and B are equal in the sampled populations. It may be noted that this t-test is equivalent to the Type III sums of squares F-test of the regression coefficient for the interaction effect being equal to zero. This is shown by the following results for BRHR*PGM reported in the Type III Model ANOVA table from Program 13.3A:

```
Source          DF Type III SS  Mean Square  F Value Pr>F
BRHR*PGM        1   462.400432   462.400432   23.99   0.0004
```

The results for the main effects PGM and BRHR from the Type III Model ANOVA table of Program 13.3A are not pertinent to Step G and therefore are not reported here but see comments in Section 13.E for a discussion as to why these results for PGM and BRHR are typically not of substantive interest.

Summary of Step G
The results of Step G indicate that the slopes of the regression lines of HR vs. BRHR are unequal between groups and thus the difference in the group means for HR is not the same over the entire range of observed values for BRHR. Therefore we should proceed to Step H.

Step H: Using an unequal slopes ANCOVA model, evaluate group differences in covariate-adjusted means of the response variable at various values of the covariate. In this step, group differences in BRHR-adjusted population means of HR are estimated by computing the difference in predicted values from each group's separate simple linear regression at various observed values of BRHR displayed in Figure 13.6. To draw inferences about the differences in mean heart rate between Groups A and B populations, adjusted for various values of BRHR, we will consider OPTIONS 1–4 for Step H.

STEP H: OPTION 1
Discrete Simulation-Based Multiplicity-Adjusted Confidence Bounds
One way to evaluate group differences in covariate-adjusted means of the response variable at a range of BRHR values is by applying the discrete simulation-based method (Westfall et al., 2011) to estimate multiplicity-adjusted confidence bounds for the difference between the predicted values of heart rate from each exercise group's simple linear regression where Y=heart rate after a six-minute run and X=baseline resting heart rate. Westfall et al. (2011, p. 268) advised that although these discrete simulation-based multiplicity-adjusted confidence bounds will underestimate the true simultaneous confidence level, this should not be a problem when the number of discrete points used in this simulation method is large. To avoid extrapolation beyond the observed data, we estimated confidence bounds for the difference in predicted values at BRHR values in the range of 58–86, a range common to both groups. These discrete simulation-based confidence bounds for the difference in predicted values from each group's simple linear regression line (shown in Figure 13.6) are displayed in Figure 13.7 (generated by Program 13.4, Section 13.G).

Evidence suggesting a difference between Exercise Programs A and B populations' mean heart rate after a six-minute run, adjusted for baseline resting heart rate (BRHR), is indicated if the estimated upper and lower confidence limits for the difference are completely below or completely above the zero reference line. Thus, from Figure 13.7 we see there is

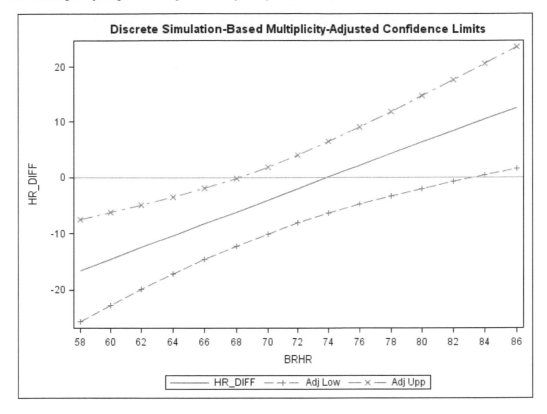

FIGURE 13.7
Discrete simulation-based multiplicity-adjusted confidence limits for HR_DIFF where HR_DIFF is the difference between Exercise Groups A and B in mean heart rate adjusted for baseline resting heart rate (BRHR) in an unequal slopes analysis.

evidence to suggest that there is a difference in BRHR-adjusted mean heart rate between Exercise Programs A and B for those individuals in the sampled population (male volunteers between the ages of 28 and 35) who have baseline resting heart rates in the range of 58–68 and in the range of 84–86. The ORTHOREG procedure, used in Program 13.4 to generate the confidence limits for the differences shown in Figure 13.7, designates Exercise Program B as the reference group (i.e., a difference is computed by subtracting the covariate-adjusted mean of Group B from that of Group A). Therefore, Figure 13.7 suggests:

- in the range of BRHR values of 58–68, where the estimated upper and lower confidence limits are completely below the zero reference line, Exercise Program B has a greater mean heart rate, adjusted for baseline resting heart rate, than that of Exercise Program A

- in the range of BRHR values of 84–86, Exercise Program A population has a greater mean heart rate, adjusted for baseline resting heart rate, than that of Exercise Program B (the estimated upper and lower confidence limits for the differences in this range of BRHR values are completely above the zero reference line).

The following results from Program 13.4 generated by the CL option on the ESTI-MATE statement of the ORTHOREG procedure give the estimated upper and lower discrete simulation-based multiplicity-adjusted confidence limits at specified values of the covariate.

These estimated upper and lower limits are displayed in Figure 13.7. For example, for individuals who have a baseline resting heart rate of 58 (Label = 58), the estimated lower limit (Adj Lower) of the discrete simulation-based multiplicity-adjusted confidence interval for the difference between exercise groups in predicted values (from the simple linear regressions of HR vs. BRHR for each exercise group) is -25.7030 and the estimated upper limit (Adj Upper) for this confidence interval is -7.4896. This confidence interval estimate (-25.7030, -7.4896) suggests a range of plausible values for the difference between the two exercise program populations in mean heart rate after a six-minute run, adjusted for a baseline resting heart rate of 58. Ordinary confidence limit estimates not adjusted for multiplicity are also reported by this CL option, e.g., at BRHR = 58, the multiplicity-unadjusted ordinary 95% confidence interval estimate is (-23.8985, -9.2942), which is as you would expect, not as wide as its multiplicity-adjusted counterpart. However, conclusions based on multiple ordinary confidence intervals in a study can be misleading and are more likely to lead to replication failure, i.e., failure to draw the same conclusions when you or other investigators replicate the study.

```
                         Estimates
            Adjustment for Multiplicity: Simulated
                                      Adj         Adj
   Label      Lower       Upper       Lower       Upper

    58      -23.8985     -9.2942    -25.7030     -7.4896
    60      -21.1446     -7.8827    -22.7833     -6.2440
    62      -18.4588     -6.4032    -19.9485     -4.9135
    64      -15.8635     -4.8332    -17.2264     -3.4703
    66      -13.3857     -3.1456    -14.6510     -1.8803
    68      -11.0543     -1.3117    -12.2582     -0.1079
    70       -8.8922      0.6914    -10.0763      1.8756
    72       -6.9074      2.8720     -8.1158      4.0804
    74       -5.0901      5.2200     -6.3640      6.4939
    76       -3.4162      7.7114     -4.7911      9.0864
    78       -1.8569     10.3175     -3.3612     11.8218
    80       -0.3854     13.0113     -2.0408     14.6667
    82        1.0200     15.7712     -0.8027     17.5939
    84        2.3759     18.5806      0.3736     20.5829
    86        3.6946     21.4273      1.5034     23.6184
```

STEP H: OPTION 2
Discrete Simulation-Based Multiplicity-Adjusted Tests

The following results from the ESTIMATE statement of the ORTHOREG procedure in Program 13.4 compare multiplicity-unadjusted (raw) p-values with discrete simulation-based multiplicity-adjusted p-values obtained from tests of the null hypothesis that there is no difference between Exercise Programs A and B in mean heart rate, adjusted for baseline resting heart rate (BRHR), at specified values of BRHR:

```
                         Estimates
            Adjustment for Multiplicity: Simulated

                      Standard
Label   Estimate       Error      DF   t Value   Pr > |t|   Adj P   Alpha

 58     -16.5963       3.3514      12    -4.95     0.0003    0.0011   0.05
```

60	-14.5137	3.0434	12	-4.77	0.0005	0.0014	0.05
62	-12.4310	2.7666	12	-4.49	0.0007	0.0023	0.05
64	-10.3483	2.5313	12	-4.09	0.0015	0.0046	0.05
66	-8.2657	2.3499	12	-3.52	0.0042	0.0123	0.05
68	-6.1830	2.2358	12	-2.77	0.0171	0.0460	0.05
70	-4.1004	2.1993	12	-1.86	0.0869	0.2012	0.05
72	-2.0177	2.2442	12	-0.90	0.3863	0.6598	0.05
74	0.06496	2.3660	12	0.03	0.9785	0.9996	0.05
76	2.1476	2.5536	12	0.84	0.4168	0.6938	0.05
78	4.2303	2.7938	12	1.51	0.1559	0.3307	0.05
80	6.3129	3.0743	12	2.05	0.0625	0.1505	0.05
82	8.3956	3.3852	12	2.48	0.0290	0.0750	0.05
84	10.4783	3.7187	12	2.82	0.0155	0.0420	0.05
86	12.5609	4.0693	12	3.09	0.0094	0.0263	0.05

The "Label" column of the preceding results from the ESTIMATE statement of the ORTHOREG procedure in Program 13.4 gives BRHR values. The "Estimate" column reports the difference in predicted values from the simple linear regressions of HR vs. BRHR for each exercise group at BRHR values from 58 to 86. For example at BRHR = 58 the difference in predicted values is -16.5963, which is computed by subtracting the predicted value for Exercise Program B (the reference group) in the simple linear regression of HR vs. BRHR at BRHR = 58 from its counterpart for Exercise Program A. As this difference in predicted values at BRHR = 58 is **negative**, this indicates that the mean heart rate after a six-minute run, adjusted for baseline resting heart rate, is **less for Exercise Program A than that for Exercise Program B in the study population** of male volunteers assigned at random to either program A or B. The discrete simulation-based multiplicity-adjusted p-value (Adj P) associated with the test of zero difference in BRHR-adjusted mean heart rate between the exercise programs at BRHR = 58 in the sampled population is 0.0011. This small p-value suggests that the data in the study are not compatible with the null hypothesis of zero difference assuming ideal inference conditions hold. As we evaluated these ideal inference conditions and found them reasonable for the data we conclude that at BRHR = 58, there is a difference between exercise programs in the mean heart rate after a six-minute run, adjusted for baseline resting heart rate of 58 in the population of male volunteers who were assigned at random to either Exercise Program A or B.

These preceding results reveal that there is evidence suggesting that there are also differences between predicted values for Exercise Programs A and B at baseline heart rate values, 60, . . . ,66, 68. We note the differences at these baseline heart rate values are negative, indicating that the mean heart rate, adjusted for baseline resting heart rate, is less for Exercise Program A than that for Exercise Program B in the study population of male volunteers.

At baseline resting heart rate values from 70 to 80, the evidence is inconclusive based on the results of the estimated differences in BRHR-adjusted mean heart rate relative to their corresponding standard errors and the relatively large multiplicity-adjusted p-values from the test of the null hypothesis of zero difference between the exercise groups.

At the baseline resting heart rate of 82, the estimated difference in the covariate-adjusted mean heart rates between the exercise groups is 8.4 with a standard error of 3.39. The p-value yielded by the discrete simulation-based multiplicity-adjusted method of the test of zero difference between groups is 0.08 suggesting that at BRHR = 82 there may be a difference in covariate-adjusted mean heart rates between the exercise groups in the sampled population. However as this discrete simulation-based method tends to be liberal, we cannot entirely rule out that a p-value of 0.08 from this method may be an underestimate of the true p-value.

Estimates of the differences between exercise groups in covariate-adjusted mean heart rate at baseline resting heart rate values of 84 and 86 respectively are 10.48 and 12.56. As these differences are **positive** this indicates that the mean heart rate after a six-minute run, adjusted for baseline resting heart rate, is **greater** for Exercise Program A than that for Exercise Program B in the study populations of male volunteers assigned at random to either program A or B. The multiplicity-adjusted p-values obtained from a test of zero difference at baseline resting heart rates of 84 and 86, respectively 0.04 and 0.03, suggest that the null hypothesis of zero difference between the exercise programs may be incompatible with the data.

Comment: Raw p-values not adjusted for multiple testing are given in the column "Pr > |t|" in the preceding results. However, drawing conclusions from these raw p-values can be misleading and may result in replication failure.

STEP H: OPTION 3
Working-Hotelling Confidence Bounds. Discrete simulation-based multiplicity-adjusted confidence bounds (OPTION 1) can be somewhat liberal, i.e., they underestimate the width of the confidence interval whereas multiplicity-adjusted confidence bounds obtained from the Working-Hotelling method can be conservative, i.e., they overestimate the width of the confidence interval (Westfall et al., 2011). In this example there is not much difference between the results obtained from the two methods. As given in the "Simulation Results" section of the output from Program 13.4 that is generated by the REPORT option in the ORTHOREG procedure, the critical value from the Working-Hotelling (Scheffé) method for the confidence bounds is 2.787577, which is only slightly larger than 2.724554, the critical value from the discrete simulation-based method for the confidence bounds. The objective of your study may indicate which method you should choose, depending whether you want to slightly err on the conservative or liberal side.

As illustrated in Program 13.5, the Working-Hotelling confidence bounds can be requested by simply replacing

/adjust=simulate(acc=0.0002 report seed=4538241) cl;
in Program 13.4
with
/adjust=scheffe cl;
as the option for the ESTIMATE statement in the ORTHOREG procedure.

Figure 13.8, which displays the Working-Hotelling confidence bounds, reveals that for this exercise example there is only a slight difference between Working-Hotelling and discrete simulation-based confidence bounds for the difference between the two regression lines of Exercise Program A and B.

STEP H: OPTION 4
Scheffé's Multiplicity-Adjusted Tests. The following results from Program 13.5 provide Scheffé's multiplicity-adjusted p-values which have been adjusted for multiple tests of the null hypothesis of zero difference between the BRHR-adjusted mean heart rate of Exercise Programs A and B at all values of BRHR from minus infinity to plus infinity.

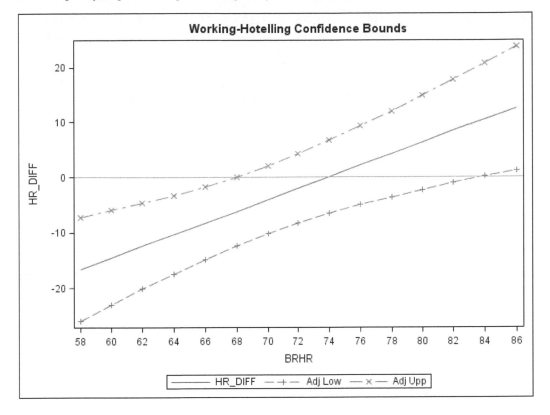

FIGURE 13.8
Working-Hotelling multiplicity-adjusted confidence bounds for HR_DIFF where HR_DIFF is the difference between Exercise Groups A and B in mean heart rate adjusted for baseline resting heart rate (BRHR) in an unequal slopes analysis.

RESULTS FROM SCHEFFÉ'S METHOD: UNEQUAL SLOPES MODEL

Exercise Program and Heart Rate Study

The ORTHOREG Procedure

Dependent Variable: HR
Estimates
Adjustment for Multiplicity: Scheffe
Exercise Program and Heart Rate Study

Label	Estimate	Standard Error	t Value	Pr > |t|	Adj P	Adj Low	Adj Upp
58	-16.5963	3.3514	-4.95	0.0003	0.0013	-25.9387	-7.2539
60	-14.5137	3.0434	-4.77	0.0005	0.0017	-22.9973	-6.0300
62	-12.4310	2.7666	-4.49	0.0007	0.0027	-20.1430	-4.7190
64	-10.3483	2.5313	-4.09	0.0015	0.0053	-17.4044	-3.2923
66	-8.2657	2.3499	-3.52	0.0042	0.0142	-14.8163	-1.7151
68	-6.1830	2.2358	-2.77	0.0171	0.0519	-12.4154	0.04936
70	-4.1004	2.1993	-1.86	0.0869	0.2173	-10.2310	2.0303

72	-2.0177	2.2442	-0.90	0.3863	0.6763	-8.2736	4.2382
74	0.06496	2.3660	0.03	0.9785	0.9996	-6.5304	6.6603
76	2.1476	2.5536	0.84	0.4168	0.7092	-4.9707	9.2659
78	4.2303	2.7938	1.51	0.1559	0.3503	-3.5577	12.0182
80	6.3129	3.0743	2.05	0.0625	0.1642	-2.2570	14.8829
82	8.3956	3.3852	2.48	0.0290	0.0835	-1.0408	17.8320
84	10.4783	3.7187	2.82	0.0155	0.0475	0.1121	20.8444
86	12.5609	4.0693	3.09	0.0094	0.0300	1.2173	23.9045

Comparison of these test results from Scheffé's method with that given previously for discrete simulation-based tests (OPTION 2) from Program 13.4 shows similarity of conclusions based on the multiplicity-adjusted p-values from the two methods.

13.9 What If an Equal Slopes Analysis Had Been Applied to the Exercise Program Data?

Suppose the researchers in the exercise experiment did not check whether an unequal slopes model was more appropriate than an equal slopes model for their ANCOVA. Suppose they fit an equal slopes model using the following SAS statements from Program 13.6 (Section 13.I):

```
proc glm data=pgm13;
class pgm;
model hr=pgm brhr;
run;
```

The following overall ANOVA and Type III sums of squares results for an equal slopes analysis were obtained by executing these preceding SAS statements in Program 13.6.

OVERALL ANOVA RESULTS FOR EQUAL SLOPES ANALYSIS

Dependent Variable: HR

Source	DF	Sum of Squares	Mean Square	F Value
Model	2	1048.053953	524.026977	9.82
Error	13	693.696047	53.361234	
Corrected Total	15	1741.750000		

TYPE III MODEL ANOVA RESULTS FOR EQUAL SLOPES ANALYSIS

Source	DF	Type III SS	Mean Square	F Value	Pr > F
PGM	1	70.616718	70.616718	1.32	0.2707
BRHR	1	1005.803953	1005.803953	18.85	0.0008

From the previous results from the equal slopes analysis we note that:

- The Mean Square Error is 53.36, i.e., more than double the Mean Square Error of 19.27, obtained when an unequal slopes analysis is applied to these data.

- The p-value for the test of the regression coefficient of PGM being equal to zero is 0.27, indicating that when an equal slopes analysis is applied to these data, there is insufficient evidence to conclude in the sampled population whether or not type of exercise program (A vs. B) has an effect on mean heart rate when adjusted for any of the baseline resting heart rates observed in the study. This is in contrast to what was concluded when an unequal slopes analysis was applied to these data.

13.9.1   What Is Going On?

What is happening is that when we specify an equal slopes model for an ANCOVA in the GLM procedure or in any other linear modelling procedure, we are specifying that we want the procedure to fit the data in each group to a simple regression line of Y, the response variable vs. the covariate, such that the regression lines for the two groups are **parallel**. Obviously whenever these two regression lines are parallel, the distance between them is the same all along the X axis. Thus, when we specify an equal slopes model for the exercise experiment we are constraining the difference in Y between groups to be identical regardless of the value of BRHR (the covariate).

We confirm this by considering the following results from Scheffé's multiplicity-adjusted method from Program 13.6 in which an equal slopes analysis is applied to the exercise program data. You will note that at each value of the covariate, BRHR (reported in the "Label" column in the output), the estimated difference between Exercise Programs A and B is identical (Estimate = -4.2094) with an identical p-value = 0.27.

RESULTS FROM SCHEFFÉ'S METHOD: EQUAL SLOPES MODEL

Estimates
Adjustment for Multiplicity: Scheffe

Label	Estimate	Standard Error	DF	t Value	Pr > |t|	Adj P	Alpha
58	-4.2094	3.6591	13	-1.15	0.2707	0.2707	0.05
60	-4.2094	3.6591	13	-1.15	0.2707	0.2707	0.05
62	-4.2094	3.6591	13	-1.15	0.2707	0.2707	0.05
64	-4.2094	3.6591	13	-1.15	0.2707	0.2707	0.05
66	-4.2094	3.6591	13	-1.15	0.2707	0.2707	0.05
68	-4.2094	3.6591	13	-1.15	0.2707	0.2707	0.05
70	-4.2094	3.6591	13	-1.15	0.2707	0.2707	0.05
72	-4.2094	3.6591	13	-1.15	0.2707	0.2707	0.05
74	-4.2094	3.6591	13	-1.15	0.2707	0.2707	0.05
76	-4.2094	3.6591	13	-1.15	0.2707	0.2707	0.05
78	-4.2094	3.6591	13	-1.15	0.2707	0.2707	0.05
80	-4.2094	3.6591	13	-1.15	0.2707	0.2707	0.05
82	-4.2094	3.6591	13	-1.15	0.2707	0.2707	0.05
84	-4.2094	3.6591	13	-1.15	0.2707	0.2707	0.05
86	-4.2094	3.6591	13	-1.15	0.2707	0.2707	0.05

When we compare these preceding results from Scheffé's tests based on an equal slopes analysis with results of Scheffé's tests based on an unequal slopes analysis (STEP H, OPTION 4, Section 13.8) we see that entirely different conclusions may be drawn from the two analyses. In contrast to the results from the equal slopes analysis, the results from the unequal slopes analysis suggest that Exercise Program B had a greater BRHR-adjusted mean heart rate than Exercise Program A at specified BRHR values in the range from 58 to 68 and that Exercise Program A had a greater BRHR-adjusted mean heart rate than Exercise Program B at BRHR values 84 and 86. In STEP G, Section 13.8, we concluded that an unequal slopes model is more appropriate than an equal slopes model for the analysis of these data. Hence the conclusion which one might make from the unequal equal slopes analysis of these data is more likely to be correct than the conclusion from the equal slopes analysis.

It is informative to note that the estimate of the common slope (regression coefficient) for BRHR that is used in the equal slopes analysis is 0.77 (0.76749633967789) as reported in the "Parameter Estimates" from the ORTHOREG procedure of Program 13.6 as follows:

Parameter	DF	Parameter Estimate	Standard Error	t Value	Pr > |t|
Intercept	1	102.600878477306	12.511462925	8.20	<.0001
(PGM='A')	1	-4.20937042459736	3.659117164	-1.15	0.2707
(PGM='B')	0	0	.	.	.
BRHR	1	0.76749633967789	0.1767797753	4.34	0.0008

Recall in Step G, Section 13.8 we found the estimated slopes for Exercise Program A and B are respectively 1.30 ($\widehat{SE}$ = 0.1699) and 0.26 ($\widehat{SE}$ = 0.12). Thus, by using the common slope of 0.77 for the regression lines of HR vs. BRHR for these exercise programs in the equal slopes analysis we may be underestimating the true slope (regression coefficient for BRHR) in Exercise Program A and overestimating this slope in Exercise Program B.

Concluding Comment

The entirely different conclusions reached when an equal slopes as compared to an unequal slopes analysis is applied to the data of the exercise program example underline the importance of always first checking whether an unequal slopes model is likely to be more appropriate for your study before applying an equal slopes analysis.

13.10   What If a One-Way ANOVA Had Been Applied to the Exercise Program Data?

What if the researchers had not adjusted for baseline resting heart rate of the subjects and had applied a one-way ANOVA rather than an unequal slopes ANCOVA to evaluate whether the two exercise programs differ in mean heart rate after a six-minute test run? The following SAS statements could be used to apply a one-way ANOVA to the exercise program data:

```
proc glm data=pgm13;
class pgm;
model hr=pgm;
means pgm;
quit;
```

The following results were obtained from the preceding SAS code found in Program 13.7 (Section 13.J):

RESULTS FROM A ONE-WAY ANOVA OF THE EXERCISE PROGRAM DATA

```
                    Exercise Program and Heart Rate Study
                              The GLM Procedure


                         Dependent Variable: HR
```

Source	DF	Sum of Squares	Mean Square	F Value	Pr > F
Model	1	42.250000	42.250000	0.35	
Error	14	1699.500000	121.392857		
Corrected Total	15	1741.750000			

Source	Pr > F
Model	0.5646
Error	
Corrected Total	

Source	DF	Type III SS	Mean Square	F Value
PGM	1	42.25000000	42.25000000	0.35

Source	Pr > F
PGM	0.5646

The conclusion from these results (p-value for PGM = 0.56) is that a one-way ANOVA analysis of the exercise program data could not detect that the mean HR for Exercise Program A was different from that for Exercise Program B in the sampled population. In contrast, the conclusion from the unequal slopes ANCOVA previously described was that there was a difference between the exercise programs in mean heart rate adjusted for BRHR at certain values of BRHR. The greatly reduced Mean Square Error of 19.275 in the unequal slopes analysis as compared to the Mean Square Error of 121.39 in the one-way ANOVA (given in the preceding overall ANOVA table) indicates that adjustment for BRHR via the unequal slopes model was greatly beneficial in this randomized trial.

13.11   Example 13.2: Effect of Study Methods on Exam Scores Adjusted for Pretest Scores

In a randomized experiment, investigators wanted to compare the effect of two study methods on student exam scores. Eight subjects were assigned completely at random to each study method. After completion of the training period in the assigned study method, a test exam was given and the score of the exam (denoted as PostScore in the analysis) was recorded for each individual.

It is typically recommended in a randomized trial of a treatment that investigators specify in their protocol one or more important pretreatment prognostic factors (nuisance variables) that they will include as covariates in the analysis phase (Senn, 1995). This eliminates the possibility of data-dredging, i.e., trying numerous prognostic factors as covariates in succession until the desired result is obtained. As it is well known that a student's prior ability can affect exam performance, the investigators tested the subjects prior to their random allocation to a study method. The investigators specified in their protocol that they would include this nuisance variable, prior test score (denoted as PreScore) as a covariate in an ANCOVA so as to:

- reduce unexplained random variation of PostScore to achieve a more powerful test of the difference between study methods and a narrower confidence interval estimate for this difference

- reduce any potential bias due to random baseline imbalance of the subjects' ability to write exams as measured by their PreScore that might occur by chance between the study method groups in their trial

The investigators were concerned about potential bias due to random baseline covariate imbalance among their treatment groups because, as illustrated by Woodward (2014), such an imbalance is likely to occur when group sample sizes are small and subjects are assigned completely at random to the treatment groups with no restrictions related to their baseline covariates (i.e., a completely randomized design). This precisely was the scenario for the investigators' experiment in Example 13.2.

The data are given in Table 13.2 and are a subset of an example given on page 362 of Milliken and Johnson (2002). It may be noted that in Table 13.2 we have relabelled Milliken and Johnson's Method 3 as Method 2 as we are considering only two study methods in Example 13.2.

Section 13.12 provides details of a step-by-step approach the researchers could implement to evaluate the effect of study method on exam score adjusting for an individual's score on a test taken before assignment to a study method. The response variable in this ANCOVA is PostScore and the two explanatory variables are Method (the group variable) and PreScore (the covariate).

13.12   A Step-by-Step Analysis of Covariance for Example 13.2

Steps A and B: Screen for data input and recording errors.

No data input or recording errors were found when Program 13.8 (Section 13.L) was applied to the data of Table 13.2. See Section 13.B in this chapter's appendix for details.

TABLE 13.2
Data for Example 13.2 where POSTSCORE is a Subject's Test Score after Participation in a Randomly Assigned Study Method and PRESCORE is the Subject's Test Score Prior to the Study Method Assignment.

| | Study Method 1 | |
Subject	Postscore	Prescore
1	59	23
2	64	31
3	60	31
4	53	15
5	53	17
6	61	11
7	59	20
8	62	15

| | Study Method 2 | |
Subject	Postscore	Prescore
09	72	37
10	62	30
11	70	25
12	68	38
13	66	20
14	61	16
15	70	36
16	60	27

Step C: Review the details of your study's sampling design study to see if there are any obvious aspects of the design which would suggest that the proposed ordinary least squares model which assumes an uncorrelated error structure is inappropriate and that a more complicated model is required for your analysis.

The following details suggest a model with a more complicated error structure is not required to reflect the sampling design of the study:

- Each individual was represented only once in the study.

- The subjects in the study were assigned completely at random to Study Method 1 or 2.

- The subjects were from a single source, viz., students from the same class in the same school.

- None of subjects were related by birth or shared the same household.

- Measurement error of the covariate, i.e., score on a pretest before assignment to a study method, was considered to be zero or negligible, as the pretest was a multiple choice exam and the assistant who marked these exams was conscientious and experienced.

Step D: Fit a simple linear regression model of Y (the response variable in the ANCOVA) vs. the covariate to the data in each group and evaluate whether

ideal inference conditions of the simple linear regression are reasonable for each group using methods described in Section 3.13.

Program 13.9 (Section 13.M) was used to implement Step D. Ideal inference conditions of the simple linear regression of PostScore vs. PreScore seem reasonable for each study method group. Details are given in Section 13.B.

Step E: Check whether any cases merit investigation due to their disproportionate influence in the simple linear regression applied to each group in the previous step using methods described in Chapter 3.

None of the cases in either group, with the exception of Case 6, merited investigation for their influence in the simple linear regression of PostScore vs. PreScore based on the values of DFBETA, DFFITS, and Cook's D output by Program 13.9 (Section 13.M). Upon investigation, nothing appeared unusual about Case 6.

Step F: Further evaluate whether ideal inference conditions of independence of errors, equality of variances, and normality are reasonable.

Independence of Errors Between Groups

We concluded that the ideal inference condition of independence of errors between groups was reasonable as this study was an experiment with a completely randomized design structure, i.e., the subjects were assigned completely at random to one of the two study group methods. There were no withdrawals from the study. Moreover, aspects of the design related to measurement of the Y variable in both groups were either held constant or assigned at random to all subjects in the experiment.

Equality of Variances Between Groups

Figure 13.9 (obtained from the last SGPLOT procedure specified in Program 13.9, Section 13.M) suggests that the ideal inference condition of equal variances between groups is reasonable.

We also investigated equality of error variances between groups by computing the ratio of the mean square errors from the simple linear regression of PostScore vs. PreScore in each group with the largest mean square error in the numerator.

The mean square error from the simple linear regression of PostScore vs. PreScore for Study Method 1 is 15.64839, as reported in the following Analysis of Variance table for this regression obtained from Program 13.9:

RESULTS FROM SIMPLE LINEAR REGRESSION FOR METHOD 1

Analysis of Variance

Source	DF	Sum of Squares	Mean Square	F Value	Pr > F
Model	1	16.98465	16.98465	1.09	0.3376
Error	6	93.89035	15.64839		
Corrected Total	7	110.87500			

The mean square error from the simple linear regression of PostScore vs. PreScore for Study Method 2 is 16.57954, as reported in the following Analysis of Variance table for this regression obtained from Program 13.9:

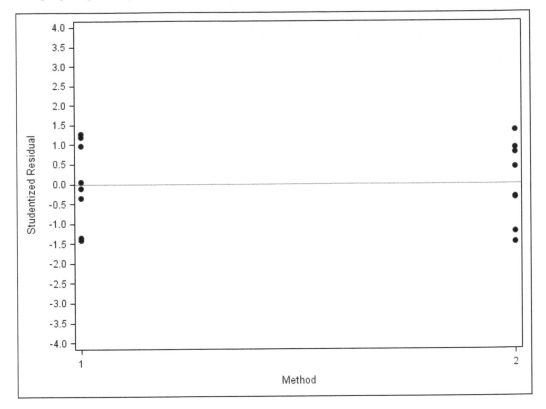

FIGURE 13.9
Between group comparison of internally studentized residuals from the simple linear regression of PostScore vs. PreScore within each group.

RESULTS FROM SIMPLE LINEAR REGRESSION FOR METHOD 2

Analysis of Variance

Source	DF	Sum of Squares	Mean Square	F Value	Pr > F
Model	1	49.39777	49.39777	2.98	0.1351
Error	6	99.47723	16.57954		
Corrected Total	7	148.87500			

The ratio of the largest mean square error to the smallest mean square error from these simple linear regressions in Study Methods 1 and 2 is $16.57954/15.64839 = 1.05950$, which suggests that these error mean squares are estimating the same population value and therefore it is not unreasonable to pool these error mean squares to form the mean square error for the analysis of covariance.

Normality
No serious departure from normality was detected in Step C when we evaluated this ideal inference condition within each group. Similarly, no serious departure from normality was detected in Figure 13.10, a normal Q-Q plot of internally studentized residuals from the unequal slopes model generated by 13.10A (Section 13.N).

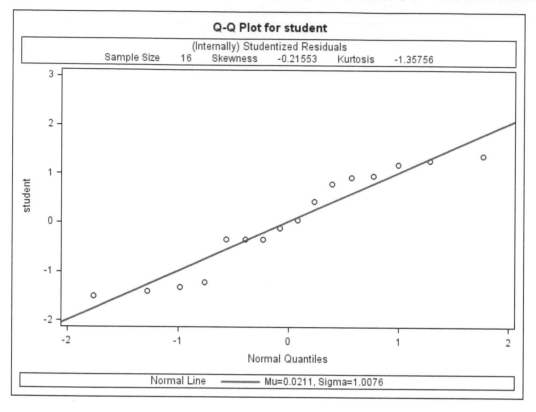

FIGURE 13.10

Normal Q-Q plot comprised by internally studentized residuals from both groups' simple linear regression of PostScore vs. PreScore.

Step G: Evaluate whether an unequal slopes rather than an equal slopes model is appropriate for your analysis of covariance.

The point estimates and their estimated standard errors (given in parentheses) of the slopes of the population simple linear regression lines of PostScore vs. PreScore are 0.209 (0.200) for Study Method 1 and 0.326 (0.189) for Study Method 2. The estimated slope of the regression line, i.e., the estimated regression coefficient for PreScore in the simple linear regression of PostScore vs. PreScore for each group, is given in the following Parameter Estimates section for each method's simple linear regression of PostScore vs. PreScore generated by Program 13.9 (Section 13.M):

Parameter Estimates For Simple Linear Regression of PostScore vs. PreScore for Method 1

Parameter Estimates

Variable	DF	Parameter Estimate	Standard Error	t Value	Pr > |t|
Intercept	1	54.62231	4.31492	12.66	<.0001
PreScore	1	0.20872	0.20034	1.04	0.3376

Parameter Estimates For Simple Linear Regression of PostScore vs. PreScore for Method 2

Parameter Estimates

Variable	DF	Parameter Estimate	Standard Error	t Value	Pr > |t|
Intercept	1	56.78389	5.59988	10.14	<.0001
PreScore	1	0.32633	0.18905	1.73	0.1351

In the preceding results for the parameter estimates from the simple linear regression for each group, it may be noted that the p-values observed for the test of the null hypothesis that the regression coefficient for PreScore = 0 are equal to 0.3376 and 0.1351 respectively for Methods 1 and 2. However, these p-values did not result in the researchers' abandoning the ANCOVA with PreScore as a covariate because on the basis of prior expert opinion, it had been specified in the protocol at the design phase of the study to adjust the outcome variable, PostScore for the variable PreScore. Moreover the size of these p-values may be related to the small sample size in each group ($n = 8$).

To continue our informal evaluation of the appropriateness of an unequal slopes model for this study, we examined Figure 13.11, which we obtained by fitting an unequal slopes multiple regression model to the combined data from Methods 1 and 2 using Program 13.10A (Section 13.N).

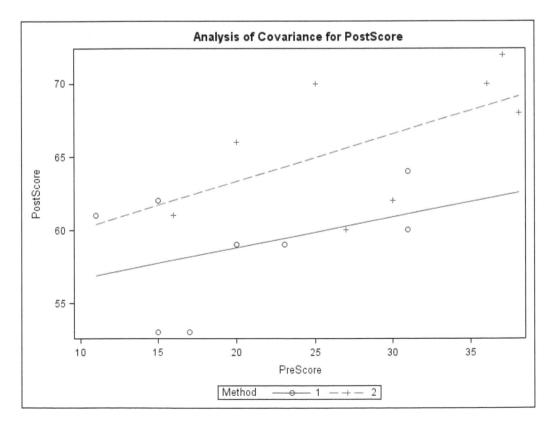

FIGURE 13.11
Fitted simple linear regression lines for PostScore vs. PreScore for Study Methods 1 and 2.

Figure 13.11 displays the fitted simple linear regression of PostScore vs. PreScore within each method. This figure suggests that an equal slopes model may be appropriate for the analysis as the simple linear regression lines for Methods 1 and 2 fitted to the sample data shown in this figure are only slightly nonparallel.

We then formally evaluated the appropriateness of an unequal slopes model by fitting this model to the data from both study methods and testing whether the interaction effect was different from zero using the GLM procedure in Program 13.10A (Section 13.N).

The following results reveal that the test of the interaction effect of the model, using Type III sums of squares in GLM, has a p-value $= 0.6774$, which suggests that with these data we are not able to detect, in the sampled populations, a difference between the study methods' slopes of the simple linear regression lines of PostScore vs. PreScore.

Source	DF	Type III SS	Mean Square	F Value	Pr > F
Method	1	1.51641441	1.51641441	0.09	0.7643
PreScore	1	60.64303561	60.64303561	3.76	0.0763
PreScore*Method	1	2.92993165	2.92993165	0.18	0.6774

Estimated limits for the 95% confidence interval for the difference between the slopes in the sampled populations provided by Program 13.10B (Section 13.O) are as follows:

Parameter	Estimate	Standard Error	t Value	Pr > |t|
Slope 1-Slope 2	-0.11760639	0.27580580	-0.43	0.6774

Parameter	95% Confidence Limits	
Slope 1-Slope 2	-0.71853560	0.48332281

From the preceding results we note that the estimated difference in slopes -0.1176 (slope of Method 1 minus slope of Method 2) is small relative to its estimated standard error of 0.2758. The 95% confidence interval estimate for this difference in population slopes (-0.71853560, 0.48332281) suggests any value in this interval, including zero, could be plausible. We therefore cannot conclude from these results whether there is or is not a difference in slopes of the regression line PostScore vs. PreScore between Method 1 and 2 in the sampled populations.

The p-value of 0.6774 reported from the t-test of the null hypothesis that the difference between slopes is equal to zero does not suggest that the slopes for Methods 1 and 2 are different in the sampled populations. As previously noted this t-test is equivalent to the Type III sums of squares F-test of the regression coefficient for the interaction effect being equal to zero given in Program 13.10A.

Summary of Step G

The results of Step G do not suggest an unequal slopes model is required for the analysis of these data. As there were no prior studies on Methods 1 and 2 to contradict this conclusion, we omitted Step H (fitting an unequal slopes model) and proceeded to Step I where we applied an equal slopes model for the analysis.

Step I: Apply an equal slopes (common slope) model to the data

Although we have spared you the details here, we did evaluate the ideal inference conditions of equality of variances and normality for the equal slopes model applied to this example before interpreting the results. We concluded these ideal inference conditions of the equal slopes model were reasonable for these data just as we had previously concluded that these ideal inference conditions of the unequal slopes model were reasonable for these data. One may anticipate that conclusions based on the pattern of studentized residuals from an equal slopes and unequal slopes models would be quite similar when the slopes of the simple regression lines based on the sample data of Methods 1 and 2 are similar to each other and therefore similar to the common slope of the equal slopes model.

We applied an equal slopes analysis to the data from Methods 1 and 2 using the following code from Program 13.11 (Section 13.P):

```
proc glm data=one;
class Method;
model PostScore= Method PreScore/ss3;
means Method;
lsmeans Method/pdiff tdiff stderr cl;
quit;
```

The following results were generated by the preceding code:

RESULTS FROM EQUAL SLOPES ANALYSIS: Program 13.11

The GLM Procedure

Dependent Variable: PostScore

Source	DF	Sum of Squares	Mean Square	F Value	Pr > F
Model	2	273.7024890	136.8512445	9.06	0.0034
Error	13	196.2975110	15.0998085		
Corrected Total	15	470.0000000			

Source	DF	Type III SS	Mean Square	F Value	Pr > F
Method	1	75.84813733	75.84813733	5.02	0.0431
PreScore	1	63.45248902	63.45248902	4.20	0.0611

The preceding results suggest that, in the sampled population, the focus variable Method has an effect on mean PostScore, when mean PostScore is adjusted for any value of PreScore in the observed range common to both methods in the study. The p-value from the Type III sums of squares F-test of the null hypothesis of no such Method effect is 0.04, suggesting that these data are not compatible with this null hypothesis, assuming an equal slopes model is appropriate for the analysis and ideal inference conditions for the test are reasonable for the data. Evaluation of these assumptions in Steps D–G indicated that they are reasonable.

The following results from fitting an equal slopes model to the data of Example 13.2 were generated by the LSMEANS statement of the GLM procedure in Program 13.11.

LSMEANS RESULTS FROM PROGRAM 13.11

The GLM Procedure
Least Squares Means

Method	PostScore LSMEAN	Standard Error	HO:LSMEAN=0 Pr > |t|	HO:LSMean1=LSMean2 t Value	Pr > |t|
1	59.9995608	1.4793316	<.0001	-2.24	0.0431
2	65.0004392	1.4793316	<.0001		

Method	PostScore LSMEAN	95% Confidence Limits	
1	59.999561	56.803659	63.195462
2	65.000439	61.804538	68.196341

Least Squares Means for Effect Method

i	j	Difference Between Means	95% Confidence Limits for LSMean(i)-LSMean(j)	
1	2	-5.000878	-9.821327	-0.180430

The preceding results report that 59.999561 and 65.000439 are respectively the estimated PostScore means for Method 1 and Method 2, adjusted for Prescore = 24.5, the average value of PreScore observed in the combined sample data of the two methods. The estimated difference (Method 1 minus Method 2) between these PostScore means is -5.0. It may be noted that because this analysis is using an equal slopes model, the difference in PostScore means between Method 1 and Method 2 is also estimated to be -5, when these PostScore means for Method 1 and 2 are adjusted for **any** value of PreScore in the observed range common to both methods in the study. This can be verified by examining results generated from multiple LSMEANS statements for the GLM procedure in Program 13.12 (Section 13.Q) such as

```
lsmeans Method /pdiff tdiff stderr cl at PreScore= 24.0;
lsmeans Method /pdiff tdiff stderr cl at PreScore= 15;
```

Further examination of results from this equal slopes analysis reveals that a p-value of 0.04 is associated with the t-test of the null hypothesis of no difference in PostScore method means, when adjusted for any value of PreScore in the observed range common to both methods in the study.[2] The 95% confidence interval estimate for this difference between Method 1 and Method 2 in PreScore-adjusted Postscore means is (-9.8, -0.18). This confidence interval estimate, which does not include zero, suggests there may be a difference between methods in the sampled populations when adjusted for PreScore. However, although the absolute value of the estimated lower confidence limit suggests the difference between methods in PreScore-adjusted PostScore means could be as large as 9.82, the absolute value of the estimated upper confidence limit suggests this difference could be as

---

[2]This t-test is equivalent to the Type III sums of squares F-test of no effect of Method on PreScore-adjusted PostScore means previously reported.

small as 0.18. Researchers may conclude from this pilot study that Method 2 may be more efficacious than Method 1 in raising test scores but more research is desirable to confirm this.

13.13   References for Other Approaches for Covariate Adjustment

We hope that understanding ANCOVA for the simple scenarios explored in this chapter where there are just two groups and a single continuous covariate will lay the foundation for understanding issues in covariate adjustment in more complex situations. For detailed descriptions of analyses of covariance involving more than two groups and more complex experimental designs using SAS, Milliken and Johnson (2002) could be used as a reference. For Koch's randomization-based nonparametric covariance analyses, requiring minimum assumptions, see Koch et al. (1998). For Rosenbaum and Rubin's approach of using propensity scores in observational studies to achieve balance of numerous observed baseline covariates in the study groups, see Rosenbaum and Rubin (1983); Faries et al. (2010); Austin (2011). Faries et al. (2010) described the instrumental variable method which has the capability of adjustment for unmeasured confounders in observational studies, provided a proper "instrument" can be identified (Murray, 2006) and validated.

13.14   Chapter Summary

This chapter:

- describes an ANCOVA model as a multiple regression model which is used to adjust group means for one or more nuisance variables

- outlines the characteristic features of a one-way ANCOVA model

- notes that when group means are adjusted for a single continuous covariate in an ANCOVA, the multiple regression model used for this covariate adjustment encompasses a single linear regression of the response variable (Y) vs. the covariate (X) for each group

- discusses two critical decisions that you have to make before embarking on an ANCOVA to adjust your group means for a single continuous covariate:

    1. You have to decide whether a linear model is appropriate for your data by evaluating if there is a linear relationship between the response variable and the covariate in each group.
    2. If you decide a linear model is appropriate, then you have to decide what form of the model to apply to your data, an unequal slopes or equal slopes model.

- illustrates how conclusions from an ANCOVA can be erroneous if a data analyst uses an equal slopes model when an unequal slopes model is the appropriate model for the data

- proposes and discusses details of a step-by-step approach to implement a one-way ANCOVA where there are two groups and a single continuous covariate

- illustrates the proposed step-by-step approach with two examples, one where an unequal slopes model is appropriate, the other where an equal slopes model appears to be appropriate

- describes, illustrating with examples, multiplicity-adjusted tests and simultaneous confidence intervals for the final step of an unequal slopes analysis where covariate-adjusted group means are compared at multiple values of the covariate

- provides SAS programs that carry out the proposed step-by-step approach to implement a one-way ANCOVA

- gives references for covariate adjustment approaches other than those described in this chapter

Appendix

13.A Details of a Nine-Step Evaluation of Ideal Inference Conditions for Simple Linear Regressions in Exercise Programs A and B in Example 13.1

This section provides results from a nine-step evaluation of ideal inference conditions for the simple linear regression applied separately to each group, A and B in the exercise program experiment where heart rate after a six-minute test run is the outcome and baseline resting heart rate is the predictor.

Steps (1) and (2) Check for data input and recording errors.

No data input or recording errors were found for Exercise Program A or B when Program 13.1 (Section 13.C) was applied to the data. Neither Figure 13.12 nor Figure 13.13 respectively reveal any contextual outliers, i.e., any cases which have an unusual combination of an X and a Y value for the population sampled in Exercise Program A and B.

Step (3) Confirm that the response is continuous.

The response variable in the simple linear regressions, heart rate after a six-minute test run, is considered to have an underlying continuous distribution (although it is recorded to the nearest whole integer).

Step (4) Review the details of the sampling design of the study to see if there are any obvious aspects of the design that would suggest that the uncorrelated error structure of the simple linear regression model is unsuitable and that a different model with a more complicated error structure is required for Group A or B.

The sampling design details reviewed in Step C (Section 13.8) suggest that a model with a more complicated error structure is not required.

Step (5) Apply a separate simple linear regression to the data in each group to obtain regression diagnostics.

This step was implemented using Program 13.2, given in Section 13.D.

Step (6) Informally evaluate the ideal inference conditions of linearity and equality of variances.

It may be noted that each group's small sample size makes evaluation of linearity and equality of variances difficult. However, bearing this in mind, we see that Figure 13.14 and Figure 13.15, plots of the internally studentized residuals against baseline resting heart rate, for Groups A and B respectively, do not reveal any systematic pattern that would suggest serious departures from linearity and equality of variances at each value of baseline resting heart rate for the simple linear regression applied separately in Groups A and B.

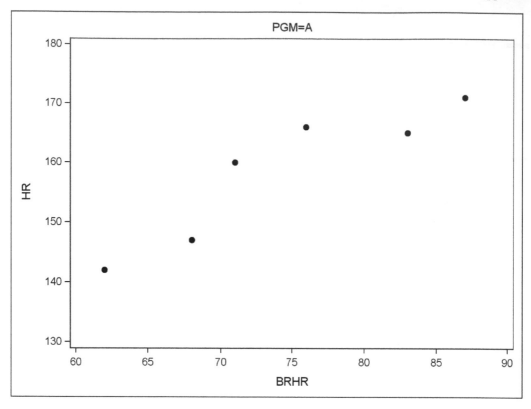

FIGURE 13.12
Scatter plot of heart rate after a six minute test run (HR) vs. baseline resting heart rate
(BRHR) for Exercise Program A (PGM = A).

**Step (7) Use graphical displays of internally studentized residuals to informally
evaluate whether the ideal inference condition of independence of errors in
the sampled population has been violated for the simple linear regression
model due to population errors being systematically related to the explanatory
variable in the model or to any other variable not included in the model
such as time, space, or some other key variable related to the details of the
implementation of the study.**
We could not apply Step (7) because we did not have sufficient information about the
study design details of this example. However, it is hoped that key variables related to
the implementation of the study that might have been associated with dependence among
the model errors were either held constant or randomized in the study, e.g.:

(i)   There was just one research assistant who followed a prescribed protocol to
        measure the subjects' heart rate after the six-minute test run and baseline
        resting heart rate.

(ii)   Each subject ran his test run in the gym solely in the company of the research
        assistant and arrangements were made to circumvent any communication among
        test subjects before and after the test run.

(iii)   The conditions under which each subject would take the six-minute test run
        were held constant: the same location, i.e., the same gym with constant tem-
        perature and lighting control; the same day of the week for the test run.

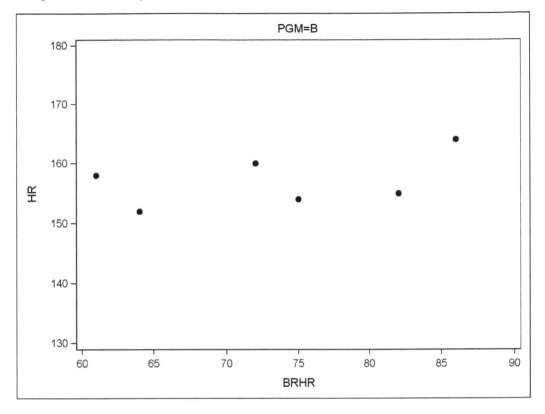

FIGURE 13.13
Scatter plot of heart rate after a six-minute test run (HR) vs. baseline resting heart rate (BRHR) for Exercise Program B (PGM = B).

(iv) The exact time of each subject's test run within a restricted time period, say from 2:00 to 4:00 pm, was assigned at random. However, even with this randomization, it would be useful to compare studentized residuals vs. time of test run to see if a pattern were evident.

Step (8) Evaluate the ideal inference condition of normality.
  Neither Figure 13.16 nor Figure 13.17, Q-Q normal plots of the internally studentized residuals from the simple linear regression model of heart rate vs. baseline resting heart rate applied separately in Exercise Program A and B respectively, reveals a major departure from normality.

Step (9) Identify and investigate model outliers using studentized deleted residuals obtained from Step (5), provided previous steps have suggested that ideal inferences conditions are reasonable.
  As previous steps suggest that ideal inference conditions are reasonable for the separate simple linear regressions in Groups A and B, studentized deleted residuals from these separate regressions were used to identify any model outliers in each group. Figures 13.18 and 13.19 respectively reveal that all of the studentized deleted residuals obtained from Step (5)'s simple linear regression for Exercise Program A and B were less than 2, suggesting there were no model outliers in either simple linear regression.

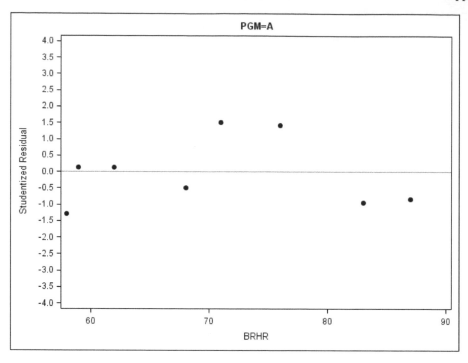

FIGURE 13.14
Internally studentized residuals vs. baseline resting heart rate (BRHR) from the simple linear regression for Exercise Program A.

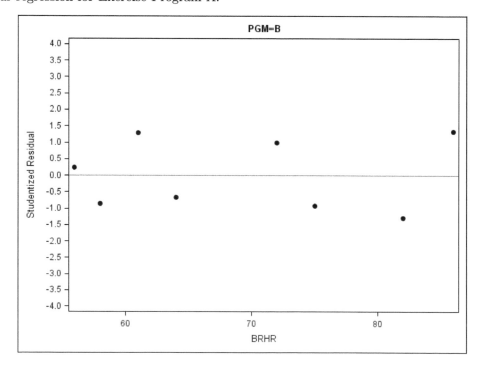

FIGURE 13.15
Internally studentized residuals vs. baseline resting heart rate (BRHR) from the simple linear regression for Exercise Program B.

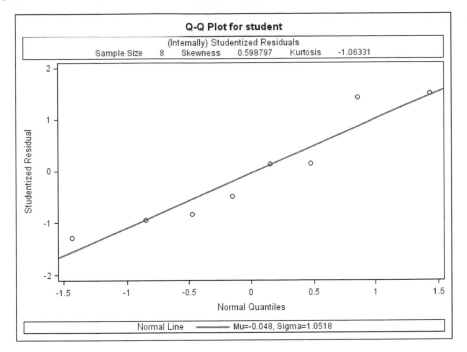

FIGURE 13.16
A normal Q-Q plot of internally studentized residuals from the simple linear regression of heart rate after a six minute test run vs. baseline resting heart rate for Exercise Program (PGM) A.

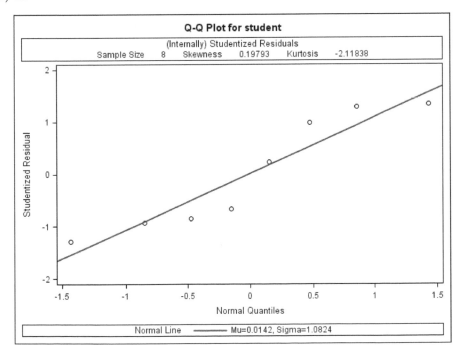

FIGURE 13.17
A Normal Q-Q plot of internally studentized residuals from the simple linear regression of heart rate vs. baseline resting heart rate for Exercise Program (PGM) B.

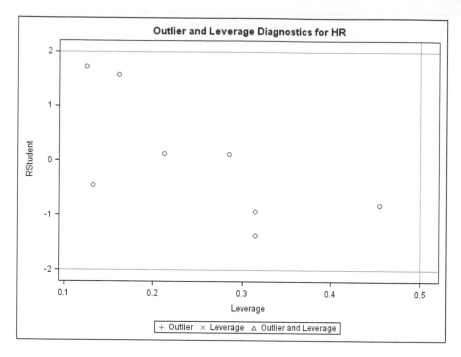

FIGURE 13.18
Studentized deleted residuals (RStudent) vs. leverage values from the simple linear regression of heart rate vs. baseline resting heart rate for Exercise Program A.

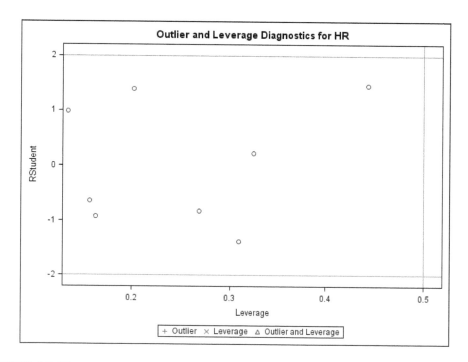

FIGURE 13.19
Studentized deleted residuals (RStudent) vs. leverage values from the simple linear regression of heart rate vs. baseline resting heart rate for Exercise Program B.

13.B Details of Evaluation of Ideal Inference Conditions for Simple Linear Regressions of Postscore vs. Prescore in Example 13.2

This section provides results from a nine-step evaluation of ideal inference conditions for a simple linear regression applied separately to each group in the study method experiment of Example 13.2 where Postscore (score on the test exam after training in a study method) is the response variable and Prescore (test score prior to training) is the predictor variable in each simple linear regression.

Steps (1) and (2) Check for data input and recording errors.
No data input or recording errors were found for Method 1 or Method 2 when Program 13.8 (Section 13.L of this chapter's appendix) was applied to the data. Neither Figure 13.20 nor Figure 13.21 (obtained from the SGPLOT procedure in Program 13.8) reveals any contextual outliers, i.e., any cases which have an unusual combination of a Prescore (X) and Postscore (Y) value for the population sampled.

Step (3) Confirm that the response is continuous.
The response variable, PostScore is continuous, rounded to the nearest integer.

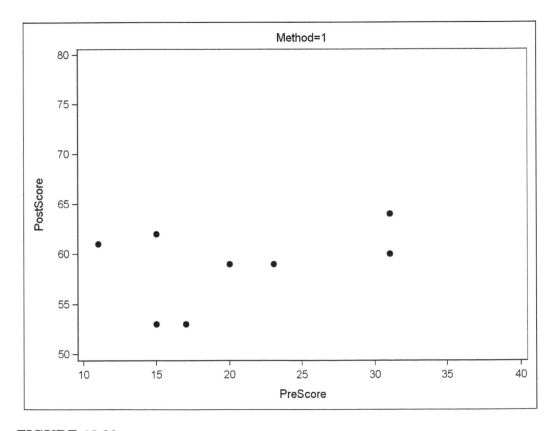

FIGURE 13.20
Scatter plot of PostScore vs. PreScore where PostScore and PreScore are respectively exam score after vs. before participation in Study Method 1.

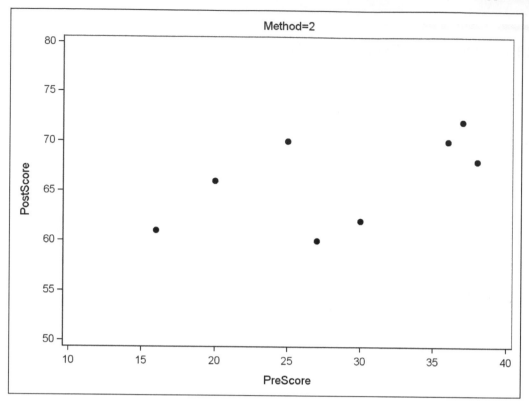

FIGURE 13.21

Scatter plot of PostScore vs. PreScore where PostScore and PreScore are respectively exam score of a subject after and before participation in Study Method 2.

Step (4). Review the details of the sampling design of the study to see if there are any obvious aspects of the design that would suggest that the uncorrelated error structure of the simple linear regression model is unsuitable and that a different model with a more complicated error structure is required.

The sampling design details have already been reviewed in Step C (Section 13.12) and it was concluded a different model with a more complicated error structure is not required.

Step (5) Apply a separate simple linear regression to the data in each study method group to obtain regression diagnostics.

This step was implemented using Program 13.9 (Section 13.M) of this chapter's appendix.

Step (6) Informally evaluate the ideal inference conditions of linearity and equality of variances.

It may be noted that the small sample size in each group makes it difficult to evaluate linearity and equality of variances. However, bearing in mind this difficulty, examination of Figures 13.22 and 13.23, respectively, plots of internally studentized residuals from the separate simple linear regression of Postscore vs. PreScore in Group 1 and 2, does not reveal any systematic pattern that would suggest serious departures from linearity and equality of variances for these group's simple linear regressions. These figures were generated by the SGPLOT procedure in Program 13.9 (Section 13.M).

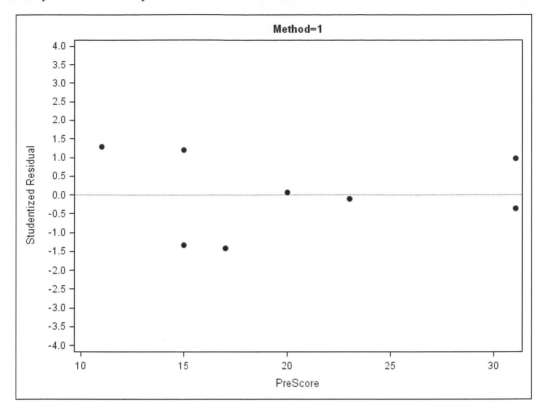

FIGURE 13.22
Internally studentized residuals from simple linear regression of PostScore vs. PreScore for Study Method 1. (PostScore and PreScore are respectively the exam score of a subject after and prior to participation in Study Method 1.)

Step (7) Use graphical displays of internally studentized residuals to informally evaluate whether the ideal inference condition of independence of errors in the sampled population has been violated for the simple linear regression due to model errors being systematically related to the explanatory variable in the model or to any other variable not included in the model such as time, space, or some other key variable related to the details of the implementation of the study.

Due to insufficient information about the study design details of this example, we could not apply Step (7). However, one could hope that key variables related to the study's implementation that might possibly be associated with dependence among model errors were either held constant or randomized in the study, for example:

(i) There was just one research assistant who administered the exams.

(ii) The pre-exam and post-exam were administered at the same time of day in the same exam room under the same environmental conditions, same lighting, same room temperature, same noise level.

(iii) The subjects were separated from each other in the exam room so that they could not see other subjects' answers.

(iv) The pre-exam and post-exam were multiple choice exams so variability in the score due to the subjectivity of the marker was not a concern.

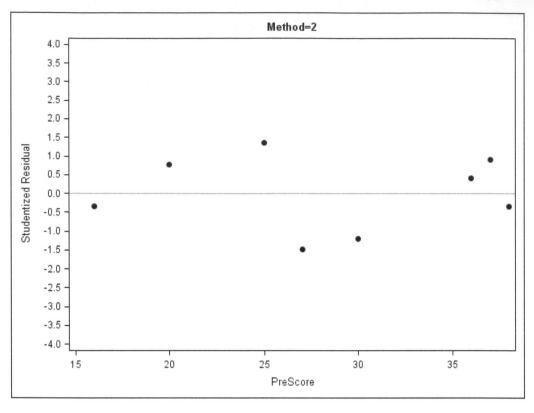

FIGURE 13.23
Internally studentized residuals from simple linear regression of PostScore vs. PreScore for Study Method 2. (PostScore and PreScore are respectively the exam score of a subject after and prior to participation in Study Method 2.)

Step (8) Evaluate the ideal inference condition of normality.
  Neither Figure 13.24 nor Figure 13.25, Q-Q plots of internally studentized residuals from the separate simple linear regressions of PostScore vs. PreScore for Methods 1 and 2 respectively, reveals a major departure from normality.

Step (9) Identify and investigate model outliers using studentized deleted residuals obtained from Step (5), provided previous steps have suggested that the ideal inferences conditions are reasonable.
  As previous steps suggest that ideal inference conditions are reasonable, studentized deleted residuals were used to identify any model outliers from the simple linear regressions of PostScore vs. PreScore for Methods 1 and 2. Figures 13.26 and 13.27 reveal that studentized deleted residuals obtained from Step 5's simple linear regressions for Method 1 and 2 were between 2 and -2 suggesting that there are no model outliers. These figures, labelled "RStudent by Leverage", are found in "Observation-wise Statistics" under "Diagnostics Plots" in the output for Methods 1 and 2 from the REG procedure of Program 13.9.

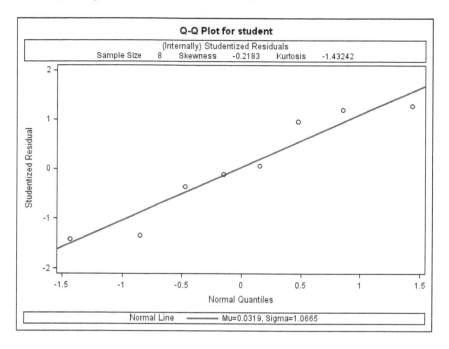

FIGURE 13.24

A normal Q-Q plot of internally studentized residuals from the simple linear regression of PostScore vs. PreScore for Study Method 1. (PostScore and PreScore are respectively the exam score of subject after and prior to participation in Study Method 1.)

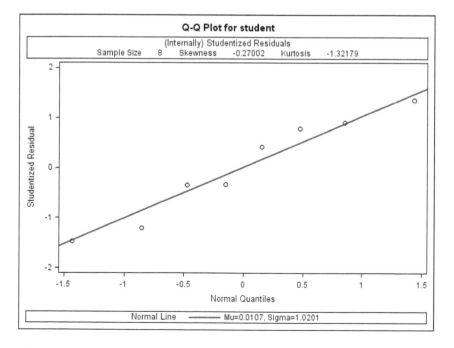

FIGURE 13.25

A normal Q-Q plot of internally studentized residuals from the simple linear regression of PostScore vs. PreScore for Study Method 2. (PostScore and PreScore are respectively the exam score of subject after and prior to participation in Study Method 2.)

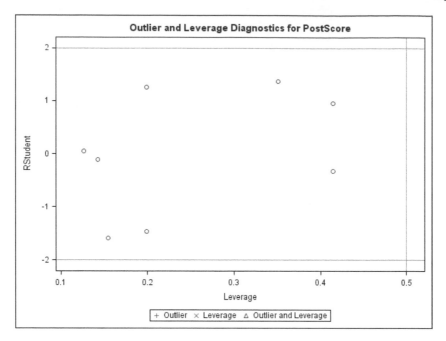

FIGURE 13.26
Studentized deleted residuals (RStudent) vs. leverage values from the simple linear regression of PostScore vs. PreScore for Study Method 1. (PostScore and PreScore are respectively a subject's exam score after and prior to participation in Study Method 1.)

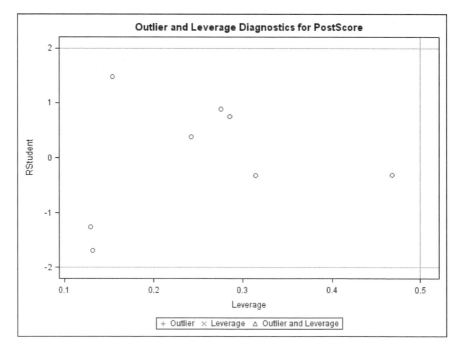

FIGURE 13.27
Studentized deleted residuals (RStudent) vs. leverage values from the simple linear regression of PostScore vs. PreScore for Study Method 2. (PostScore and PreScore are respectively a subject's exam score after and prior to participation in Study Method 2.)

13.C   Program 13.1

Program 13.1 reads in the exercise experiment data of Example 13.1 given in Table 13.1 and stores it in the temporary SAS data set PGM13 for the duration of a SAS session in most operating environments. This program checks for data input and data recording errors in the exercise experiment data and is similar to data checking programs presented in previous chapters.

```
* Program 13.1;
data pgm13;
   input IDNUM PGM $ HR BRHR;
   datalines;
 1 A 131 58
 2 A 138 59
 3 A 142 62
 4 A 147 68
 5 A 160 71
 6 A 166 76
 7 A 165 83
 8 A 171 87
 9 B 153 56
10 B 150 58
11 B 158 61
12 B 152 64
13 B 160 72
14 B 154 75
15 B 155 82
16 B 164 86
 ;

*STEP A;
proc print data=pgm13;
run;
*STEP B;
proc univariate data=pgm13 plots;
   var hr brhr;
   by pgm;
   ods select ExtremeValues Frequencies;
   ods exclude TestsForLocation;
   id idnum;
run;
/*The following plots can be examined for contextual,
predictor and response outliers that merit investigation for recording
errors*/
    ods listing style=journal2 image_dpi=200;
proc sgplot data=pgm13;
   scatter x=BRHR y=HR /markerattrs=(symbol=circlefilled color=black);;
   yaxis values =(130 to 180 by 10);
   xaxis values = (60 to 90 by 5);
   by pgm;
run;
```

13.D   Program 13.2

Program 13.2 applies a simple linear regression in each exercise group of Example 13.1 and computes regression diagnostics and measures of influence for each group's simple linear regression. The response variable is heart rate after a six-minute test run and the predictor is baseline resting heart rate. Program 13.2 uses as input the **temporary** SAS data set PGM13 created by Program 13.1.

Please note: If you attempt to execute Program 13.2 after having logged out from the SAS session in which you executed Program 13.1, you will get error messages in most operating environments. One way to avoid this is by executing Program 13.1 in your current SAS session prior to running Program 13.2 or alternatively by copying the DATA step statements and data from Program 13.1 and pasting them at the beginning of Program 13.2.

```
*Program 13.2;
/*This program applies a simple linear regression
of Y vs. the covariate within each group
and provides regression diagnostics and measures of influence
for each group*/
 title;
*STEP A;
proc print data=pgm13;
   by pgm;
run;
proc reg data=pgm13
    plots(only label)=(cooksd dffits dfbetas RStudentByLeverage);
    model hr= brhr / r p clb influence;by pgm;
    output out=diag
    r=r rstudent=rstudent student=student p=p dffits=dffits cookd=cookd;
run;


*Evaluate Ideal Inference Conditions for each study method;
/*Obtain plots of internally studentized residuals vs. covariate, PRESCORE,
 */
 proc sgplot data=diag;
    refline 0/ axis=Y;
    scatter x=brhr y=student /markerattrs=(symbol=circlefilled color=black);
    yaxis values= (-4 to 4 by .5);
    by pgm;
 run;
* Evaluate normality;
* Obtain a Q-Q normal plot for internally studentised residuals;
proc univariate data =diag;
    by pgm;
    var student;
    qqplot student  /normal (mu=est sigma=est);
    inset n='Sample Size' skewness kurtosis
    /pos=tm header='(Internally) Studentized Residuals';
run;
/* Get a listing of variables of interest for cases identified as
```

```
      regression outliers based on a specified cutoff value for
      the studentized deleted residual (rstudent) */

proc print data=diag;
   where abs(rstudent) gt 2.0;
run;
*STEP H;
*Check for influential cases;
/*CALCULATE SIZE-ADJUSTED CUT-OFF VALUES
for DFBETAS: Belsley et al 1980
for DFFITS Belsley et al 1980,Staudte and Sheather 1990*/

data cutvalues;
   /*parm=number of parameters in model including the intercept*/
   /*n = equal number of cases in each simple linear regression */
   /*dffitscutbels is cutoff value recommended by Belsley et al 1980*/
   /*dffitscutss is cutoff value recommended by Staudte and Sheather 1990*/
   parm=2;
   n=8;
   nminusp=n-parm;
   dfbetacut=2/(sqrt(n));
   dffitscutbels=2*(sqrt(parm/n));
   dffitscutss=1.5*(sqrt(parm/n));
run;
proc print data=cutvalues;
   title Size-Adjusted Cutoff Values for DFBETAS and DFFITS;
run;
 data influence;
    if _n_=1 then set cutvalues;
    set diag;
    absdffits=abs(dffits);
run;
proc print data=influence;
   title;
run;
proc print data=influence;
   where cookd gt 0.5 ;
   title Cases with COOKD Value > 0.5 ;
   var idnum cookd pgm hr brhr p r student rstudent  dffits;
run;
proc print data=influence;
   where absdffits gt 1.00;
   title Cases with DFFITS Absolute Value > 1.00;
   var idnum dffits pgm hr brhr p r student rstudent cookd ;
run;
proc print data=influence;
   where absdffits gt dffitscutbels;
   title Cases with Abs. DFFITS Value > Belsley et al 1980 Cutoff Value;
   var idnum dffits pgm hr brhr p r student rstudent cookd ;
run;
proc print data=influence;
```

```
      where absdffits gt dffitscutss;
      title Cases with Abs. DFFITS Value > Staudte and Sheather 1990 Cutoff
      Value;
      var idnum dffits pgm brhr p r student rstudent cookd ;
run;


/*Obtain plot where STUDENT (internally studentized residuals from
the simple linear regression of HR vs. BRHR) is
on the Y axis and Exercise Regime (PGM) is on the X axis.*/
/*Obtain plot of internally studentized residuals vs. covariate, BRHR,
 */
 title;
proc sgplot data=diag;
   refline 0/ axis=Y;
   scatter x=PGM y=STUDENT /markerattrs=(symbol=circle);
   yaxis values= (-4 to 4 by .5);
   xaxis type=discrete;
run;
```

Comments about Program 13.2

The statements in Program 13.2 have been explained previously as this program is an amalgamation of Programs 3.1A, 3.1B, and 3.1C (Sections 3.A.1, 3.B.1, and 3.C.1 respectively).

13.E   Program 13.3A

Program 13.3A uses the GLM procedure to fit an unequal slopes multiple regression model to the exercise experiment data in the **temporary** SAS data set PGM13. The GLM procedure of this program provides an ANCOVA plot for an informal visual evaluation of an unequal slopes model's fit to the exercise experiment data. In the ANCOVA plot, fitted simple linear regression lines for each group, based on an unequal slopes model specified on the MODEL statement of the GLM procedure, are displayed along with the original data points for each group.

Please note: If you attempt to execute Program 13.3A after having logged out from the SAS session in which you executed Program 13.1, you will get error messages in most operating environments. One way to avoid this is by executing Program 13.1 in your current SAS session prior to running Program 13.3A or alternatively by copying the DATA step statements and data from Program 13.1 and pasting them at the beginning of Program 13.3A.

```
*Program 13.3A;
proc glm data=pgm13;
class pgm;
model hr=pgm brhr brhr*pgm/ss3;
output out=diag  student =student r=r star=star h=h
cookd=cookd rstudent=rstudent;
```

```
run;
proc print data=diag;
run;
proc univariate data =diag;
   var student;
   qqplot student /normal (mu=est sigma=est);
   inset n='Sample Size' skewness kurtosis
   /pos=tm header='(Internally) Studentized Residuals ';
run;
```

Comments about Program 13.3A

- Program 13.3A illustrates how the GLM procedure makes it easier to represent class (qualitative) variables and interaction terms in a model as compared to the REG procedure. This is because the GLM procedure, unlike the REG procedure, does not require the data analyst to create interaction terms or indicator variables representing categories of a class (qualitative) predictor in a preceding data step.

- The CLASS statement in the GLM procedure in Program 13.3A specifies that PGM should be treated as a qualitative (classification) variable in the analysis. The CLASS statement should always be placed before the MODEL statement in your program.

- The SS3 option on the MODEL statement of a GLM procedure specifies that Type III sums of squares hypothesis tests should be carried out.

- For those interested in technical details underpinning the MODEL statement and associated Type III sums of squares hypothesis tests, a related discussion may be found on page 27 of Milliken and Johnson (2002) and in SAS's online support documentation (support.sas.com/documentation) under the topic "Parameterization of Proc GLM Models".

13.F   Program 13.3B

To enable a formal evaluation of the unequal slopes model's fit to the exercise experiment data stored in the **temporary** SAS dataset, PGM13, the GLM procedure in Program 13.3B computes:

- an unbiased estimate of the slope and standard error of the regression line for Y vs. the covariate for each group

- an estimate of the difference between these slopes and associated standard error

- a 95% confidence interval estimate for the difference between slopes

- a hypothesis test of the difference between slopes being equal to zero

Please note: If you attempt to execute Program 13.3B after having logged out from the SAS session in which you executed Program 13.1, you will get error messages in most operating environments. One way to avoid this is by executing Program 13.1 in your current SAS session prior to running Program 13.3B or alternatively by copying the DATA step statements and data from Program 13.1 and pasting them at the beginning of Program 13.3B.

```
*Program 13.3B;
/*The following SAS code gives estimates of the slopes of the simple linear
regression of HR (Y) vs. BRRH (the covariate) for each group (PGM)
and also the estimated difference between these simple slopes
with estimated 95% confidence limits for this difference */
proc glm data=pgm13;
title Evaluation of Unequal Slopes Covariance Model;
class pgm;
model hr=pgm brhr*pgm/noint clparm solution ss3;
estimate 'Slope A-Slope B' brhr*pgm 1 -1;
ods select ParameterEstimates Estimates;
quit;
```

Comments about Program 13.3B

- The CLPARM option on the MODEL statement for a GLM procedure requests confidence limits for the parameter estimates (if the SOLUTION option is also specified) and for the results of all ESTIMATE statements.

- The ESTIMATE statement, "**estimate 'Slope A-Slope B' brhr*pgm 1 -1;**" of this program requests that the difference between the slopes of the simple linear regression of HR vs. BRH R in each exercise group be estimated along with the standard error of the difference, as well as a hypothesis test that the difference in slopes is equal to zero. The estimated difference is computed by subtracting the estimated slope of exercise protocol B from that of exercise protocol A as specified by **brhr*pgm 1 -1** in the ESTIMATE statement. The 'Slope A- Slope B' is an arbitrary label given by the data analyst.

- In the "Solution" output produced by Program 13.3.B, the parameter estimates labelled PGM and BRHR*PGM for A and B are respectively unbiased estimates of the intercept and slope for the simple linear regression of HR vs. BRHR in each exercise group. These estimates are identical to those obtained from each group's separate simple linear regression in Program 13.2. The intercept and slope estimates for the simple linear regression are respectively 60.49 and 1.30 for Exercise Group A and 137.48 and 0.26 for Exercise Group B. However it may be noted that the standard errors of these intercept and slope estimates from Program 13.2 and 13.3 are different. This is because the standard errors of these estimates from Program 13.3B are based on the mean square errors of an unequal slopes multiple regression model and not on the mean square error of each group's simple linear regression as in Program 13.2. The mean square error of the unequal slopes multiple regression model is a weighted average of the mean square errors from each group's simple linear regression, where the weight for each group is the degrees of freedom associated with each group's mean square error.

13.G Program 13.4

Program 13.4 implements Step H, Option 1, for the exercise study data in the **temporary** SAS data set PGM13, i.e., this program computes and provides a graphical display of discrete simulation-based multiplicity-adjusted confidence limits for the difference between each group's simple linear regression line, HR vs. BRHR, at specified values of BRHR in the range of 58 to 84. The program also provides discrete simulation-based multiplicity-adjusted hypothesis tests of zero difference between Exercise Programs A and B in mean

HR, adjusted for the covariate, BRHR, at specified values of BRHR in the range of 58 to 84
(Step H, Option 2). In other words, using discrete simulation-based multiplicity-adjusted
confidence limits and hypothesis tests, this program evaluates whether the difference in
predicted values obtained from each group's simple linear regression of HR vs. BRHR is
zero at specified values of BRHR.

Please note: If you attempt to execute Program 13.4 after having logged out from the
SAS session in which you executed Program 13.1, you will get error messages in most
operating environments. One way to avoid this is by executing Program 13.1 in your current
SAS session prior to running Program 13.4 or alternatively by copying the DATA step
statements and data from Program 13.1 and pasting them at the beginning of Program
13.4.

```
*Program 13.4;
title Unequal Slopes Covariance Model;
title Discrete Simulation-Based Multiplicity-Adjusted Confidence Limits
and Tests;
%macro MakeDiffs;
%do i=58 %to 84 %by 2;
"&i" PGM 1 -1 PGM*BRHR &i -&i,
%end;
"&i" PGM 1 -1 PGM*BRHR &i -&i
%mend;
proc orthoreg data=pgm13;
/*apply an unequal slopes model*/
class pgm;
model HR= PGM BRHR PGM*BRHR;
estimate %MakeDiffs
/adjust=simulate(acc=0.0002 report seed=4538241) cl;
ods output Estimates=Estimates;
run;
proc print data=Estimates noobs label;
var Label Estimate StdErr tValue probt Adjp AdjLower AdjUpper;
run;
proc sgplot data=Estimates(rename=(Estimate=HR_DIFF label=BRHR));
series x=BRHR Y=HR_DIFF;
series x=BRHR Y=AdjLower/ markers;
series x=BRHR Y=AdjUpper/ markers;
refline 0;
title 'Discrete Simulation-Based Multiplicity-Adjusted
 Confidence Limits';
run;
title;
```

Comments about Program 13.4

- Program 13.4 is based on Program 9.10 in Section 9.3.3 of Westfall et al. (2011), which
  utilizes PROC ORTHOREG, a SAS procedure that is able to implement a covariance
  analysis and can accommodate the SAS macro "MakeDiffs". Entire books have been
  devoted to SAS macros, which are generally considered an advanced topic so we will not
  attempt to describe macro programming here, other than to say a SAS macro can save
  you time and effort as it allows you to reuse code over and over again in the same or in a

different program. An introductory paper on SAS macros (Slaughter and Delwiche, 2004) is available at

http://www2.sas.com/proceedings/sugi29/243-29.pdf

- The following macro "MakeDiffs" of Program 13.4 is used by the ESTIMATE statement of the ORTHOREG procedure to specify the covariate (BRHR) values from **58 to 84 in units of 2**, at which to estimate discrete simulation-based confidence limits and tests for the difference between the regression lines of the two exercise programs, as shown in the output for Step H, Options 1 and 2 in Section 13.8:

```
%macro MakeDiffs;
%do i=58 %to 84 %by 2;
"&i" PGM 1 -1 PGM*BRHR &i -&i,
%end;
"&i" PGM 1 -1 PGM*BRHR &i -&i
%mend;
```

- The specifications of the ESTIMATE statement in this program are beyond the scope of this book, as they are based on the design matrix construction used by the ORTHOREG procedure to represent a linear model that includes a classification variable as a predictor. For those interested in the details of design matrix construction in various SAS procedures including Proc ORTHOREG, a useful reference is

https://support.sas.com/documentation/onlinedoc/stat/131/introcom.pdf

- The CLASS statement in the ORTHOREG procedure has the same purpose as we described for the CLASS statement in the GLM procedure, i.e., to specify which variables should be treated as classification (qualitative) predictors in the analysis.

13.H   Program 13.5

Program 13.5 implements Step H, Option 3 for the exercise study data in the **temporary** SAS data set PGM13, i.e., this program provides a graphical display of Working-Hotelling multiplicity-adjusted confidence limits for the difference between the simple linear regression lines where Y is HR (heart rate after a six-minute test run) and X, the covariate is BRHR (baseline resting heart rate). The program also provides Working-Hotelling multiplicity-adjusted hypothesis tests of zero difference between Exercise Programs A and B in mean HR, adjusted for the covariate, BRHR (Step H, Option 4).

Please note: If you attempt to execute Program 13.5 after having logged out from the SAS session in which you executed Program 13.1, you will get error messages in most operating environments. One way to avoid this is by executing Program 13.1 in your current SAS session prior to running Program 13.5 or alternatively by copying the DATA step statements and data from Program 13.1 and pasting them at the beginning of Program 13.5.

```
*Program 13.5;
title Unequal Slopes Covariance Model;
```

```
/* This rogram provides Working-Hotelling Multiplicity-Adjusted Confidence
Limits and Scheffe Tests*/
%macro MakeDiffs;
%do i=58 %to 84 %by 2;
"&i" PGM 1 -1 PGM*BRHR &i -&i,
%end;
"&i" PGM 1 -1 PGM*BRHR &i -&i
%mend;
proc orthoreg data=pgm13;
class pgm;
model HR= PGM BRHR PGM*BRHR;
estimate %MakeDiffs
/adjust=scheffe cl;
ods output Estimates=Estimates;
run;
proc print data=Estimates noobs label;
var Label Estimate StdErr tValue probt Adjp AdjLower AdjUpper;
run;
proc sgplot data=Estimates(rename=(Estimate=HR_DIFF label=BRHR));
series x=BRHR Y=HR_DIFF;
series x=BRHR Y=AdjLower/ markers;
series x=BRHR Y=AdjUpper/ markers;
refline 0;
title Working-Hotelling Confidence Bounds;
run;
```

Comments on Program 13.5

The statements in Program 13.5 are identical to Program 13.4, except for the OPTION "/adjust=scheffe cl;" specified for the ESTIMATE statement in Program 13.5 which requests Working-Hotelling confidence limits and multiplicity-adjusted tests based on Scheffé's method.

13.I   Program 13.6

Program 13.6 fits an equal slopes model to the exercise study data in the **temporary** SAS data set PGM13 and applies Scheffé's tests and Working-Hotelling confidence bounds for the difference in group covariate-adjusted means at specified covariate values. This program is identical to Program 13.5, except an equal slopes model is specified in Program 13.6.

Please note: If you attempt to execute Program 13.6 after having logged out from the SAS session in which you executed Program 13.1, you will get error messages in most operating environments. One way to avoid this is by executing Program 13.1 in your current SAS session prior to running Program 13.6 or alternatively by copying the DATA step statements and data from Program 13.1 and pasting them at the beginning of Program 13.6.

```
* Program 13.6;
/* This program applies an equal slopes model to evaluate
BRHR-adjusted mean differences between Exercise Program A and B
```

```
and provides Working-Hotelling Multiplicity-Adjusted Confidence Limits
and Scheffe Tests*/
proc glm data=pgm13;
class pgm;
model hr=pgm brhr;
run;
%macro MakeDiffs;
%do i=58 %to 84 %by 2;
"&i" PGM 1 -1,
%end;
"&i" PGM 1 -1
%mend;
proc orthoreg data=pgm13;
class pgm;
model HR= PGM BRHR ;
estimate %MakeDiffs
/adjust=scheffe cl;
ods output Estimates=Estimates;
run;
proc print data=Estimates noobs label;
var Label Estimate StdErr tValue probt Adjp AdjLower AdjUpper;
run;
proc sgplot data=Estimates(rename=(Estimate=HR_DIFF label=BRHR));
series x=BRHR Y=HR_DIFF;
series x=BRHR Y=AdjLower/ markers;
series x=BRHR Y=AdjUpper/ markers;
refline 0;
title 'Working-Hotelling Confidence Bounds for Difference
in Mean Heart Rate (HR_DIFF) Between Exercise Programs A and B';
run;
```

13.J Program 13.7

Program 13.7 applies a one-way ANOVA to the exercise study data in the **temporary** SAS data set PGM13. See Section 13.10 for a discussion of the results from this one-way ANOVA which does not adjust each group's mean heart rate for the nuisance variable baseline resting heart rate.

Please note: If you attempt to execute Program 13.7 after having logged out from the SAS session in which you executed Program 13.1, you will get error messages in most operating environments. One way to avoid this is by executing Program 13.1 in your current SAS session prior to running Program 13.7 or alternatively by copying the DATA step statements and data from Program 13.1 and pasting them at the beginning of Program 13.7.

```
* Program 13.7;
title One-Way ANOVA of Exercise Program Data;
 /* Fit a oneway ANOVA model */;
```

```
proc glm data=pgm13;
   class pgm;
   model hr=pgm;
   means pgm;
quit;
```

13.K   Summary of SAS Programs for Example 13.2

Programs 13.8 to 13.10B (Sections 13.L to 13.O) are equivalent to Programs 13.1 to 13.3B except for changes related to adapting the programs for the data of Example 13.2. Program 13.11 fits an equal slopes model which was determined to be appropriate for Example 13.2 based on evaluation of results from Programs 13.10A and 13.10B.

13.L   Program 13.8

Program 13.8 stores the data of Example 13.2 in the **temporary** SAS data set ONE for the duration of a SAS session in most operating environments. This program checks for data input and data recording errors in the data and is similar to data checking programs presented in previous chapters.

```
*Program 13.8;
data one;
input Idnum Method PostScore PreScore;
datalines;
1 2 72 37
2 2 62 30
3 2 70 25
4 2 68 38
5 2 66 20
6 2 61 16
7 2 70 36
8 2 60 27
9 1 59 23
10 1 64 31
11 1 60 31
12 1 53 15
13 1 53 17
14 1 61 11
15 1 59 20
16 1 62 15
;
proc sort data=one presorted out=onesort; by method;
*STEP A;
proc print data=onesort;by method;
run;
```

```
*STEP B;
proc univariate data=onesort plots;
var PostScore PreScore; by method;
ods exclude TestsForLocation;
id idnum;
run;
/*The following plots can be examined for contextual,
predictor and response outliers that merit
investigation for recording errors*/
ods listing style=journal2 image_dpi=200;

proc sgplot data=onesort;
   scatter x=PreScore y=PostScore/markerattrs=(symbol=circlefilled
   color=black);
   yaxis values=(50 to 80 by 5);
   xaxis values=(10 to 40 by 5);
   by method;
run;
```

13.M   Program 13.9

Program 13.9 applies a simple linear regression of Y vs. the covariate within each group for Example 13.2, using as input the data stored in the **temporary** SAS data set ONE created by Program 13.8. Regression diagnostics and measures of influence for each group's simple linear regression are provided by Program 13.9.

Please note: If you attempt to execute Program 13.9 after having logged out from the SAS session in which you executed Program 13.8, you will get error messages in most operating environments. One way to avoid this is by executing Program 13.8 in your current SAS session prior to running Program 13.9 or alternatively by copying the DATA step statements and data from Program 13.8 and pasting them at the beginning of Program 13.9.

```
*Program 13.9;
proc sort data=one presorted out=onesort; by method;
*STEP A;
proc print data=onesort;by method;
run;
proc reg data=onesort
   plots(only label)=(cooksd dffits dfbetas RStudentByLeverage);
   model PostScore = PreScore  / r p clb influence;by method;
   output out=diag
  r=r rstudent=rstudent student=student p=p dffits=dffits cookd=cookd;
run;

*Evaluate Ideal Inference Conditions for each study method;
/*Obtain plots of internally studentized residuals vs. covariate, PRESCORE,
 */
 proc sgplot data=diag;
```

```
      refline 0/ axis=Y;
      scatter x=prescore y=student /markerattrs=(symbol=circlefilled
      color=black);
      yaxis values= (-4 to 4 by .5);
      by method;
run;
* Evaluate normality;
* Obtain a Q-Q normal plot for internally studentized residuals;
proc univariate data =diag;
      by method;
      var student;
      qqplot student  /normal (mu=est sigma=est);
      inset n='Sample Size' skewness kurtosis
      /pos=tm header='(Internally) Studentized Residuals';
run;
/* Get a listing of variables of interest for cases identified as
   regression outliers based on a specified cutoff value for
   the studentized deleted residual (rstudent) */

proc print data=diag; where abs(rstudent) gt 2;
run;
*STEP H;
*Check for influential cases;
/*CALCULATE SIZE-ADJUSTED CUT-OFF VALUES
for DFBETAS: Belsley et al 1980
for DFFITS Belsley et al 1980,Staudte and Sheather 1990*/

data cutvalues;
 /*parm=number of parameters in model including the intercept*/
 /*n = equal number of cases in each simple linear regression */
/*dffitscutbels is cutoff value recommended by Belsley et al 1980*/
/*dffitscutss is cutoff value recommended by Staudte and Sheather 1990*/
      parm=2;
      n=8;
      nminusp=n-parm;
      dfbetacut=2/(sqrt(n));
      dffitscutbels=2*(sqrt(parm/n));
      dffitscutss=1.5*(sqrt(parm/n));
run;
proc print data=cutvalues;
 title Size-Adjusted Cutoff Values for DFBETAS and DFFITS;
run;
data influence;
      if _n_=1 then set cutvalues;
      set diag;
      absdffits=abs(dffits);
run;
proc print data=influence;
title;
run;
proc print data=influence; where cookd gt 0.5 ;
```

```
    title Cases with COOKD Value > 0.5 ;
    var idnum cookd method postscore prescore p r student rstudent  dffits;
run;
proc print data=influence; where absdffits gt 1.00;
    title Cases with DFFITS Absolute Value > 1.00;
    var  idnum dffits method postscore prescore p r student rstudent cookd;
run;
proc print data=influence; where absdffits gt dffitscutbels;
    title Cases with Abs. DFFITS Value > Belsley et al 1980 Cutoff Value;
    var  idnum dffits method postscore prescore p r student rstudent cookd ;
run;
proc print data=influence; where absdffits gt dffitscutss;
    title Cases with Abs. DFFITS Value > Staudte and Sheather 1990 Cutoff
    Value;
    var idnum dffits method postscore prescore p r student rstudent cookd;
run;
 title;
 proc sgplot data=diag;
   refline 0/ axis=Y;
   scatter x=METHOD y=STUDENT /markerattrs=(symbol=circlefilled
   color=black);
   yaxis values= (-4 to 4 by .5);
   xaxis type=discrete;
 run;
```

13.N   Program 13.10A

Program 13.10A uses the GLM procedure to fit an unequal slopes multiple regression model to the data of Example 13.2 stored in the **temporary** SAS data set ONE created in Program 13.8.

Please note: If you attempt to execute Program 13.10A after having logged out from the SAS session in which you executed Program 13.8, you will get error messages in most operating environments. One way to avoid this is by executing Program 13.8 in your current SAS session prior to running Program 13.10A or alternatively by copying the DATA step statements and data from Program 13.8 and pasting them at the beginning of Program 13.10A.

```
*Program 13.10A;
*Unequal Slopes Model;
proc glm data=one;
   class method;
   model postscore=method prescore method*prescore/ss3;
   output out=diag  student =student;
run;
*Evaluation of Normality;
proc univariate data =diag;
   var student;
   qqplot student  /normal (mu=est sigma=est);
```

```
   inset n='Sample Size' skewness kurtosis
   /pos=tm header='(Internally) Studentized Residuals ';
run;
```

13.O Program 13.10B

To enable an evaluation of the unequal slopes model's fit to the data of Example 13.2 stored in the **temporary** SAS data set ONE created by Program 13.8, Program 13.10B gives unbiased estimates of the slopes of the simple linear regression of PostScore (Y) vs. PreScore (the covariate) for each group (Method) and also the estimated difference between these simple slopes with estimated 95% confidence limits for the difference in simple slopes in the sampled populations.

Please note: If you attempt to execute Program 13.10B after having logged out from the SAS session in which you executed Program 13.8, you will get error messages in most operating environments. One way to avoid this is by executing Program 13.8 in your current SAS session prior to running Program 13.10B or alternatively by copying the DATA step statements and data from Program 13.8 and pasting them at the beginning of Program 13.10B.

```
*Program 13.10B;
proc glm data=one;
   class method;
   model postscore=method method*prescore/noint solution clparm ss3;
   estimate 'Slope 1-Slope 2' method*prescore 1 -1;
   ods select ParameterEstimates Estimates;
quit;
```

13.P Program 13.11

Program 13.11 applies an ANCOVA (equal slopes model) to the data of Example 13.2 stored in the **temporary** SAS data set ONE created by Program 13.8.

Please note: If you attempt to execute Program 13.11 after having logged out from the SAS session in which you executed Program 13.8, you will get error messages in most operating environments. One way to avoid this is by executing Program 13.8 in your current SAS session prior to running Program 13.11 or alternatively by copying the DATA step statements and data from Program 13.8 and pasting them at the beginning of Program 13.11.

```
*Program 13.11;
title Example 13.2 Equal Slopes Model;
proc glm data=one;
class Method;
model PostScore= Method PreScore/ss3;
means Method;
lsmeans Method/pdiff tdiff stderr cl;
quit;
```

13.Q   Program 13.12

Program 13.12 applies an ANCOVA (equal slopes model) to the data of Example 13.2 stored in the **temporary** SAS data set ONE created by Program 13.8. The results from the LSMEANS statements in this program illustrate that, because an equal slopes model is used, the difference in PostScore means between Method 1 and Method 2 is estimated to be -5 when PostScore means for Methods 1 and 2 are adjusted for different values of PreScore in the observed range common to both methods in the study.

Please note: If you attempt to execute Program 13.12 after having logged out from the SAS session in which you executed Program 13.8, you will get error messages in most operating environments. One way to avoid this is by executing Program 13.8 in your current SAS session prior to running Program 13.12 or alternatively by copying the DATA step statements and data from Program 13.8 and pasting them at the beginning of Program 13.12.

```
*Program 13.12;
proc glm data=one;
   class Method;
   model PostScore= Method PreScore/ss3;
   means Method;
   lsmeans Method/pdiff tdiff stderr cl;
   lsmeans Method /pdiff tdiff stderr cl at PreScore= 24.5 ;
   lsmeans Method /pdiff tdiff stderr cl at PreScore= 24.0 ;
   lsmeans Method /pdiff tdiff stderr cl at PreScore= 15;
quit;
```

14

Alternative Approaches If Ideal Inference Conditions Are Not Satisfied

---

14.1 Introduction

The methods presented in previous chapters – simple linear regression, multiple linear regression, one-way analysis of variance (ANOVA), and analysis of covariance (ANCOVA) – are all based on an ordinary least squares linear regression model which has a single error term and which has an outcome variable that is continuous. These methods share the following ideal inference conditions:

- random sampling

- independence of model errors

- equality of model error variances in the sampled population(s)

- normal distribution of model errors in the sampled population(s)

In simple and multiple linear regression where predictor variables are continuous, there is the additional ideal inference condition that there is a linear relationship between the outcome and the continuous predictor(s). Moreover, in multiple regression, absence of harmful collinearity among predictor variables is also a condition for ideal inference.

In this chapter we provide an overview of some general alternative approaches that can be considered when ideal inference conditions of simple linear regression and multiple linear regression including one-way ANOVA and ANCOVA are not satisfied in practice. Although a number of these alternative approaches are beyond the scope of this book, we nevertheless briefly introduce them, so our readers will be aware of these alternatives and of their accessibility in SAS.

In real life, violation of more than one ideal inference condition may occur concurrently in a researcher's data set. However, for clarity of presentation, we initially consider violation of each ideal inference condition separately.

14.2 When Random Sampling Is Not an Option

Random sampling is an ideal method of ensuring your research is based on a sample that is representative of your population of interest. In the real world, however, it is not always possible to first specify a population of interest and then randomly sample from that specified population. For example, sometimes all the subjects used in an investigation are those who volunteered for the study. Obviously these subjects are not a random sample from a target population of interest. In these situations where random sampling is not feasible, it is critical that investigators summarize the characteristics of their sample that are known to effect or be associated with the outcome and predictor variables in the study (e.g., age, gender, socioeconomic status) so that subject matter expertise can be applied to decide whether the results obtained in the study are applicable to the population of interest (Moodie and Craig, 1986).

14.3 When Errors Are Not All Independent

14.3.1 Introduction

Recall that the model error, which is usually simply referred to as the error, is that part of an individual's value on the outcome variable Y that is unexplained by the model. Errors are not all independent if the error of one observation is in some way related to the error of one or more observations in the analysis. Violation of independence of errors, resulting in invalid hypothesis tests and confidence intervals, always occurs in a statistical analysis which ignores any correlation of errors present in the data. The following examples illustrate scenarios where violation of independence of errors occurred.

Repeated Measurements Example

Independence of errors was violated in a simple linear regression investigating the relationship between systolic blood pressure (Y) and age (X) in a study in which two systolic blood pressure measurements were taken five minutes apart on the same arm of each subject. There were ten subjects in the study but the investigators used the repeated measurements on each subject as independent observations in the simple linear regression so that they had 20 observations (two measurements per subject) in their analysis. Such an analysis is referred to as a **naive pooled analysis** (Burton et al., 1998). Dependence of errors occurred in this analysis because the observations for each subject tended to be more alike than observations on different individuals. The data in this example are **clustered data**, i.e., data comprised by clusters where observations within a cluster tend to be more similar than observations between clusters. In this study each individual is considered to be a cluster.

Alternative approaches to circumvent dependence of errors in this example include:

- **methods based on fitting a model for correlated data**, which are beyond the scope of this book but references are provided in subsection 14.3.2

- **the single summary statistic approach**, which involves computing an appropriate single summary statistic (such as the mean of the two systolic blood pressure measurements) for each cluster (subject) and using these summary statistic values (means) as observations for the Y variable in a statistical analysis that assumes independence (such as simple linear regression as in this example). It is important to note that this approach is not valid when there is an unequal number of repeated observations on each subject. A disadvantage of the single summary statistic approach is that information about the variability of blood pressure measurements within a subject would be disregarded.

Nested (Hierarchical) Design Example

Consider the data in Table 14.1, provided by Sokal and Rohlf (1981) from unpublished records of a study by J. A. Weir, who wanted to compare blood pH in two strains of male mice where the strains had been selected for genetically high and low blood pH.

TABLE 14.1

Blood pH measurements for two strains of male mice.

Strain	Litter	Blood pH measurements on individual mice
High pH	A	7.43 7.38 7.49 7.49
	B	7.39 7.46 7.50 7.55
	C	7.53 7.50 7.63 7.47
	D	7.39 7.39 7.44 7.55
	E	7.48 7.43 7.47 7.44
	F	7.43 7.55 7.44 7.50
	G	7.49 7.49 7.51 7.54
Low pH	H	7.40 7.46 7.43 7.42
	I	7.35 7.40 7.46 7.38
	J	7.51 7.39 7.42 7.43
	K	7.46 7.53 7.49 7.45
	L	7.48 7.53 7.52 7.43
	M	7.43 7.40 7.48 7.47
	N	7.53 7.47 7.50 7.53

If a one-way ANOVA for two groups had been used to compare the two mouse strains for mean blood pH, violation of independence of errors would have been an issue because of the study design. In this study the subjects were selected from litters that had at least four males. If a litter had more than four males, four were selected at random from this litter. Thus, the data are clustered data where each litter is considered to be a cluster and the individual mice are nested within litters. The design of this study is a nested design, also known as a hierarchical or a multilevel design.

Alternative approaches that would circumvent dependence of errors in this example include methods based on fitting a model for correlated data (subsection 14.3.2). The single summary statistic approach is not appropriate in this genetics study where the investigator had been expressly interested in the variation among individual mice within a litter. It would have been helpful for Dr. Weir if at the design phase of his study, he had known about methods based on correlated data (subsection 14.3.2) that can handle unequal sample sizes, which would have allowed him to study all the males in each litter rather than selecting only four males at random from litters that had more than four males.

14.3.2 Fitting a Model for Correlated Data

When a data set in an analysis is comprised by groups (clusters) with correlated errors, a recommended approach, which gleans the most information from the data, is to fit a model to the data that takes the correlation structure of the errors into account (Kutner et al., 2005; Ryan, 2009; Milliken and Johnson, 2009). Methods which involve modelling correlated errors are beyond the scope of this book. Such methods include generalized estimating equations (Diggle et al., 2002; Hanley et al., 2003) and mixed (multilevel) models (Burton et al., 1998; Milliken and Johnson, 2002; Littell et al., 2006; Zhu, 2014). Burton et al. (1998) provided a practical comparison of generalized estimating equations and mixed effects modelling approaches. References discussing methods that address serial correlation (autocorrelation) in time series data include Freund and Littell (2000), Kutner et al. (2005) and Greene (2008).

14.4   Transformations

In this section we introduce the topic of data transformation, an alternative approach, which may be considered as a remedial measure for a violation of linearity, equal variances, and/or normality assumptions.

A transformation of data values is the re-expression of the original data on a different scale of measurement, e.g., the log to the base 10.

If linearity, equal variances, and/or normality of a conventional linear model method (e.g., ordinary least squares linear regression, ANOVA) are not satisfied for the original data, then transforming these data and reapplying the conventional method to the transformed data may fix one or more of these ideal inference condition violations. It is important to note, however, that violation of independence of errors can never be remediated by transforming your data.

Once you have applied a particular transformation and refit your linear model to the transformed data, it is extremely important to evaluate, using the model checking methods of Chapter 3, whether linearity, equal variances, and normality are more reasonable when that particular transformation has been applied. If that particular transformation appears to make assumption violations worse, another transformation may be tried and evaluated. Ideally, however, knowledge of subject matter should guide the choice of a particular transformation or alternatively the choice of a different remedial approach other than transformations.

14.4.1   Advantages and Disadvantages of a Data Transformation Approach

Advantages of a data transformation include:

* It is easy to implement.

* A transformation may sometimes remediate more than one ideal inference condition violation simultaneously (e.g., non-linearity and heterogeneity of variances).

* Model parameter estimation and hypothesis testing are straightforward, provided ideal inference conditions of the particular conventional linear model applied are plausible upon data transformation.

* Confidence limits in transformed units can be transformed back into the original units.

  Disadvantages of a data transformation include:

* A complex relationship between an outcome variable and a predictor may be oversimplified.

* Sometimes applying a transformation while fixing the violation of one ideal inference condition may simultaneously result in violations of other ideal inference conditions.

* It may be difficult to interpret the model parameters obtained from fitting a model to transformed data.

* Estimates of standard errors obtained from an analysis of transformed data should not be directly "back transformed" into the original units.

Comments

Tukey (1977) advised that it is not useful to apply transformations to differences (e.g., pretreatment Y minus post-treatment Y) although it may sometimes be helpful to transform variables with which the difference is computed.

Hoaglin et al. (1982) suggested a transformation is likely to be helpful only if ratio of the largest X value to the smallest X value is "say greater than 20" and is unlikely to be helpful if this ratio is "say less than 2". They note this rule of thumb does not apply if there is no "natural zero" inherent in the data, as for example occurs with temperature data values where a different ratio will be obtained depending on whether the temperature is measured in Fahrenheit or Celsius. The rule of thumb may be applied when there is no natural zero if the ratio is calculated as

$$\frac{\text{largest data value - natural boundary}}{\text{smallest data value - natural boundary}} \qquad (14.1)$$

where, for example, the natural boundary for temperatures recorded in Fahrenheit would be 32 whereas the natural boundary for temperatures recorded in Celsius is zero.

14.4.2 Power Transformations

A power transformation is a transformation which raises a variable to a particular power, e.g., X^2, $X^{1/2}$.

Tukey's Ladder of Power Transformations

Tukey (1977) organized power transformations into a graduated series, which he called a ladder, to help data analysts understand the relative effect of various transformations on the data. This understanding can help in the choice of an appropriate transformation for the values of X or Y or both so that ideal inferences conditions become more reasonable on the transformed scale as compared to the original scale. Tukey's ladder of power transformations is organized as follows:

- The top rungs of the power transformation ladder consist of transformations which raise the variable to a power greater than one.

- The lower rungs of the ladder consist of transformations which raise the variable to a power less than one.

- The raw untransformed data is raised to the power one.

- The log transformation, by convention, is assigned its position on the ladder as if it were a transformation with power equal to zero.

The following illustrates Tukey's ladder of powers for typically used power transformations where p, using Tukey's (1977) notation, denotes the power to which a variable is raised and Y denotes any random variable, not necessarily an outcome variable:

Y^3 $(p = 3)$
Y^2 $(p = 2)$
Y^1 $(p = 1$, raw data no transformation)
$Y^{.5}$ $(p = .5$, square root of $Y)$ defined only for non-negative values
$\log Y$ $(p = 0$ by convention) defined only for positive values
$1/Y$ $(p = -1$, reciprocal of $Y)$

356   *Alternative Approaches*

Comments

When a power transformation with $p < 0$ is applied, the order of the original data values is not preserved, in fact the order is reversed. For example, when a reciprocal transformation is applied (i.e., $p = -1$), the raw data values 2, 3, 4 become 1/2, 1/3, 1/4, i.e., the original order of the raw data values is reversed (1/2 is greater than 1/3, which is greater than 1/4). Tukey (1977) and Mosteller and Tukey (1977) suggested that data analysts may find it convenient to use the negative reciprocal rather than the reciprocal, if they want to preserve the order of their original data values. The original order of 2, 3, 4 is preserved if the negative reciprocal transformation is applied (-1/2 is less than -1/3, which is less than -1/4).

We will refer to Tukey's ladder of powers in subsequent sections when we discuss transformations as a remedial measure for linearity, equal variances, and/or normality violations.

14.4.3   Applying Power Transformations in SAS

Power transformations can be applied in a SAS DATA step and the transformed data values, saved in a SAS data set, can be input to SAS procedures. The following program illustrates how common power transformations can be applied in a DATA step and saved in a temporary SAS data set arbitrarily named ONE. The data set ONE is input to the PRINT procedure.

```
data one;
input Y@@;
YCUBE  =Y**3  ;           /* Y cubed */
YSQUARE = Y**2;        /* Y squared */
YSQROOT = sqrt(Y);     /* square root of Y */
LNY    = log(Y);       /* where log(Y) = natural logarithm (base e) */
LOGY   = log10(Y);     /* where log10(Y)= log to base 10            */
RECIP_Y=Y**-1  ;       /* where Y**-1 is the reciprocal transformation*/
NEG_RECIP_Y =  -(Y**-1) ;    /* where-(Y**-1) is the negative
of the reciprocal transformation    */
datalines;
 4 9 6 10 15 35
 ;
run;
proc print data=one;
run;
```

The results from this program are:

Obs	Y	YCUBE	YSQUARE	YSQROOT	LNY	LOGY	RECIP_Y	NEG_RECIP_Y
1	4	64	16	2.00000	1.38629	0.60206	0.25000	-0.25000
2	9	729	81	3.00000	2.19722	0.95424	0.11111	-0.11111
3	6	216	36	2.44949	1.79176	0.77815	0.16667	-0.16667
4	10	1000	100	3.16228	2.30259	1.00000	0.10000	-0.10000
5	15	3375	225	3.87298	2.70805	1.17609	0.06667	-0.06667
6	35	42875	1225	5.91608	3.55535	1.54407	0.02857	-0.02857

Comment

In SAS, the **log** function calculates the natural log whereas the function **log10** calculates the log base 10.

14.4.4 Log Transformations

The log transformation is one of the most commonly used transformation when linearity, equal variances and/or normality are not satisfied. Applying a log transformation to a set of data values has the effect of bringing large data values closer together and spreading out small data values. This effect is achieved regardless of the base of the log used. Therefore it does not matter whether you use log base 10 or log base e. We will outline how this effect of log-transforming data can specifically help remediate nonlinearity, unequal variances, and non-normality when we discuss these ideal inference condition violations individually in subsequent sections.

Log transformations are defined only for positive numbers. Therefore, if you want to apply a log transformation to a variable and there are zeros in your data, you must add an arbitrary constant to that variable's values in your data before applying a log transformation. This constant could be a very small number such as 0.5 or 1.0. You can add a constant to the data values for a particular variable, say ORIGY, in a SAS data step, as follows:

```
data one;
input ORIGY@@;
/* Create a new variable  denoted as NEWY
by adding a small constant 0.5 to the original Y variable input ORIGY */
NEWY =ORIGY+0.5;
LN_NEWY   = log(NEWY);     /* where LN_NEWY is the
natural log transformed NEWY */
datalines;
 0 9 6 10 15 35
 ;
run;
proc print data=one;
run;
```

The results from this program are:

Obs	ORIGY	NEWY	LN_NEWY
1	0	0.5	-0.69315
2	9	9.5	2.25129
3	6	6.5	1.87180
4	10	10.5	2.35138
5	15	15.5	2.74084
6	35	35.5	3.56953

Log Transformations and Interpretation of Regression Coefficients

Vittinghoff et al. (2012) provided formulas for interpreting regression coefficients obtained from applying a linear regression to log-transformed data. These various formulas depend on whether the X variable, or the Y variable or both are log-transformed.

Interpretation When Only a Predictor, X is Log-Transformed

When only a predictor, X (not the response variable, Y) is log-transformed, regardless of the base of the log, the estimated regression coefficient for log-transformed X multiplied by log (1.01) is interpretable as the estimate of the change in the average value of Y for every 1% **increase** in X in the population sampled.

Vittinghoff et al. (2012) illustrated with an example of a linear regression where Y, the response variable, systolic blood pressure in mm Hg is in the original scale, i.e., not transformed, and X, the predictor variable, weight is transformed using a natural log (ln) transformation. The estimated regression coefficient for ln(weight) from this analysis is 3.004517. Therefore, the interpretation of this regression analysis is that the estimated average systolic blood pressure in the sampled population **increases** by approximately 0.03 mm Hg (i.e., 3.004517 multiplied by ln(1.01)) for every 1% increase in weight of the individuals in the population sampled.

Usually we might be more interested in the change in average SBP associated with a larger percentage weight gain, for example, a 10% weight gain. If we want to obtain an estimate of the change in average Y for each **10% increase** in X, we multiply the estimated regression coefficient for ln(X) by ln(1.10). For example, if we multiply 3.00451, the estimated regression coefficient for ln(weight) from Vittinghoff et al.'s regression by ln(1.10), we obtain the estimate that the average SBP increases by approximately 0.29 mmHg for each 10% increase in weight (3.004517 × ln(1.10) = 0.29). You can verify this result using Program 14.A (Section 14.A). The output from this program is 0.28636.

Similarly if we want to obtain an estimate of the change in average Y for each **20% increase** in X, we multiply the estimated regression coefficient for ln(X) by ln(1.20).

Interpretation When Only the Response Variable, Y is Log-Transformed

When only the response variable Y is log-transformed in a linear analysis, the following interpretations can be made:

- When a natural log is used to transform the Y values, then $100(e^{\hat{\beta}}-1)$ is interpretable as the estimated **percentage** change in the average value of Y per unit increase in X in the population sampled.

- When log to the base 10 is used for transforming Y, then $100(10^{\hat{\beta}}-1)$ is interpretable as the estimated **percentage** change in the average value of Y per unit increase in X in the population sampled.

Interpretation When Both X and Y are Log-Transformed

When both X and Y are log-transformed in a linear analysis, the following interpretations can be made:

- When a natural log is used to transform the X and Y values, then $100(e^{\hat{\beta} ln(1.01)} - 1)$ is interpretable as the estimated **percentage** change in the average value of Y per 1% increase in X.

- When log to the base 10 is used for transforming X and Y, then $100(10^{\hat{\beta} log_{10}(1.01)} - 1)$ is interpretable as the estimated **percentage** change in the average value of Y per 1% increase in X.

14.5 Alternative Approaches When Linearity Is Not Satisfied

14.5.1 Introduction

In simple and multiple regression, when a linear relationship between the outcome variable and a continuous predictor does not seem reasonable in the range of the predictor values observed in the data, alternative approaches include:

- **adding one or more additional predictor variables** to the original linear model (subsection 14.5.2)

- **applying a transformation to the predictor(s) and/or** Y and fitting a linear model to the transformed data (subsection 14.5.3)

- **rank regression** (subsection 14.5.4)

- **polynomial regression** (subsection 14.5.5)

- **fitting a nonlinear model** using nonlinear estimation techniques (subsection 14.5.6)

14.5.2 Adding One or More Predictor Variable(s) to the Model to Achieve Linearity

It may be reasonable to model Y with a linear function when one or more additional predictors are included in your original model. When a plot of internally studentized residuals from your original linear regression against a prospective candidate variable suggests a linear trend, you can try fitting a new linear model with that candidate variable added to your original model. You should then check if this new model appears to be a better fit for your data. However, if too many additional predictor variables are added to the original model there is a potential danger of over-fitting, i.e., we fit the model to our sample data so well that it will not be a good model for predicting Y in a sample with slightly different data values, even though this slightly different sample has also been randomly sampled from the same population.

14.5.3 Transformations and the Linearity Assumption

Transformations are sometimes applied to the predictor(s) or the outcome variable Y or both in an attempt to achieve the linearity assumption. If, however, model checking methods on the original untransformed data have revealed that the ideal inference conditions of normality and equal variances appear reasonable, then a transformation on only the predictor(s) should be considered because a linear transformation on Y in this situation may achieve linearity but at the expense of introducing non-normality and/or inequality of variances. Ideally, knowledge of subject matter being investigated should guide your choice of a linearizing transformation.

It is important to note that a power transformation, such as a log transformation, may be useful only if in the sampled population, a nonlinear relationship between X and Y in the original scale of measurement is **entirely monotonic**, i.e., where an increase in X always involves an increase in Y or where a decrease in X always involves a decrease in Y. An entirely monotonic curve is also referred to as a **simple monotonic** curve.

Mosquito Study Example: Applying a Log Transformation When the Relationship between X and Y is Nonlinear

Researchers wanted to study the density of mosquitoes as a function of the amount of water that is available for breeding. They collected data from 40 different city blocks, approximately 800 square meters in size. Their predictor variable X was the surface area of water in square meters available in a block, and their outcome variable Y was the total weight, in grams, of mosquitoes captured in a trap over a certain specified time period for a block. Previous studies have shown that total mosquito weight in grams is highly correlated with the number of mosquitoes. The data are given in Table 14.2.

Figure 14.1 reveals a curvilinear relationship between mosquito weight (Y) vs. block water surface area (X), which indicates that a simple linear (straight line) model would not be a good fit for these data expressed on the original scale. The researchers preferred the simplicity of a simple linear regression analysis to that of a nonlinear analysis so they tried applying a log transformation to mosquito weight (the Y variable) to see if re-expressing it on a log scale would straighten out the curve. They chose a log transformation because it was compatible with prior research on mosquito population size which has been shown to increase exponentially provided suitable resources are available.

Figure 14.2 is a scatter plot of natural log-transformed mosquito weight vs. surface water, which reveals that the transformation appears to have the desired effect of linearizing the relationship. Thus, a simple linear regression applied to these data on a log-transformed

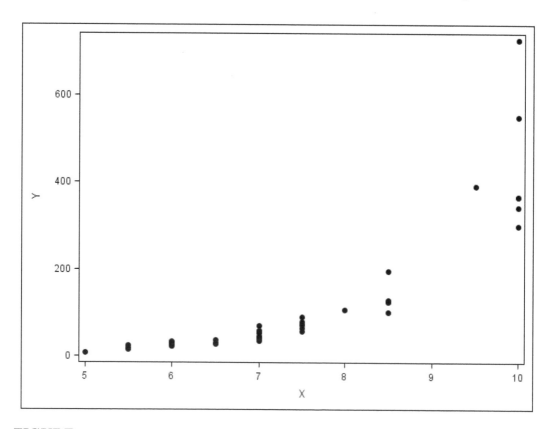

FIGURE 14.1

Scatter plot of Y = mosquito weight (gm) caught in block trap vs. X = block surface water (sq m).

TABLE 14.2

Data from study investigating mosquito prevalence as a function of neighbourhood surface water.

Block	Surface Water (sq m)	Mosquito Weight (gms)	Ln (Mosquito Weight)
1	7.5	74.0	4.30407
2	5.5	14.5	2.67415
3	6.0	22.9	3.13114
4	10.0	300.8	5.70645
5	5.0	8.4	2.12823
6	8.5	125.7	4.83390
7	6.0	34.0	3.52636
8	8.5	125.7	4.83390
9	7.5	90.5	4.50535
10	9.5	391.4	5.96973
11	7.0	69.9	4.24707
12	7.0	42.2	3.74242
13	7.5	79.6	4.37701
14	7.0	55.0	4.00733
15	8.5	130.3	4.86984
16	5.5	24.6	3.20275
17	8.5	101.6	4.62104
18	7.0	53.0	3.97029
19	6.0	26.5	3.27714
20	7.0	46.3	3.83514
21	8.5	196.8	5.28219
22	6.0	27.6	3.31782
23	7.0	34.8	3.54962
24	10.0	727.9	6.59016
25	6.0	32.0	3.46574
26	7.5	73.3	4.29456
27	10.0	367.4	5.90645
28	7.5	58.5	4.06903
29	5.5	18.1	2.89591
30	10.0	342.9	5.83744
31	7.5	65.8	4.18662
32	7.0	39.5	3.67630
33	6.5	27.9	3.32863
34	10.0	551.0	6.31173
35	6.5	30.6	3.42100
36	6.5	37.5	3.62434
37	10.0	367.4	5.90645
38	7.0	59.8	4.09101
39	7.0	42.4	3.74715
40	8.0	107.1	4.67376

scale may be appropriate, provided the other ideal inference conditions of normality and equal variances at each X were reasonable when the y-values (mosquito weight) were re-expressed as natural logs.

In Program 14.B (Section 14.B), we generated regression diagnostics (discussed in Chapter 3) to evaluate whether ideal inference conditions are reasonable when a simple linear regression is applied to the data with mosquito weight log-transformed. Graphical displays of studentized residuals from this regression do not suggest any serious departures from

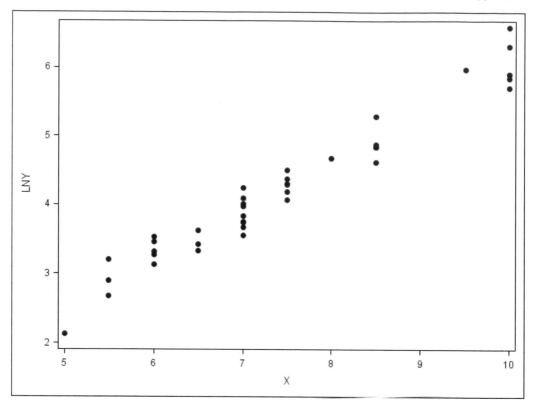

FIGURE 14.2
Scatter plot of LNY = natural log of mosquito weight (gm) caught in block trap vs. X = block surface water (sq m).

equality of population error variances (Figure 14.3) and normality of the population errors (Figure 14.4). Summary statistics from the UNIVARIATE procedure of Program 14.B regarding the distribution of these internally studentized residuals do not suggest a drastic departure from a symmetrical distribution of the model errors, as the mean (0.00) and median (-0.09) of these internally studentized residuals are not remarkably different in value. These summary statistics, the normal QQ plot (Figure 14.4), and the sample size of 40 suggest that normality assumption may not be unreasonable when the mosquito weight is log-transformed.

The estimated regression coefficient for block surface water for the regression where mosquito weight (Y) is natural log-transformed is 0.71018, as given by the parameter estimate for X from the REG procedure in Program 14.B.

Parameter Estimates

Variable	DF	Parameter Estimate	Standard Error	t Value	Pr > |t|
Intercept	1	-1.06010	0.19374	-5.47	<.0001
X	1	0.71018	0.02546	27.90	<.0001

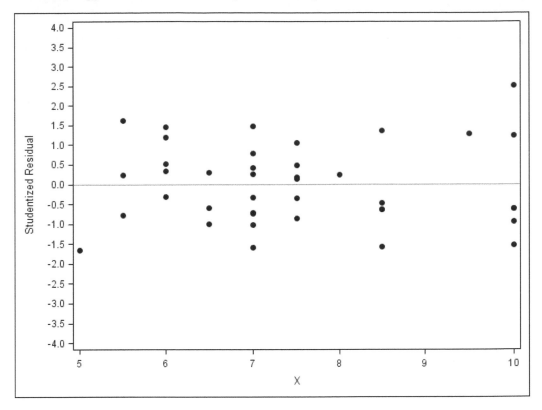

FIGURE 14.3
Scatter plot of studentized residuals vs. block surface water (sq m) from regression of natural log of mosquito weight (gm) caught in block trap vs. block surface water (sq m).

```
                    Parameter Estimates

        Variable      DF        95% Confidence Limits

        Intercept      1        -1.45231      -0.66789
        X              1         0.65864       0.76172
```

Program 14.C (Section 14.C) provides calculations for interpreting the estimated regression coefficient for X (0.71018 for block water surface area in sq m) and its estimated 95% confidence limits (0.65864, 0.76172), when only Y is natural log-transformed. The results from Program 14.C, where the variable, "interpret" is $100(e^{\hat{\beta}}-1)$

```
        Obs      obs      estbeta      interpret

         1        1       0.71018       103.436
         2        2       0.65864        93.216
         3        3       0.76172       114.196
```

inform us that the estimated regression coefficient for X (estbeta=0.71018)) in the mosquito study where only Y is natural log-transformed may be interpreted as follows:

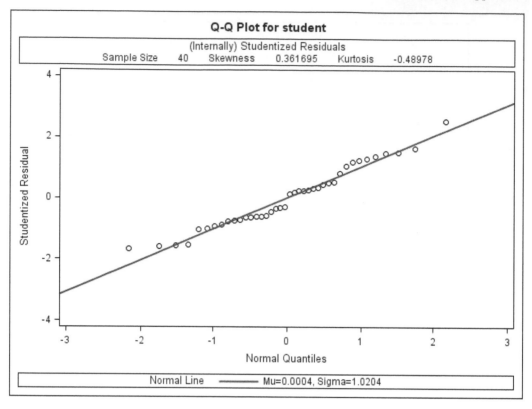

FIGURE 14.4

Normal QQ plot of internally studentized residuals from regression of natural log of mosquito weight (gm) caught in block trap vs. block surface water (sq m).

- the estimated percentage increase in the average value of mosquito weight in grams captured in a trap in a block over a certain specified time period is 103% per unit increase in square meters of available block surface water

- the 95% confidence interval estimate for the estimated percentage increase in the average value of Y per increase in each square meter of available block surface water is (93–114%).

14.5.4   Overview of Rank Regression

Rank regression is a nonparametric method proposed by Iman and Conover (1979) for situations where the relationship between the response variable and a predictor is entirely monotonic, either entirely monotonically increasing or entirely monotonically decreasing. Their procedure performs well whether the monotonic relationship between X and Y is linear or nonlinear. Iman and Conover (1979) wrote that "if one is interested in a model that predicts well over a monotonic nonlinear surface then rank regression will work well". The key advantage of rank regression is its simplicity, as it involves replacing the original data values by their ranks for each variable or average ranks in case of ties and then fitting an ordinary least squares linear regression model to the ranks (Conover, 2012). Applications of rank regression to simple linear and additive multiple linear regressions have been discussed in Conover (2012).

The key disadvantage of rank regression is that it cannot provide standard errors and hence confidence interval estimates and hypothesis tests either for the Y intercept or the slope. Moreover if the relationship between X and Y is not monotonic, rank regression does not perform well for simple or multiple linear regression problems.

14.5.5   Polynomial Regression and Nonlinearity

A polynomial regression model is a linear model (i.e., linear in the parameters), which can be useful when subject matter expertise suggests that a curvilinear relationship between X and Y exists in the population sampled. A polynomial model is also useful as an approximation to an unknown nonlinear model (Cook and Weisberg, 1999; Montgomery et al., 2012). However, not all nonlinear relationships can be approximated by a polynomial, for example, polynomials are not useful for modelling a threshold effect. A threshold effect occurs when there is a particular relationship between X and Y above a threshold value of X but not below it such as the relationship between occupational dust exposure and chronic bronchitic reactions (Ulm, 1991).

A polynomial regression model for a predictor variable X is a special case of the linear model which includes X as well as higher powers of X, e.g., X^2, X^3, as linear terms. A polynomial model of degree 2, known as a quadratic model in X, has X^2 and X as predictors in the linear model and is given by

$$Y = \beta_0 + \beta_1 X + \beta_2 X^2 + \epsilon \tag{14.2}$$

A polynomial model of degree 3, known as a cubic model in X is given by

$$Y = \beta_0 + \beta_1 X + \beta_2 X^2 + \beta_3 X^3 + \epsilon \tag{14.3}$$

A polynomial model is called a hierarchical model because it always includes all the lower orders of X as predictors, e.g., a polynomial model of degree 3 in addition to X^3 also includes the lower orders of X, i.e., X^2 and X as predictors, regardless of their significance.

You may recall from high school math that a quadratic model defines a parabola, i.e., a curve that is symmetric about a single minimum or a maximum. The plot of y1 vs. x and y2 vs. x in Figure 14.5 are examples of data points generated by quadratic models. However, data generated by a quadratic model can look like y3 vs. x or like y4 vs. x in Figure 14.5 when the minimum or maximum of the response variable Y is outside the range of the observed X. The last two plots in Figure 14.5 illustrate Cook and Weisberg's suggestion (1999) that a portion of a quadratic curve can be a useful approximation for a relationship where Y increases at an increasing rate as X increases or where Y has a decreasing rate of change with X.

Polynomials with degrees greater than 2 are more flexible in that they can describe relationships that are not symmetric about a minimum or maximum or relationships where there are more than one minima or maxima. However, polynomials with degrees greater than 3 are usually not fit by data analysts. Polynomial regression is a special case of multiple regression, as it has more than one predictor, for example, a polynomial model of degree 2 has three predictors β_0, X, and X^2. An example of a polynomial regression of degree 2 for a single variable, X, is given in Montgomery et al. (2012).

A potential problem with polynomial models, especially with higher degree polynomials, is that the fit of data in one region can be seriously affected by data in other regions. This potential problem can be circumvented by fitting **piece-wise polynomials**, which involves dividing the data into intervals (pieces) on the X axis and fitting a polynomial function to the data in each interval.

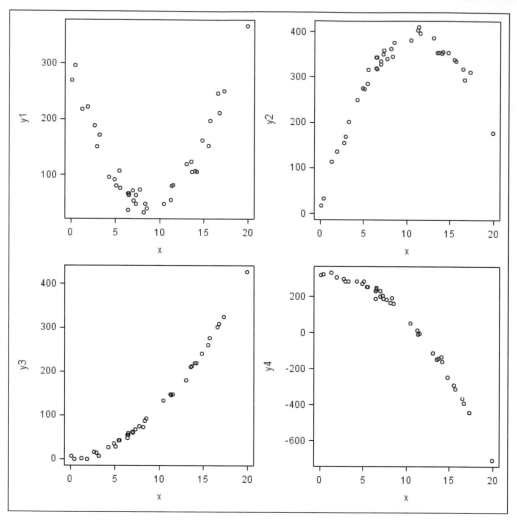

FIGURE 14.5
Plots of simulated data from four different quadratic models.

Spline-Based Approach of Fitting Piece-Wise Polynomials

One approach that can be used when fitting piece-wise polynomials is the spline-based approach in which regression splines are used to fit a polynomial function to the data in each interval on the X axis with the constraint that the polynomial function has to join at the adjacent "knots", i.e., at the value of X where two pieces join together. Harrell (2001) is an enthusiastic advocate of regression splines as an approach for dealing with nonlinear relationships when subject matter expertise does not suggest a particular function. He suggests **cubic splines** , i.e., **a polynomial function of degree 3** (with X, X^2, and X^3) have "nice properties with good ability to fit sharply curving shapes". Piece-wise polynomials can be fitted using regression splines in the TRANSREG procedure. Details are given in the reference "Using Knots and Splines" for the TRANSREG procedure available on-line at

```
https://support.sas.com/documentation/cdl/en/statug/63033/HTML
/default/viewer.htm#statug_transreg_sect016.htm
```

This documentation shows how easily a data analyst can fit spline functions using the TRANSREG procedure. For example the following code fits cubic spline functions (the default in TRANSREG) with nine knots to the data in the data set named "a":

```
proc transreg data=a;
    model identity(y)=spline(x / nknots=9);
run;
```

14.5.6   Overview of Fitting a Nonlinear Model

A nonlinear model is any model that is not linear in the parameters. In other words, a nonlinear model is a model that does not have the basic general form of a model linear in the parameters as described in Section 2.5. An example of a nonlinear model is an exponential model.

$$Y_i = \alpha e^{\beta X_i} + \epsilon_i \qquad (14.4)$$

Exponential models are often used to describe phenomena related to growth within a limited time period when availability of resources is not a limiting factor. For example, subject matter expertise and empirical evidence (as given in the mosquito data displayed in Figure 14.1) indicate an exponential model could be used to describe the relationship between mosquito density and available surface water. Other examples of nonlinear models are discussed in Rawlings et al. (1998) and Draper and Smith (1998).

An important advantage of nonlinear models is that they can fit a broad range of functions that cannot be fit by linear models. Other advantages of nonlinear models, as compared to competing models such as polynomials, are parsimony (fewer parameters have to be estimated), interpretability of the parameters, and robust prediction (Bates and Watts, 2007).

Wakefield (2004) cites the following challenges associated with fitting a complex nonlinear model:

- The model's statistical properties are often not completely understood.

- There is potential loss of precision in estimating the parameters of interest.

- The potential problem of over-fitting (being data-driven) may be greater with increased model complexity.

SAS offers a powerful procedure, PROC NLIN, that can fit nonlinear models. Details of this procedure can be found in SAS online support documentation.

14.6   Alternative Approaches When Variances Are Not All Equal

14.6.1   Introduction

The effect of unequal population error variances on the efficiency of a simple linear regression analysis (which affects results of hypothesis tests and confidence limits estimates) depends on a number of factors which include:

- sample size

- how unequal the error variances are at each value of the predictor, X

- the relationship between the error variances and the x-values

The effect of unequal population variances becomes even more troublesome to predict where there are multiple predictors. It is therefore impossible to make general conclusions as to when an unequal variances violation would cause serious harm. However, some rough guidelines have been offered:

- Fox (2016) suggested, as a conservative rule of thumb, that one typically does not have to worry about unequal variances in regression unless the ratio of the largest to smallest error variance exceeds four. As a liberal "rough rule", he suggested not worrying unless the largest error variance is about 10 times greater than the smallest.

- Carroll and Ruppert (1988) suggested that corrective action for unequal variances in regression is not crucial if the ratio of the largest to smallest variance is 2.25:1, assuming you have sufficient data to compute reliable estimates of the error variances at each X. If, however, the ratio of the largest to smallest variance is 9:1, they suggested that corrective action for unequal variances should be considered.

- The effect of moderately unequal variances in a one-way ANOVA is typically not considered harmful when there are equal sample sizes in the groups. However, the combination of unequal variances and unequal group sample sizes can either increase Type I or Type II error rates depending on whether the group samples sizes are negatively or positively related to the size of the variances. Section 11.8 provides details and references.

Alternative approaches that can deal with unequal variances include:

- **adding one or more additional predictors** to the model. Unequal variances can sometimes be mitigated or eliminated by adding to the model, one or more indicator (dummy) variable(s) representing a classification (qualitative) variable such as gender.

- **applying a transformation** and fitting a linear model to the transformed data (subsection 14.6.2)

- **weighted least squares** (subsection 14.6.3)

- **fitting a model that accommodates unequal error variances**, an approach which is beyond the scope of this book. Examples and details of this approach in SAS have been given for: ANCOVA applications in Chapter 14 of Milliken and Johnson (2002); a regression application in Chapter 9 of Littell et al. (2006); one-way ANOVA applications in Chapter 2 of Milliken and Johnson (2009).

- **applying a robust statistics method** (Section 14.8)

- **applying a bootstrapping method** (Section 14.9)

14.6.2   Transforming Data to Achieve Equality of Variances

Equality of variances can sometimes be induced by applying a transformation to the outcome variable, Y. However, caution is advised, as it is possible that transforming Y to achieve equality of variances may result in non-normality. Moreover, transforming Y may change the relationship between Y and a continuous predictor(s) so that a linear model is no longer a good fit when the Y values are transformed. In some situations where the data in the original units are a good fit to the linear regression model, transforming the Y values, while remediating the unequal variances problem, may result in a worse fit of the transformed data to the model. One possible way to circumvent or mitigate this problem in regression is to apply the same transformation to the X variable as applied to the Y variable, which is known as the **"transform both sides"** or **"TBS"** approach (Carroll and Ruppert, 1988).

Variance-Stabilizing Transformations

Tukey's Ladder of Power transformations may be considered when the variance of Y given X is related to the value of X. In many data sets, the variance of Y given X increases as X increases. In these situations, applying a power transformation to Y where the power, p, is less than 1 may be helpful in reducing the unequal variances at higher X values. For this reason, a log-transformation applied to Y (p close to zero) is commonly used as it compresses the larger Y values.

The arcsine transformation may be considered as a remedial measure for unequal variances in situations where Y has values that range between 0 and 1 and when unequal variances occur as a result of the size of the variance of Y given the predictor(s) being related to the mean of Y given the predictor(s). An example of this would be a data set where Y values are proportions. The arcsine transformation can be applied using the ARSIN function in a SAS DATA step.

A data analyst should never transform Y to stabilize variances if it causes serious departure from normality, e.g., it causes the errors to have a skewed distribution. Fortunately, it has been found in some real-life applications that the same transformation on Y which remediates unequal variances also has the effect of remediating non-normality. However, a data analyst cannot count on this. Each time a transformation is tried, the model assumptions have to be checked again using diagnostics obtained from applying the model to the transformed data.

14.6.3 Overview of Weighted Least Squares for Unequal Variances

Weighted least squares, a special case of generalized linear modelling, adjusts for unequal variances by assigning a weight to every case in the sample. The basic idea underlying this procedure is that the weight for each case is equal or proportional to the reciprocal of the error variance associated with that case, i.e., the weight for case i, denoted as w_i is given by

$$w_i \propto \frac{1}{\text{VAR}(e_i)} \tag{14.5}$$

where
    $\text{VAR}(e_i)$ is the population error variance associated with the ith case.

Thus, a case that is less reliable because it has a larger error variance is given a smaller weight in the analysis. A case that is more reliable because it has a smaller error variance is assigned a larger weight.

When there are unequal error variances and the true weights (the true w_i) are known (as opposed to estimated) for each case, the standard errors for the model parameters will typically be smaller using the method of weighted least squares as compared to that of ordinary least squares (OLS), which assumes equal population variances. Thus, if the true weights are known, more powerful hypothesis tests and narrower confidence interval estimates will be obtained using the weighted least squares method when variances are unequal as compared to using OLS. An example where weights are known is the situation where the response value, Y_i for each case is a proportion based on n_i observations. Therefore the true weight, w_i, is n_i (the reciprocal of $1/n_i$) because n_i is proportional to the variance of a proportion. (Recall from introductory statistics that the variance of a proportion is pq/n where p = the proportion and q = 1 − p.) However, weights typically are not known and have to be estimated. In situations when there are unequal error variances and the weights are poorly estimated due to limited available information, the method of weighted least squares can produce larger standard errors for the model parameters and therefore less powerful hypothesis tests and wider confidence intervals than an ordinary least squares analysis (Myers, 1990; Carroll and Ruppert, 1988). Non-statistical researchers are well advised to seek help from a professional statistician who can recommend whether weighted least squares or an alternative remedial measure or no corrective action should be considered when an unequal variances violation occurs in a conventional ordinary least squares analysis.

14.7 Alternative Approaches If Model Errors Are Not Normal

14.7.1 Introduction

When population model errors are not normally distributed, particularly if the error distribution is non-symmetrical (i.e., skewed to the left or to the right as is often encountered with real life data) or if symmetrical the error distribution is long-tailed (subsection 3.9.3), it is possible that an alternative approach, resistant to non-normality, may be more efficient than applying ordinary least squares linear regression to the original data.

Alternative approaches that may be considered include:

- applying a **transformation** to Y (subsection 14.7.2)

- applying a **classical nonparametric method**, e.g., rank regression (subsection 14.5.4)

- applying a **robust statistics method** when the non-normality is associated with outliers (Section 14.8)

- applying a **bootstrapping** method (Section 14.9)

14.7.2 Reducing Skewness with Power Transformations

The family of power transformations can be applied to reduce a skewed distribution of Y thereby mitigating an important source of non-normality of model errors in simple and multiple linear regression and ANOVA. Let Y^p denote a power transformation for Y. Possible choices for the power p and the associated effect of the transformation on the distributional shape of Y are:

- $p > 1$ reduces negative skewness

- $p = 1$ no change in distributional shape of Y

- $p < 1$ reduces positive skewness

As one moves up Tukey's ladder of power transformations (i.e., increases the value of p away from 1), the effect of the transformation in reducing negative skewness is greater. For example, the effect of the transformation in reducing negative skewness is greater when Y is raised to the third power ($p = 3$) as compared to squaring the Y values ($p = 2$).

As one moves down the ladder of powers (i.e., decreases the value of p away from 1), the effect of reducing positive skewness is greater. For example, moving down the ladder and taking the log of Y (p = zero by convention) reduces positive skewness to a greater extent than taking the square root of Y ($p = 0.5$).

Optimally, when choosing a value of p to transform Y, researchers should base their decision on subject matter knowledge. In the absence of this knowledge, sometimes a transformation is selected by trial and error. A more formal approach is the Box-Cox method

(Box and Cox, 1964), which selects a value of p to transform Y with the objective to make the model errors as close to normally distributed as possible. The Box-Cox procedure is available in the TRANSREG procedure. A disadvantage of the Box-Cox procedure is that it is sensitive to outliers and heterogeneous variances. Sakia (1992) provided a review of some modifications of the Box-Cox method that take heterogeneous variances into account. Regardless of the method of selecting a normalizing transformation, it is always important to check if **all** ideal inference conditions are reasonable on the transformed scale via model checking methods discussed in Chapter 3.

14.8 Robust Statistics

Robust statistics refer to methods that are stable when there are minor deviations from conventional Gaussian model assumptions associated with one or more outliers. The term stable here means that **both the level** (the true probability of a Type I error) **and the efficiency** (which relates to the power to detect a true effect) of a classical parametric test (e.g., simple or multiple linear regression) are unaffected by minor deviations from the test's assumptions. Some of the commonly used robust statistics methods are provided by the ROBUSTREG procedure and an overview of these methods may be found in SAS online documentation for this procedure.

The choice of a robust statistics method is critical as some methods are stable when only Y outliers are present in the data and are not stable when X as well as Y outliers are present. Moreover, certain robust statistics methods will fail to reveal violation of linearity in the data and will fit the wrong model when the relationship between a predictor and a response is curvilinear (Cook et al., 1992). Huber (2009), one of the founding fathers of robust statistics, cautioned that efficiency, hence, the ability to detect a true effect, greatly varies among robust statistics methods. He noted that loss of efficiency may sometimes be related to a method's high breakdown point (i.e., the fraction of outliers to which a method is stable) when X as well Y outliers are considered. Huber warned against utilizing methods which belong to "the Souped-up Car Syndrome" where one optimal property of a method (such as a large breakdown point) has been achieved at the expense of efficiency, a key optimal property of robust statistics. In view of these concerns, researchers in other disciplines are strongly recommended to seek the advice of a statistician before venturing into robust statistics methodology.

14.9 Bootstrapping

Bradley Efron's 1979 article on bootstrap procedures heralded a new era in statistics, which has eased the burden (or if you prefer, the challenges) for users of statistical methods. Bootstrap procedures are intuitively simple, computer-based methods which "can routinely answer questions that are far too complicated for traditional statistical analysis" (Efron and Tibshirani, 1986). The complexity of the analysis may arise from the complicated nature of the model per se or from a lack of knowledge about the distribution of a statistic when model assumptions do not hold.

The basic idea underpinning bootstrapping is that the researcher's original sample is considered to be the source population and a large number of bootstrap samples are independently generated by repeated random sampling with replacement from the researcher's original sample. It has been shown (Efron, 1979) that information obtained from these bootstrap samples can be used to make inferences about the population of interest from which the researcher's original sample was drawn.

For example, in simple linear regression we can consider using bootstrapping to estimate the standard error of $\hat{\beta}_1$, the regression coefficient statistic for X when the ideal inference condition of equal variances or normality or both are not reasonable and thus the conventional approach for computing the standard error of $\hat{\beta}_1$ (Section 4.4) may not be reliable.

14.9.1   Case Resampling

Case resampling is one of the basic resampling approaches used to obtain bootstrap samples in regression. It is a model-free resampling approach as it involves a random sampling with replacement of **cases** in the researchers' original sample without consideration of any model. In the context of simple linear regression, **case resampling** means that the x, y values of each case are considered as the unit in the sampling scheme. Case resampling is sometimes called **pairs resampling in simple linear regression** (usually in economics) as a pair of values (the x and y values) for each case is sampled as a unit. Case resampling is also referred to as **vector resampling** in regression. Program 14.E (Section 14.E) illustrates how case resampling may be implemented in SAS.

Case resampling is typically used when there is concern about the terms of the model not being correctly specified (e.g., wrong functional form, wrong predictors) or if the model's assumptions are not reasonable for the data. However, the nonparametric simple bootstrap method involving case resampling is not, in general, recommended for dependent data (Davison and Hinkley, 1997).

We will not include **residual resampling**, another basic bootstrap resampling method in our discussion of alternative approaches for situations when ideal inference conditions are not satisfied because residual resampling is model-dependent and more sensitive to violation of model assumptions than case resampling (Efron and Tibshirani, 1993).

14.9.2   The Bootstrap Estimate of Standard Error

The **nonparametric simple bootstrap** method, sometimes referred to as **the naive bootstrap** method, can be used to estimate the standard error of virtually any statistic. This method is algorithm-based and requires no complicated theoretical calculations. As an example we describe how the nonparametric simple bootstrap method provides an estimate of the standard error of the regression coefficient for X in simple linear regression via the following three steps:

1. Generate a large number of bootstrap samples, say N bootstrap samples, where each bootstrap sample consists of n cases obtained by resampling (random sampling with replacement) of cases from the researcher's original sample of n cases.

2. Fit a simple linear regression model to each bootstrap sample to get the bootstrap estimate of β_1 for that bootstrap sample, which we denote as $\hat{\beta}_1{}^*$.

3. Compute the bootstrap standard error, by computing the standard deviation of the $\hat{\beta}_1{}^*$ values from the N bootstrap samples. The bootstrap standard error which you compute in this step is the simple bootstrap estimate of the standard error of $\hat{\beta}_1{}^*$ in simple linear regression.

Program 14.E illustrates how these steps can be readily implemented in SAS. An example of a bootstrap standard error is given in subsection 14.10.1.

14.9.3 The Bootstrap Percentile Confidence Interval

The bootstrap percentile interval (Efron, 1982) is an approximate confidence interval for a population parameter, which we arbitrarily denote as θ. This popular bootstrap confidence interval is the interval between 100α and $100(1 - \alpha)$ percentiles of the distribution of the corresponding bootstrap statistic, which we arbitrarily denote as $\hat{\theta}^*$. As illustration we consider how the bootstrap percentile method provides an approximate 95% confidence interval for β_1, the population regression coefficient for X in simple linear regression via the following four steps:

1. Generate a large number of bootstrap samples, say N bootstrap samples, where each bootstrap sample consists of n cases obtained by case resampling from your original sample of n cases.

2. Fit a simple linear regression model to each bootstrap sample to get a bootstrap estimate of β_1 for every bootstrap sample, which we denote as $\hat{\beta_1}^*$.

3. Order the N values of the $\hat{\beta_1}^*$ obtained in Step 2 from smallest to largest.

4. Obtain the 2.5th and 97.5th percentiles of the $\hat{\beta_1}^*$. These percentile values constitute the lower and upper limits of the 95% bootstrap percentile confidence interval for β_1. The basic idea is that you are choosing an interval that **includes 95%** of the values of the $\hat{\beta_1}^*$ **by excluding 5% of the values**, i.e., by excluding the bottom 2.5% and the top 2.5%. Therefore 2.5th and 97.5th percentiles identify the lower and upper limits of the 95% bootstrap percentile confidence interval.

These four steps can be readily executed in SAS, as illustrated by Program 14.E. An example of a bootstrap percentile confidence interval is provided in subsection 14.10.1.

Advantages of the bootstrap percentile confidence interval include its simplicity and its ability to automatically incorporate an appropriate transformation of the statistic, θ, to improve the effectiveness of the interval. Efron and Tibshirani (1993) provided a theoretical explanation of this second advantage. A disadvantage of the bootstrap percentile confidence interval is that the coverage error is often quite large unless the distribution of $\hat{\theta}$ is nearly symmetrical. Recall the coverage error is the proportion of intervals computed that do not include the true value of the parameter being estimated. A refinement to the percentile method was developed (Efron, 1982) to adjust for non-symmetric distributions of $\hat{\theta}$ and is known as the bias-corrected method or BC method. A further refinement which adjusts for non-symmetry of $\hat{\theta}$ as well as adjusts for the possibility of its changing shape and skewness as θ changes (Efron, 1987) is known as the BC_a method. A review of the percentile, the BC, and the BC_a confidence intervals as well as a number of other bootstrap confidence intervals was given by Carpenter and Bithell (2000).

14.10 When Is n Too Small And What Is An Adequate Number Of Bootstrap Samples?

When n, the size of the bootstrap sample is less than 10, Chernick and LaBudde (2014) concluded that the bootstrap procedure may not be reliable. However, they suggested that an n as small as 20 may be adequate for bootstrapping.

A concern with small n is that true variability of the parameter being estimated will be underrepresented. Random sampling n observations with replacement from the original sample consisting of n observations may result in selecting bootstrap samples that are identical. However, Hall (1992) has shown that when $n = 20$ and N the number of bootstrap samples is 2000, the probability is greater than 0.95 that none of the bootstrap samples will consist of identical observations.

Efron and Tibshirani (1993, p. 52) stated, that based on their experience, even a small number of bootstrap samples, say $N = 25$, is usually informative and $N = 50$ is often sufficient to give a good estimate of a standard error. Very seldom are more than 200 bootstrap samples needed for estimating a standard error. However they cautioned (p. 275) that a much larger N is needed for bootstrap confidence intervals, say ≥ 500 or 1000 for constructing bootstrap percentile confidence intervals.

Hesterberg (2014, p. 96) argued that a large N (e.g., $N = 15,000$) is needed for a bootstrap percentile interval, especially in clinical trial applications where reproducibility of results is crucial. In this era of modern computers where calculations can be rapid and inexpensive, Chernick (2008) reported that an N of 100,000 or more is commonplace for simple problems.

14.10.1   Example of a Bootstrapping Application

In this section we consider a data set from Littell et al. (Chapter 9, 2006) which they provided as an example of unequal variances in simple linear regression. In Program 14.D we have generated regression diagnostics for fitting a simple linear regression model to these data.

Summary of Regression Diagnostics for a Simple Linear Regression applied to Littell et al.'s Unequal Variances Example. Figure 14.6 suggests that the model errors are not independent of X when a simple linear regression model is fitted to these data.

Summary statistics from the UNIVARIATE procedure regarding the distribution of internally studentized residuals obtained from applying a simple linear regression model to these data suggest the model errors have approximately a symmetrical distribution. The skewness statistic is -0.30, i.e., not far from zero and the mean (0.000810) and median (0.031132) of the internally studentized residuals are both close to zero.

Figure 14.7 from the UNIVARIATE procedure does not suggest any serious departure of the model errors from the normality assumption.

Results from Program 14.D: Estimated Parameters and Associated Standard Errors for a Simple Linear Regression Applied to Unequal Variances Data of Littell et al.

Parameter Estimates

Variable	DF	Parameter Estimate	Standard Error	t Value	Pr > |t|
Intercept	1	4.49757	4.49986	1.00	0.3324
X	1	7.78248	2.17297	3.58	0.0025

Parameter Estimates

Variable	DF	95% Confidence Limits	
Intercept	1	-5.04170	14.03685
X	1	3.17598	12.38898

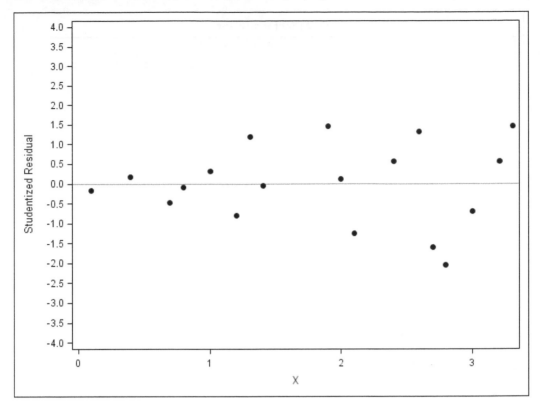

FIGURE 14.6
Plot of internally studentized residuals vs. X: simple linear regression applied to data of
Littell et al. (2006).

As the model errors may not be independent of X in the sampled population when a
simple linear regression model is fitted to these data, there is a concern whether the preceding estimated standard error for $\hat{\beta}_1$ (2.17) as well as the estimated 95% confidence limits for
β_1 (3.17, 12.39) are reliable. This concern arises because these estimates of standard error
and confidence interval limits for $\hat{\beta}_1$ are based on the Mean Square Error from ordinary
least squares simple linear regression, which assumes equality of variances of the model errors within the range of observed X. We therefore use this example to illustrate calculation
of a nonparametric bootstrap estimate of the standard error of $\hat{\beta}_1$ and a 95% bootstrap
percentile confidence interval for β_1 in Program 14.E.

**Results from Program 14.E giving a Bootstrap Standard Error Estimate for the
Regression Coefficient for X**
The following results generated by Program 14.E are summary statistics of the bootstrap
distribution of the $\hat{\beta}_1^*$ based on case resampling 100,000 bootstrap samples from Littell
et al.'s data. The standard deviation of this distribution for the $\hat{\beta}_1^*$ is 2.00691602, i.e.,
the bootstrap standard error estimate for the regression coefficient for X in the simple
linear regression is 2.01. This bootstrap standard error estimate, 2.01, is similar to 2.17,
the standard error for the regression coefficient for X obtained from applying ordinary
least squares simple linear regression analysis assuming equal variances to these data in
Program 14.D.

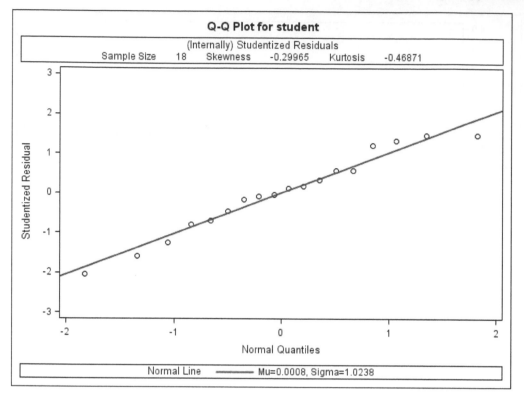

FIGURE 14.7
Normal probability plot of internally studentized residuals from simple linear regression applied to Littell et al.'s unequal variances example.

```
                    The UNIVARIATE Procedure
                          Variable:  X


                              Moments

   N                       100000    Sum Weights              100000
   Mean                7.67408897    Sum Observations     767408.897
   Std Deviation       2.00691602    Variance              4.0277119
   Skewness            -0.3580989    Kurtosis             0.33357927
   Uncorrected SS      6291935.34    Corrected SS          402771.19
   Coeff Variation     26.1518472    Std Error Mean                .
```

In subsection 14.E.1 we explain why the $\hat{\beta}_1^*$ bootstrap regression coefficient for X in simple linear regression is labelled as X in the preceding UNIVARIATE procedure output.

95% Bootstrap Percentile Confidence Interval for β_1 given in Output from the PRINT procedure in Program 14.E

```
          95% Bootstrap Percentile Confidence Interval


            Obs        ci2_5        ci97_5


             1        3.39937       11.2734
```

The preceding result (3.39937, 11.2734) obtained via the bootstrap percentile method is an approximate 95% confidence interval for β_1, the regression coefficient for X in the population from which the researcher's original sample was drawn. This bootstrap percentile 95% confidence interval (3.39937, 11.2734) is somewhat similar to (3.17598, 12.38898), the 95% confidence interval estimated in Program 14.D when a simple linear regression analysis was applied to the researcher's original sample.

Comment

The values of the bootstrap standard error estimate for the regression coefficient for X (1.9958) and the bootstrap percentile 95% confidence interval (3.43, 11.27) based on 50,000 bootstrap samples are remarkably similar to these results based on 100,000 bootstrap samples. If you would like to verify these values based on 50,000 bootstrap samples, simply specify rep = 50000 in the PROC SURVEYSELECT statement in Program 14.E.

14.11   Alternative Approaches If Harmful Collinearity Is Detected

If harmful collinearity is detected in a multiple regression analysis, we suggest using exploratory factor analysis to create new uncorrelated variables from the original predictor variables. These new uncorrelated variables can be then used as predictors to replace the original predictors in the multiple regression model. If the primary objective of the regression analysis is to understand the underlying relationship between the outcome and predictor variables, a discussion of caveats (Johnson, 1998) regarding subjectivity of interpretation of new variables created in factor analysis may be of interest. References for applying factor analysis using SAS include Johnson (1998); O'Rourke and Hatcher (2013); and Osborne and Banjanovic (2016).

Adding new cases to the sample has been suggested as a possible remedial measure for harmful collinearity. However, this option is not always practical at the analysis phase of research. Moreover, this option will not necessarily be helpful to disentangle the effects of confounding if the pattern of collinearity in the sample is representative of that in the sampled population and simple random sampling is used to select new subjects.

Dropping one or several of the collinear predictor variables from the model is another approach suggested as a remedial measure for harmful collinearity. This approach, however, can lead to misleading conclusions because although a variable is no longer included in the model, it is still intrinsic to the subjects being analysed. For example, consider a multiple regression analysis of occupational asbestos exposure where "years of smoking" and "years of occupational asbestos exposure" are predictor variables that are highly correlated because most asbestos workers in the sample are long-time smokers. Suppose a data analyst were to drop "years of smoking" and retain only "years of occupational asbestos exposure" as a predictor in the multiple regression model. This would not circumvent the problem that the subjects in the analysis who are occupational asbestos workers are also mostly smokers. Thus, the effect of years of occupational asbestos exposure on the outcome variable would be still confounded by the effect of years of smoking. The analysis may yield a misleading substantive conclusion as the magnitude of the regression coefficient for years of occupational asbestos exposure could be unduly large as its effect might be confounded with but not adjusted for the effect of years of smoking in the analysis.

Belsley (1991) referred to the option of "tossing out collinear variates" from the model as the "nonsolution" of "Occam's hatchet". He provided a detailed numerical example to illustrate why this option is "quite generally neither a good nor a recommended solution to the collinearity problem" and stated, "If collinearity can be shown to be adversely affecting the estimate of the coefficient of a variate of interest or rendering some a priori important effect insignificant, the appropriate conclusion is that the data lack the information needed to accomplish the statistical task at hand with precision, not that the model must be molded in a form that looks good relative to the data". He advised that under these circumstances "one must either get better data or introduce appropriate prior information", the latter approach is beyond the scope of our discussion but see Chapter 10 in Belsley (1991) for a detailed example.

Another approach proposed to mitigate harmful collinearity is ridge regression, a procedure where the method of least squares is modified by allowing biased estimation of the predictor variables' regression coefficients in order to reduce the standard error of these coefficients. A major limitation of ridge regression is that the choice of how much bias that should be introduced in estimating the regression coefficients is typically left up to the judgement of the analyst. Although methods have been developed to formalize this choice, these methods have their own limitations (Kutner et al., 2005). Another limitation of ridge regression is that conventional inference procedures (hypothesis tests and confidence interval estimates) cannot be applied to evaluate the precision of regression coefficients obtained from ridge regression as exact distributional properties of these coefficients are unknown. Bootstrapping can be employed to assess the precision of ridge regression coefficients. For a technical discussion of ridge regression, see Kutner et al. (2005).

14.12 Chapter Summary

This chapter:

- underlines the commonality of ideal inference conditions that are shared by the methods presented in this book

- provides overviews of some general alternative approaches that may be considered when an ideal inference condition is not satisfied

- strongly recommends, when random sampling is not an option, that researchers describe characteristics of their sample which may affect or be associated with the outcome and/or predictor variables, so that expert opinion can determine whether the sample results are applicable to a population of interest

- emphasizes that violation of independence of errors in an analysis should never be ignored as it can lead to substantively incorrect conclusions

- describes the single summary statistic approach for circumventing dependence of errors and the disadvantages of this approach

- provides references for methods that take the correlation structure of errors into account thereby avoiding violation of independence of errors

- provides a general discussion of transformations as an alternative approach when one or more ideal inference conditions are not satisfied

- provides interpretations of regression coefficients when X or Y or both X and Y are log-transformed

- gives an overview of alternative approaches that may be considered in simple and multiple linear regression when the relationship between the outcome variable and a continuous predictor is nonlinear

- cites approximate guidelines regarding when unequal variances may cause serious harm and gives an overview of alternative approaches that may be considered when unequal variances occur

- provides an overview of alternative approaches that may be considered for situations where the ideal inference condition of normality is not reasonable

- briefly discusses robust statistics methodology

- introduces the concept of bootstrapping, using as illustration an example where the equal variances assumption for a simple linear regression model may not be reasonable

- provides an overview of alternative approaches when harmful collinearity occurs in multiple regression

Appendix

14.A Program 14.A

Program 14.A calculates the variable "est_change_avgY", the estimated change in the average value of Y per **10%** increase in X, based on an estimated regression coefficient for $\ln(X)$ from a linear regression where X but not Y is log-transformed.

```
data beta;
/*betahat denotes the estimated regression coefficient for
the natural log of X */
/*In this program 3.004517 is input as the value for betahat*/
/* The log function in SAS denotes the natural log */
/* The * symbol in SAS means multiply*/
input betahat;
est_change_avgY=betahat *log(1.10);
datalines;
3.004517
;
run;
proc print data=beta;
run;
```

Comment

Program 14.A can easily be modified. For example, to calculate the estimated change in average Y per 20% increase in X when only the predictor X is natural log-transformed, replace log(1.10) by log(1.20) in Program 14.A.

14.B Program 14.B

Program 14.B applies a natural log-transformation to the Y values in the mosquito example. It then fits a simple linear regression model and generates regression diagnostics so that ideal inference conditions using these transformed Y values may be evaluated.

```
data nipper;
input ID X Y;
LNY=log(y);
datalines;
        1       7.5       74.0
        2       5.5       14.5
        3       6.0       22.9
```

```
        4      10.0     300.8
        5       5.0       8.4
        6       8.5     125.7
        7       6.0      34.0
        8       8.5     125.7
        9       7.5      90.5
       10       9.5     391.4
       11       7.0      69.9
       12       7.0      42.2
       13       7.5      79.6
       14       7.0      55.0
       15       8.5     130.3
       16       5.5      24.6
       17       8.5     101.6
       18       7.0      53.0
       19       6.0      26.5
       20       7.0      46.3
       21       8.5     196.8
       22       6.0      27.6
       23       7.0      34.8
       24      10.0     727.9
       25       6.0      32.0
       26       7.5      73.3
       27      10.0     367.4
       28       7.5      58.5
       29       5.5      18.1
       30      10.0     342.9
       31       7.5      65.8
       32       7.0      39.5
       33       6.5      27.9
       34      10.0     551.0
       35       6.5      30.6
       36       6.5      37.5
       37      10.0     367.4
       38       7.0      59.8
       39       7.0      42.4
       40       8.0     107.1
    ;
title;
run;
proc print data=nipper;
run;
proc sgplot data=nipper;
scatter X=X Y=Y/markerattrs=(symbol=circlefilled color=black);
 run;
proc sgplot data=nipper;
scatter x=X y=LNY/markerattrs=(symbol=circlefilled color=black);
run;
* Regression diagnostics for ln Y vs. X;
proc reg data=nipper;
   model lny= x / r p clb;
```

```
    output out=diag
    student=student p=p stdi=stdi stdp=stdp;
run;
proc print data=diag;
 run;
*Evaluate Ideal Inference Conditions;;
/*Obtain plots of internally studentized residuals vs. X*/
proc sgplot data=diag;
  refline 0/ axis=Y;
  scatter X=X y=student/markerattrs=(symbol=circlefilled color=black);
  yaxis values= (-4 to 4 by .5);
run;
/*Obtain plots of internally studentized residuals vs. predicted value*/
proc sgplot data=diag;
  refline 0/ axis=Y;
  scatter x=p y=student/markerattrs=(symbol=circlefilled color=black);
 run;
* Evaluate normality;
* Obtain a Q-Q normal plot for internally studentised residuals;
proc univariate data =diag;
var student;
qqplot student  /normal (mu=est sigma=est);
inset n='Sample Size' skewness kurtosis
    /pos=tm header='(Internally) Studentized Residuals';
run;
```

14.C   Program 14.C

Program 14.C provides calculations for interpreting the estimated regression coefficient for X (0.71018 for block water surface area in sq m) and its estimated 95% confidence limits (0.65864, 0.76172) when only Y is natural log-transformed.

```
data betas;
input id estbeta;
/*estbeta is estimated regression coefficient for X
obtained from regression when only Y
is natural log transformed*/
interpret=100*(exp(estbeta)-1);
datalines;
1 0.71018
2 0.65864
3 0.76172
;
run;
proc print data=betas;
run;
```

14.D   Program 14.D

Program 14.D fits a simple linear regression model and provides regression diagnostics for an example, which Littell et al. (2006) provided to illustrate unequal variances.

```
data one;
input IDNUM X Y;
datalines;
1 0.1   4
2 0.4   9
3 0.7   6
4 0.8   10
5 1.0   15
6 1.2   7
7 1.3   25
8 1.4   15
9 1.9   32
10 2.0   21
11 2.1   10
12 2.4   28
13 2.6   36
14 2.7   12
15 2.8   9
16 3.0   22
17 3.2   34
18 3.3   42
;
title;
proc reg data=one
plots(only label)=(cooksd dffits dfbetas RStudentByLeverage);
model y= x / r p clb influence;
output out=diag
r=r rstudent=rstudent student=student p=p h=h stdi=stdi stdp=stdp
dffits=dffits cookd=cookd;
run;
proc print data=diag;
 run;


*Evaluate Ideal Inference Conditions;;
/*Obtain plots of internally studentized residuals vs. X*/
   proc sgplot data=diag;
  refline 0/ axis=Y;
  scatter x=x y=student/markerattrs=(symbol=circlefilled color=black);
  yaxis values= (-4 to 4 by .5);
   run;
/*Obtain plots of internally studentized residuals vs. predicted value*/
proc sgplot data=diag;
   refline 0/ axis=Y;
   scatter x=p y=student/markerattrs=(symbol=circlefilled color=black);
     run;
```

```
* Evaluate normality;
* Obtain a Q-Q normal plot for internally studentised residuals;
*Apply significance tests for normality;

proc univariate data =diag normal;
var student;
qqplot student  /normal (mu=est sigma=est);
inset n='Sample Size' skewness kurtosis
/pos=tm header='(Internally) Studentized Residuals';
run;
```

14.E   Program 14.E

Program 14.E uses as input the **temporary** SAS data set ONE created in Program 14.D which contains the data provided by Littell et al. (2006) to illustrate unequal variances in simple linear regression. Program 14.E:

- implements case resampling to obtain 100,000 bootstrap samples, each of which has a sample size of 18 (the same sample size as the researchers' actual original sample)

- estimates a standard error and an approximate confidence interval for β_1 in simple linear regression via the simple bootstrap and percentile method

This program is based on helpful information given in Wicklin (2013) and Cassell (2007).

```
title;
sasfile one load;
proc surveyselect data=one
seed= 59020 method =urs samprate=1 outhits
rep=100000 out=boots;
run;
sasfile one close;
ods listing close;
 /* suppress output to ODS destinations */
ods exclude all;
ods noresults;
proc reg data=boots outest=estreg (drop=_:)
plots=none;
model y=x;
by replicate;
quit;
ods listing;
ods exclude none;
ods results;
proc univariate data=estreg vardef=N;
var x ;
output out=final pctlpts=2.5, 97.5 pctlpre=ci;
run;
proc print data=final;
```

```
title 95% Bootstrap Percentile Confidence Interval;
run;
```

14.E.1   Explanation of SAS Statements in Program 14.E

title;
This title command tells SAS that there is to be no title displayed until a title command with an actual specified title is encountered.

sasfile one load;
This statement loads the data set ONE into RAM (random access memory). This allows the program to run faster, as SAS reads the values in the original data set called ONE only once and stores them in RAM. If you do not specify "sasfile one load;", the SAS data set ONE will be read by the SURVEYSELECT procedure every time a bootstrap sample is generated, typically a huge number of times, 100,000 times in this example.

proc surveyselect data=one
seed= 59020 method =urs samprate=1
outhits rep=100000 out=boots;

Explanation of Options Specified in the Preceding PROC SURVEYSELECT Statement

data=one specifies that the input data set is called ONE.

seed=59020
By using the SEED= *any positive integer* option to specify an initial value for the pseudo-random number generator, you will be able to exactly reproduce your results should you or someone else need to run the program at another time. If you omit this option or specify SEED=0, SAS will use the time of day from the computer's clock as an initial seed and your results will be slightly different each time your program is executed.

method=urs specifies you want **u**nrestricted **r**andom **s**ampling, i.e., simple random sampling with replacement from data set ONE.

samprate=1 informs SAS that you want each sample generated by the SURVEYSE-LECT procedure to be the same size as the original data set. An alternative way of specifying this is to specify **samprate=100**.

outhits is a keyword which ensures that the procedure generates an output record every time it randomly selects a record, even if the procedure has selected the same record previously. In other words the keyword **outhits** allows the same record to be included twice in a bootstrap sample which is what is required in bootstrap random sampling **with replacement**.

rep=100000 specifies that 100,000 samples should be generated by the SURVEYSE-LECT procedure. When you specify this option it automatically creates a variable called REPLICATE which starts at 1 to identify the first sample generated by the procedure and increases by 1 for each sample subsequently generated. Thus, in this program the last bootstrap sample generated is labelled as REPLICATE 100000.

out=boots specifies that the data set output by the SURVEYSELECT procedure should be called BOOTS.

The next statements in this program
sasfile one close;
ods listing close;
improve the efficiency of the program as it closes the sasfile ONE which contains the original data as these data are no longer needed. The program from this point on will be working with the bootstrap samples that were generated by the SURVEYSELECT procedure and stored in the SAS dataset BOOTS.

The following statements also improve the efficiency of the program:

ods exclude all; which excludes all tables from open ODS destinations such as HTML or LISTING

ods noresults; which prevents an entry in the results window for each of the subsequent PROC REG steps that are generated. This is a considerable saving of resources as PROC REG will apply a simple linear regression to each one of the 100,000 bootstrap samples.

Explanation of Options Specified in the Statement proc reg data=boots plots=none outest=estreg (drop=_:);

data=boots specifies that the input data set is called BOOTS.

plots=none disables ODS Graphics just for the current REG procedure.

outest=estreg produces a SAS TYPE=EST output data set which contains estimates and optional statistics from the regression model specified in the model statement of the REG procedure.

(drop=_:) informs SAS to drop variables that have names preceded by _ which saves computer resources. When you specify this option the following four variables generated automatically by the REG procedure, viz., _MODEL_ _TYPE_ _DEPVAR_ _RMSE_ will not be stored for any of the 100,000 bootstrap replicates, which is efficient as you do not need these variables in this example.

model y=x;
This model statement directs SAS to fit an ordinary least squares simple linear regression model to the input data where Y is the response and X is the predictor variable.

by replicate;
This BY statement directs the REG procedure to apply a separate regression analysis to each bootstrap sample. Each bootstrap sample is identified by the value of variable, REPLICATE, where REPLICATE 1 is the first bootstrap sample generated by the SURVEYSELECT procedure and REPLICATE 100000 is the last bootstrap sample generated by this procedure. The combination of specifying this "by replicate;" statement and the "outest=estreg" option of the PROC REG statement means that summary regression statistics are output for each bootstrap sample in the output data set called ESTREG. Thus, 100,000 estimates of the $\hat{\beta}_1^*$, the bootstrap estimates of β_1 obtained when a simple linear regression model was fit to each bootstrap sample, will be stored in the output data set ESTREG.

The next statements in this program re-enable ODS output so that results generated by subsequent procedures will be displayed.
ods listing;
ods exclude none;
ods results;

proc univariate data=estreg vardef=N;
var x
These statements direct SAS to generate univariate summary statistics for the $\hat{\beta}_1^*$, the bootstrap estimates of β_1 generated in the previous REG procedure. The option "vardef=N" requests that N be used for degrees of freedom rather than the default N-1 in computing summary statistics where N is the number of bootstrap samples generated. However, either N or N-1 is acceptable for degrees of freedom (Chernick, 2008), especially when a large number of bootstrap samples have been generated. In the input data set ESTREG, each bootstrap estimate of β_1 is labelled as "x". Therefore to request summary statistics for the $\hat{\beta}_1^*$, we specify "x" in the VAR statement.

output out=final pctlpts=2.5, 97.5 pctlpre=ci;

This OUTPUT statement directs the UNIVARIATE procedure to compute the 2.5 and 97.5 percentiles of the $\hat{\beta}_1^*$, which are the upper and lower limits for the 95% confidence interval for the population regression coefficient for X, as estimated by the bootstrap percentile procedure. The values of these percentiles are to be output in the data set called FINAL. The 2.5 and 97.5 percentiles are requested via the "pctlpts= " option. The 2.5 and 97.5 percentiles are labelled as "ci2_5" and "ci97_5" in the output via the "pctlpre=ci" option. This latter option tells SAS to use "ci" as a prefix to create the variable names for the percentiles created via the pctlpts= option.

This OUTPUT statement with PCTLPTS= and PCTLPRE= options is required to direct the UNIVARIATE procedure to output the 2.5 and 97.5 percentiles. Without this OUTPUT statement only the 1st, 5th, 10th, 25th, 50th, 75th, 90th, 95th, and 99th percentiles (quantiles) are automatically given in the output of the UNIVARIATE procedure.

The following statements direct SAS to print the data set FINAL which contains the limits of the bootstrap percentile confidence interval for the regression coefficient for the predictor X in this simple linear regression example:

proc print data=final;
title Bootstrap Percentile Confidence Interval;
run;

References

Abbott, D. (2014). *Applied predictive analytics: Principles and techniques for the professional data analyst*. John Wiley and Sons.

Aiken, L. S. and S. G. West (1991). *Multiple regression: Testing and interpreting interactions*. Sage Publications Inc.

Akaike, H. (1973). Information theory and an extension of the maximum likelihood principle. In B. Petrov and F. Csaki (Eds.), *Proceedings of the Second International Symposium on Information Theory*, pp. 267–281.

Algina, J., S. Olejnik, and R. Ocanto (1989). Type I error rates and power estimates for selected two-sample tests of scale. *Journal of Educational Statistics 14*, 373–384.

Andrews, D. F. (1979). The robustness of residual displays. In R. Launer and G. Wilkinson (Eds.), *Robustness in statistics*, pp. 19–32. Academic Press.

Austin, P. (2011). An introduction to propensity score methods for reducing the effects of confounding in observational studies. *Multivariate Behavioral Research 46*, 399–424.

Bachoc, F., H. Leeb, and B. Pötscher (2019). Valid confidence intervals for post-model-selection predictors. *The Annals of Statistics 47*, 1475–1504.

Bartlett, M. S. (1937). Properties of sufficiency and statistical tests. *Proceedings of the Royal Statistical Society, A 160*, 268–282.

Bates, D. M. and D. G. Watts (2007). *Nonlinear regression analysis and its applications*. John Wiley and Sons.

Belsley, D. A. (1991). *Conditioning diagnostics, collinearity and weak data in regression*. John Wiley and Sons.

Belsley, D. A., E. Kuh, and R. E. Welsch (1980). *Regression diagnostics: Identifying influential data and sources of collinearity*. John Wiley and Sons.

Bender, R. and S. Lange (2001). Adjusting for multiple testing-when and how? *Journal of Clinical Epidemiology 54*, 343–349.

Benjamini, Y., R. Deveaux, B. Efron, S. Evans, M. Glickman, B. I. Graubard, X. He, X.-L. Meng, N. Reid, S. M. Stigler, S. B. Vardeman, C. K. Wikle, T. Wright, L. J. Young, and K. Kafadar (2021). The ASA president's task force statement on statistical significance and replicability. *The Annals of Applied Statistics 15*, 1084–1085.

Benjamini, Y. and Y. Hochberg (1995). Controlling the false discovery rate: A practical and powerful approach to multiple testing. *Journal of the Royal Statistical Society: B 57*, 289–300.

Benjamini, Y. and D. Yekutieli (2001). The control of the false discovery rate in multiple testing under dependency. *The Annals of Statistics 29*, 1165–1188.

Berglund, P. and S. Heeringa (2014). *Multiple imputation of missing data using SAS*. SAS Institute Inc.

Berk, R., L. Brown, A. Buja, K. Zhang, and L. Zhao (2013). Valid post-selection inference. *The Annals of Statistics 41*, 802–837.

Bolland, M., P. Barber, R. Doughty, B. Mason, A. Horne, R. Ames, G. D. Gamble, A. Grey, and I. R. Reid (2008). Vascular events in healthy older women receiving calcium supplementation: Randomised controlled trial. *BMJ 336*, 262–266.

Boneau, C. A. (1960). The effects of violations of assumptions underlying the t-Test. *Psychological Bulletin 57*, 49–64.

Box, G. E. P. (1953). Non-normality and tests on variances. *Biometrika 40*, 318–335.

Box, G. E. P. (1954). Some theorems on quadratic forms applied in the study of analysis of variance problems, I: Effect of inequality of variance in the one-way classification. *Annals of Mathematical Statistics 25*, 290–302.

Box, G. E. P. and D. R. Cox (1964). An analysis of transformations. *Journal of the Royal Statistical Society: Series B 26*, 211–243 (discussion pp. 244–252).

Bradley, J. V. (1978). Robustness? *British Journal of Mathematical and Statistical Psychology 31*, 144–152.

Brown, M. B. and A. B. Forsythe (1974a). Robust tests for the equality of variances. *Journal of the American Statistical Association 69*, 364–367.

Brown, M. B. and A. B. Forsythe (1974b). The small sample behavior of some statistics which test the equality of several means. *Technometrics 16*, 129–132.

Brynjolfsson, E., Y. Hu, and D. Simester (2011). Goodbye Pareto Principle, hello long tail: The effect of search costs on the concentration of product sales. *Management Science 57*, 1373–1386.

Brynjolfsson, E., Y. J. Hu, and M. D. Smith (2010). The longer tail: The changing shape of Amazon's sales distribution curve. https://ssrn.com/abstract=1679991.

Burnham, K. P. and D. R. Anderson (2002). *Model selection and multimodel inference: A practical information-theoretic approach*. Springer.

Burnham, K. P. and R. Anderson (2004). Multimodel inference: Understanding AIC and BIC in model selection. *Sociological Methods and Research 33*, 261–304.

Burton, P., L. Gurrin, and P. Sly (1998). Extending the simple linear regression model to account for correlated responses: An introduction to generalized estimating equations and multi-level mixed modelling. *Statistics in Medicine 17*, 1261–1291.

Carpenter, J. and J. Bithell (2000). Bootstrap confidence intervals: When, which, what? A practical guide for medical statisticians. *Statistics in Medicine 19*, 1141–1164.

Cassell, D. L. (2007). Don't be loopy: Re-sampling and simulation the SAS® way. SAS Global Forum Proceedings, Paper 183.

Carroll, R. J. and D. Ruppert (1988). *Transformation and weighting in regression*. Chapman and Hall.

Chernick, M. R. (2008). *Bootstrap methods: A Guide for practioners and researchers.* John Wiley and Sons.

Chernick, M. R. and R. A. LaBudde (2014). *An Introduction to Bootstrap Methods with Applications to R.* John Wiley and Sons.

Clarke, B. R. (2000). An adaptive method of estimation and outlier detection in regression applicable for small to moderate sample sizes. *Discussiones Mathematicae Probability and Statistics 20,* 25–50.

Clinch, J. J. and H. J. Keselman (1982). Parametric alternatives to the analysis of variance. *Journal of Educational Statistics 7,* 207–214.

Cochran, W. G. (1947). Some consequences when the assumptions for the analysis of variance are not satisfied. *Biometrics 3,* 22–38.

Cohen, J., S. G. West, L. Aiken, and P. Cohen (2003). *Applied multiple regression/correlation analysis for the behavioral sciences, 3rd ed.* Routledge.

Cohen, R. A. (2006). *Introducing the GLMSELECT Procedure for Model Selection.* Paper 207-31, SAS Institute Inc.

Cohen, R. A. (2009). *Applications of the GLMSELECT Procedure for Megamodel Selection.* Paper 259-2009, SAS Institute Inc.

Conover, W. J. (2012). The rank transformation – an easy and intuitive way to connect many nonparametric methods to their parametric counterparts for seamless teaching introductory statistics courses. *WIREs Computational Statistics 4,* 432–438.

Cook, R. D. (1977). Detection of influential observations in linear regression. *Technometrics 19,* 15–18.

Cook, R. D., D. M. Hawkins, and S. Weisberg (1992). Comparison of model misspecification diagnostics using residuals from least mean of squares and least median of squares fits. *Journal of the American Statistical Association 87,* 419–424.

Cook, R. D. and S. Weisberg (1999). *Applied Regression Including Computing and Graphics.* John Wiley and Sons.

D'Agostino Jr., R. B. and R. B. D'Agostino Sr. (2007). Estimating treatment effects using observational data. *JAMA 297,* 314–316.

David, F. N. and N. L. Johnson (1951). The effect of non-normality on the power function of the F-test in the analysis of variance. *Biometrika 38,* 43–57.

Davison, A. C. and D. V. Hinkley (1997). *Boostrap methods and their applications.* Cambridge University Press.

Diggle, P., P. Heagerty, K.-Y. Liang, and S. Zeger (2002). *Analysis of longitudinal data, 2nd ed.* Oxford.

Dmitrienko, A., A. C. Tamhane, and F. Bretz (2010). *Multiple testing problems in pharmaceutical statistics.* Chapman and Hall/CRC.

Draper, N. R., I. Guttman, and H. Kanemasu (1971). The distribution of certain regression statistics. *Biometrika 58,* 295–298.

Draper, N. R. and H. Smith (1998). *Applied regression analysis, 3rd ed.* John Wiley and Sons.

Dunnett, C. W. (1955). A multiple comparison procedure for comparing several treatments with a control. *Journal of the American Statistical Association 50*, 1096–1121.

Efron, B. (1979). Bootstrap methods: Another look at the jackknife. *Annals of Statistics 7*, 1–26.

Efron, B. (1982). *The jacknife, the bootstrap and other resampling plans*, Volume 38. Society for Industrial and Applied Mathematics.

Efron, B. (1987). Better bootstrap confidence intervals. *Journal of the American Statistical Association 82*, 171–185.

Efron, B. (2007). Size, power and false discovery rates. *Annals of Statistics 35*, 1351–1377.

Efron, B., T. Hastie, I. Johnstone, and R.Tibshirani (2004). Least angle regression. *The Annals of Statistics 32*, 407–499.

Efron, B. and R. Tibshirani (1986). Bootstrap methods for standard errors, confidence intervals, and other measures of statistical accuracy. *Statistical Science 1*, 54–75.

Efron, B. and R. J. Tibshirani (1993). *An introduction to the bootstrap.* Chapman and Hall.

Efroymson, M. A. (1960). Multiple regression analysis. In A. Ralston and H. S. Wilf (Eds.), *Mathematical Methods for Digital Computers*, pp. 191–203, John Wiley and Sons.

Faries, D. E., A. C. Leon, J. M. Haro, and R. L. Obenchain (2010). *Analysis of observational health care data using SAS.* SAS Institute, Inc.

Fisher, R. A. (1935). *The design of experiments.* Macmillan.

Fithian, W., D. Sun, and J. Taylor (2014). Optimal inference after model selection. arXiv:1410.2597.

Foster, D. P. and R. A. Stine (2004). Variable selection in data mining: Building a predictive model for bankruptcy. *Journal of the American Statistical Association 99*, 303–313.

Fox, J. (2016). *Applied regression analysis and generalized linear models.* SAGE.

Freund, R. J. and R. C. Littell (2000). *SAS system for regression, 3rd ed.* SAS Institute, Inc.

Ganguli, B., S. S. Roy, M. Naskar, E. J. Malloy, and E. A. Eisen (2016). Deletion diagnostics for the generalised linear mixed model with independent random effects. *Statistics in Medicine 35*, 1488–1501.

Glass, G. V., P. D. Peckham, and J. R. Sanders (1972). Consequences of failure to meet assumptions underlying the fixed effects analyses of variance and covariance. *Review of Educational Research 42*, 237–288.

Gnanadesikan, R. (1997). *Methods for statistical data analysis of multivariate observations.* John Wiley and Sons.

Goad, C. L. (2020). *SAS programming for elementary statistics, getting started.* Chapman and Hall/CRC.

Goeman, J. J. and A. Solari (2014). Multiple hypothesis testing in genomics. *Statistics in Medicine 33*, 1946–1978.

Goodman, S. N. (2008). A dirty dozen: Twelve p-value misconceptions. *Seminars in Hematology 45*, 135–140.

Graybill, F. A. and H. K. Iyer (1994). *Regression analysis: Concepts and applications.* Duxbury Press.

Greene, W. H. (2008). *Econometric analysis.* Pearson Prentice Hall.

Greenhouse, S. W. and S. Geisser (1959). On methods in the analysis of profile data. *Psychometrika 24*, 95–112.

Greenland, S., S. J. Senn, J. Rothman, J. B. Carlin, C. Poole, S. N. Goodman, and D. G. Altman (2016). Statistical tests, p values, confidence intervals, and power: A guide to misinterpretations. *European Journal of Epidemiology 31*, 337–350.

Gunes, F. (2005). *Penalized regression methods for linear models in SAS/STAT.* SAS Institute Inc.

Gustafson, P. (2004). *Measurement error and misclassification in statistics and epidemiology: Impacts and Bayesian adjustments.* Chapman and Hall/CRC.

Hall, P. (1992). *The bootstrap and Edgeworth expansion.* Springer.

Hanley, J. A., A. Negassa, M. deB Edwardes, and J. E. Forrester (2003). Statistical analysis of correlated data using generalized estimating equations: An orientation. *American Journal of Epidemiology 157*, 364–375.

Harrell, F. E. (2001). *Regression modeling strategies: With applications to linear models, logistic regression, and survival analysis.* Springer.

Hartley, H. O. (1950). The maximum F-ratio as a short-cut test for heterogeneity of variance. *Biometrika 37*, 308–312.

Harwell, M. R., E. N. Rubinstein, W. S. Hayes, and C. C. Olds (1992). Summarizing Monte Carlo results in methodological research: The one- and two-factor fixed effects ANOVA cases. *Journal of Educational Statistics 17*, 315–339.

Hastie, T., R. Tibshirani, and J. Friedman (2009). *The elements of statistical learning, data mining, inference, and prediction, 2nd ed.* Springer.

Hayter, A. J. (1984). A proof of the conjecture that the Tukey-Kramer multiple comparisons procedure is conservative. *Annals of Statistics 12*, 61–75.

Heller, R. (2010). Comment: Correlated z-values and the accuracy of large-scale statistical estimates. *Journal of the American Statistical Association 105*, 1059–1063.

Hesterberg, T. (2014). *Bootstrap in methods and applications of statistics in clinical trials, Volume 2 : Planning, analysis, and inferential methods.* John Wiley and Sons.

Hoaglin, D. C., F. Mosteller, and J. W. Tukey (1982). *Understanding robust and exploratory data analysis.* John Wiley and Sons.

Hochberg, Y. and A. C. Tamhane (1987). *Multiple comparison procedures.* John Wiley and Sons.

Hocking, R. R. (2013). *Methods and applications of linear models: Regression and the analysis of variance, 3rd ed.* John Wiley and Sons.

Holm, S. (1979). A simple sequentially rejective multiple test procedure. *Scandinavian Journal of Statistics 6*, 65–70.

Horsnell, G. (1953). The effect of unequal-group variances on the F-test for the homogeneity of group means. *Biometrika 40*, 128–136.

Huber, P. (2009). On the non-optimality of optimal procedures. *Institute of Mathemtical Statistics Lecture Notes-Monograph Series 57*, 31–46.

Hurvich, C. M. and C.-L. Tsai (1989). Regression and time series model selection in small samples. *Biometrika 76*, 297–307.

Iman, R. and W. Conover (1979). The use of rank transform in regression. *Technometrics 21*, 499–509.

Johnson, D. E. (1998). *Applied multivariate methods for data analysts.* Duxbury Press.

Kabaila, P. (2009). The coverage properties of confidence regions after model selection. *International Statistical Review 77*, 405–414.

Kendall, M. and A. Stuart (1976). *Applied regression including computing and graphics, 3rd ed.*, Volume 3. Hafner Press.

Keselman, H. J., R. R. Wilcox, J. Algina, A. R. Othman, and K. Fradette (2008). A comparative study of robust tests for spread: asymmetric trimming strategies. *British Journal of Mathematical and Statistical Psychology 61*, 235–253.

Keyes, T. K. and M. S. Levy (1997). Analysis of Levene's test under design imbalance. *Journal of Educational and Behavioral Statistics 22*, 227–236.

Kim, K. I. and M. A. van de Wiel (2008). Effects of dependence in high-dimensional multiple testing problems. *BMC Bioinformatics 9*. https://doi.org/10.1186/1471-2105-9-114.

Koch, G. G., S. M. Davis, and R. L. Anderson (1998). Methodological advances and plans for improving regulatory success for confirmatory studies. *Statistics in Medicine 17*, 1675–1690.

Kramer, C. Y. (1956). Extension of multiple range tests to group means with unequal numbers of replications. *Biometrics 12*, 307–310.

Kuhn, M. and K. Johnson (2013). *applied predictive modeling.* Springer.

Kullback, S. and R. A. Leibler (1951). On information and sufficiency. *Annals of Mathematical Statistics 22*, 79–86.

Kutner, M. H., C. J. Nachtsheim, J. Neter, and W. Li (2005). *Applied linear statistical models, 5th ed.* McGraw-Hill.

Lana, R. E. and A. Lubin (1963). The effect of correlation on the repeated measures design. *Educational and Psychological Measurement 23*, 729–739.

Lee, J., D. Sun, Y. Sun, and J. Taylor (2016). Exact post-selection inference, with application to the Lasso. *The Annals of Statistics 44*, 907–927.

Leeb, H. and B. M. Pötscher (2005). Model selection and inference: Facts and fiction. *Econometric Theory 21*, 21–59.

Levene, H. (1960). Robust tests for the equality of variances. In I. Olkin, S. G. Churye, W. Hoeffding, W. G. Madow, and H. B. Mann (Eds.), *Contributions to Probability and Statistics: Essays in Honor of Harold Hotelling*, pp. 278–292. Stanford University Press.

Lindquist, E. F. (1953). *Design and analysis of experiments in education and psychology.* Houghton Mifflin.

Littell, C., G. A. Milliken, W. W. Stroup, R. D. Wolfinger, and O. Schabenberger (2006). *SAS system for mixed models, 2nd ed.* SAS Institute, Inc.

Littell, R. C., W. W. Stroup, and R. J. Freund (2002). *SAS for linear models, 4th ed.* SAS Institute, Inc.

Lumley, T., P. Diehr, S. Emerson, and L. Chen (2002). The importance of the normality assumption in large public health data sets. *Annual Review of Public Health 23*, 151–169.

Mallows, C. L. (1973). Some comments on Cp. *Technometrics 15*, 661–675.

Meier, U. (2006). A note on the power of Fisher's least significant difference procedure. *Pharmaceutical Statistics 5*, 253–263.

Mickey, R. M., O. J. Dunn, and V. A. Clark (2004). *Applied statistics: Analysis of variance and regression, 3rd ed.* John Wiley and Sons.

Miller, A. (2002). *Subset selection in regression, 2nd ed.* Chapman and Hall/CRC.

Miller, R. G. (1968). Jackknifing variances. *The Annals of Mathematical Statistics 39*, 567–582.

Miller, R. G. (1981). *Simultaneous statistical inference, 2nd ed.* Springer.

Miller, R. G. (1998). *Beyond ANOVA: Basics of applied statistics, 1st ed.* Chapman and Hall/CRC.

Milliken, G. A. and D. E. Johnson (2002). *Analysis of messy data. Volume 3. Analysis of Covariance.* CRC Press.

Milliken, G. A. and D. E. Johnson (2009). *Analysis of messy data, 2nd ed. Volume 1. Designed Experiments.* CRC Press.

Montgomery, D. C., E. A. Peck, and G. G. Vining (2012). *Introduction to linear regression analysis, 5th ed.* John Wiley and Sons.

Moodie, G. E. E. (1985). Gill raker variation and the feeding niche of some temperate and tropical freshwater fishes. *Environmental Biology of Fishes 13*, 71–76.

Moodie, P. F. and D. B. Craig (1986). Invited special article. Experimental design and statistical analysis. *Journal of the Canadian Anaesthetists' Society 33*, 63–65.

Mosteller, F. and J. W. Tukey (1977). *Data analysis and regression: A second course in statistics.* Addison-Wesley.

Muller, K. E. and B. A. Fetterman (2012). *An integrated approach using SAS software.* SAS Institute Inc.

Murray, M. P. (2006). Avoiding invalid instruments and coping with weak instruments. *Journal of Economic Perspectives 20*, 111–112.

Myers, R. H. (1990). *Classical and modern regression with applications, 2nd ed.* PWS-Kent Publishing Company.

Nelson, N. A. and P. F. Moodie (1985). Application of a multivariate analysis of variance: A nested design. *Biometrics 40*, 1195.

Nelson, N. A. and P. F. Moodie (1988). Robust multivariate analysis of variability. *Communications in Statistics - Simulation and Computation 17*, 1409–1430.

Neuhaus, K.-L., R. Von Essen, U. Tebbe, A. Vogt, M. Roth, M. Reiss, W. Niederer, F. Forycki, A. Wirtzfeld, W. Maeurer, P. Limbourg, W. Merx, and K. Haerten (1992). Improved thrombolysis in acute myocardial infarction with front-loaded administration of alteplase: Results of the rt-PA-APSAC patency study (TAPS). *Journal of the American College of Cardiology 19*, 885–891.

Neyman, J. and E. S. Pearson (1928). On the use and interpretation of certain test criteria for purposes of statistical inference: Part I. *Biometrika 20A*, 175–240.

Norton, D. W. (1952). An empirical investigation of the effects of nonnormality and heterogeneity upon the F-test of analysis of variance. Unpublished doctoral dissertation, State University of Iowa.

O'Brien, R. G. (1978). Robust techniques for testing heterogeneity of variance effects in factorial designs. *Psychometrika 43*, 327–344.

O'Brien, R. G. (1979). A general ANOVA method for robust tests of additive models for variances. *Journal of the American Statistical Association 74*, 877–880.

O'Brien, R. G. (1981). A simple test for variance effects in experimental designs. *Psychological Bulletin 89*, 570–574.

O'Brien, R. M. (2007). A caution regarding rules of thumb for variance inflation factors. *Quality & Quantity 41*, 673–690.

O'Rourke, N. and L. Hatcher (2013). *A step-by-step approach to using SAS for factor analysis and structural equation modeling, 2nd ed.* SAS Institute Inc.

Osborne, J. W. and E. S. Banjanovic (2016). *Exploratory factor analysis with SAS.* SAS Institute, Inc.

Pearson, E. S. (1931). The analysis of variance in cases of non-normal variation. *Biometrika 23*, 114–133.

Pletcher, M. J., J. A. Tice, M. Pignone, and W. S. Browner (2004). Using the coronary artery calcium score to predict coronary heart disease events: A systematic review and meta-analysis. *Archives of Internal Medicine 164*, 1285–1292.

Ploner, A., S. Calza, A. Gusnanto, and Y. Pawitan (2006). Multidimensional local false discovery rate for microarray studies. *Bioinformatics 22*, 556–565.

Pope, P. T. and J. T. Webster (1972). The use of an F-statistic in stepwise regression procedures. *Technometrics 14*, 327–340.

Ranney, G. B. and C. C. Thigpen (1981). The sample coefficient of determination in simple linear regression. *The American Statistician 35*, 152–153.

Rawlings, J. O., S. G. Pantula, and D. A. Dickey (1998). *Applied regression analysis: A research tool, 2nd ed.* Springer.

Reid, I. R., B. A. Schooler, S. Hannan, and H. K. Ibbertson (1986). The acute biochemical effects of four proprietary calcium preparations. *Australian and New Zealand Journal of Medicine 16*, 193–197.

Reiner-Benaim, A. (2007). FDR control by the BH procedure for two-sided correlated tests with implications to gene expression data analysis. *Biometrical Journal 49*, 107–126.

Rodriguez, R. N. (2016). *Statistical model building for large, complex data: Five new directions in SAS/STAT software.* Paper SAS4900-2016, SAS Institute Inc.

Rogers, J. A. and J. C. Hsu (2001). Multiple comparisons of biodiversity. *Biometrical Journal 43*, 617–625.

Romano, J. P., A. M. Shaikh, and M. Wolf (2008). Control of the false discovery rate under dependence using the bootstrap and subsampling. *TEST 17*. https://doi.org/10.1007/s11749-008-0126-6.

Rosenbaum, P. R. and D. B. Rubin (1983). The central role of the propensity score in observational studies for causal effects. *Biometrika 70*, 41–55.

Rubin, D. R. (1976). Inference and missing data. *Biometrika 63*, 581–592.

Ryan, T. P. (2009). *Modern regression methods, 2nd ed.* John Wiley and Sons.

Sakia, R. M. (1992). The Box-Cox transformation technique: A review. *The Statistician 41*, 169–178.

Sarkar, S. (2004). FDR-controlling stepwise procedures and their false negatives rates. *Journal of Statistical Planning and Inference 125*, 119–137.

Saville, D. J. (1990). Multiple comparison procedures: The practical solution. *The American Statistician 44*, 174–180.

Scheffé, H. (1953). A method for judging all contrasts in the analysis of variance. *Biometrika 40*, 87–110.

Scheffé, H. (1959). *The analysis of variance.* John Wiley and Sons.

Scheffé, H. (1969). A method for judging all contrasts in the analysis of variance. *Biometrika 56*, 229.

Schwarz, G. (1978). Estimating the dimension of a model. *Annals of Statistics 6*, 461–464.

Senn, S. J. (1995). In defence of analysis of covariance: A reply to Chambless and Roeback. *Statistics in Medicine 14*, 2283–2285.

Sharma, D. and B. M. Kibria (2013). On some test statistics for testing homogeneity of variances: A comparative study. *Journal of Statistical Computation and Simulation 83*, 1944–1963.

Simon, S. D. and J. P. Lesage (1988). Benchmarking numerical accuracy of statistical algorithms. *Computational Statistics & Data Analysis 7*, 197–209.

Slaughter, S. J. and L. D. Delwiche (2004). SAS macro programming for beginners. In *Proceedings of the Twenty-Ninth Annual SAS Users Group International Conference.* SAS Institute Inc. Paper 243–29.

Sokal, R. R. and F. J. Rohlf (1981). *Biometry: The principles and practice of statistics in biological research, 2nd ed.* Freeman.

Srivastava, A. B. L. (1959). Effect of non-normality on the power of the analysis of variance test. *Biometrika 46*, 114–122.

Staudte, R. G. and S. J. Sheather (1990). *Robust estimation and testing.* John Wiley and Sons.

Stokes, M. E., C. S. Davis, and G. G. Koch (2012). *Categorical data analysis using SAS, 3rd ed.* SAS Institute.

Sugiura, N. (1978). Further analysis of the data by Akaike's information criterion and corrections. *Communications in Statistics, Theory and Methods 7*, 13–26.

Tibshirani, R. (1996). Regression shrinkage and selection via the Lasso. *Journal of the Royal Statistical Society B 58*, 267–288.

Tibshirani, R. J., J. Taylor, R. Lockhart, and R. Tibshirani (2016). Exact post-selection inference for sequential regression procedures. *Journal of the American Statistical Association 111*, 600–620.

Tiku, M. L. (1971). Student's distribution under nonnormal situations. *Australian Journal of Statistics 13*, 142–148.

Tukey, J. W. (1953). *The problem of multiple comparisons in: The Collected Works of John W. Tukey VIII. Multiple Comparisons:194819831300.* Chapman and Hall.

Tukey, J. W. (1977). *Exploratory data analysis.* Addison-Wesley.

Ulm, K. (1991). A statistical method for assessing a threshold in epidemiological studies. *Statistics in Medicine 10*, 341–349.

van Belle, G., L. D. Fisher, P. J. Heagerty, and T. Lumley (2004). *Biostatistics: A methodology for the health sciences, 2nd ed.* John Wiley and Sons.

van Buuren, S. (2012). *Flexible imputation of missing data.* Chapman and Hall.

van der Vaart, H. R. (1961). On the robustness of Wilcoxon's two-sample test. In H. de Jonge (Ed.), *Quantitative Methods in Pharmacology*, pp. 140–158. Wiley Interscience.

Van Valen, L. (1978). The statistics of variation. *Evolutionary Theory 4*, 33–43. (Erratum in Evolutionary Theory 4, p. 202).

Vittinghoff, E., D. Glidden, S. Shiboski, and C. E. McCulloch (2012). *Regression methods in biostatistics, linear, logistic, survival, and repeated measures models, 2nd ed.* Springer.

Wakefield, J. C. (2004). Non-linear regression modelling and inference. In *Methods and Models in Statistics: In Honour of Professor John Nelder, FRS*, pp. 119–153. Imperial College Press.

Wasserstein, R. L. and N. Lazar (2010). The ASA statement on p-values: Context, process, and purpose. *The American Statistician 70*, 129–133.

Wasserstein, R. L., A. L. Schirm, and N. Lazar (2019). Moving to a world beyond p < 0.05. *The American Statistician 73*, 1–19.

Welch, B. L. (1951). On the comparison of several mean values: An alternative approach. *Biometrika 38*, 330–336.

Westfall, P. H., R. D. Tobias, and R. D. Wolfinger (2011). *Multiple comparisons and multiple tests using SAS, 2nd ed.* SAS Institute Inc.

Wicklin, R. (2013). *Simulating data with SAS*. SAS Institute Inc.

Wicklin, R. (2014). Fat-tailed and long-tailed distributions. https://blogs.sas.com/content/iml/2014/10/13/fat-tailed-and-long-tailed-distributions.html.

Wiegand, R. E. (2010). Performance of using multiple stepwise algorithms for variable selection. *Statistics in Medicine 29*, 1647–1659.

Woodward, M. (2014). *Epidemiology: Study design and data analysis, 3rd ed.* Chapman and Hall/CRC.

Working, H. and H. Hotelling (1929). Applications of the theory of error to the interpretation of trends. *Journal of the American Statistical Association 24*, 73–85.

Yekutieli, D. (2008). Hierarchical false discovery rate-controlling methodology. *Journal of the American Statistical Association 103*, 309–316.

Zhang, P. (1992). Inference after variable selection in linear regression models. *Biometrika 79*, 741–746.

Zhu, M. (2014). Analyzing multilevel models with the GLIMMIX procedure. *Cary, NC: SAS Institute Inc.* Paper SAS0262014.

Zou, H. (2006). The adaptive Lasso and its oracle properties. *Journal of the Amerian Statistical Association 101*, 1418–1429.

Zou, H. and T. Hastie (2005). Regularization and variable selection via the elastic net. *Journal of The Royal Statistical Society Series B 67*, 301–320.

Index